Zootechnologies

Zootechnologies

A Media History of Swarm Research

Sebastian Vehlken

Translated by Valentine A. Pakis

Routledge
Taylor & Francis Group
LONDON AND NEW YORK

This publication is funded by MECS Institute for Advanced Study on Media Cultures of Computer Simulation, Leuphana University Lüneburg (German Research Foundation Project KFOR 1927).

Already published as: *Zootechnologien. Eine Mediengeschichte der Schwarmforschung,* Sebastian Vehlken. Copyright 2012, Diaphanes, Zürich-Berlin.

First published in 2019 by Amsterdam University Press Ltd.

Published 2025 by Routledge
4 Park Square, Milton Park, Abingdon, Oxon OX14 4RN
605 Third Avenue, New York, NY 10158

Routledge is an imprint of the Taylor & Francis Group, an informa business

ISBN: 9789462986206 (hbk)
ISBN: 9781041191001 (pbk)
ISBN: 9781003710011 (ebk)
NUR 670

Cover design: Suzan Beijer

DOI: 10.5117/9789462986206

"The whole is greater than the sum of its parts. As one of my colleagues once remarked: 'Can't the numbskulls even add?'" – Heinz von Foerster

"Among many techniques, this strange science called media history would do well to focus on those which themselves read or write." – Friedrich Kittler

Table of Contents

Acknowledgements

This book is the revised, updated, and in large part re-arranged translation of *Zootechnologien: Eine Mediengeschichte der Schwarmforschung*, published by the Diaphanes Verlag (Berlin, Zurich) in 2012. The publication is funded by MECS Institute for Advanced Study on Media Cultures of Computer Simulation, Leuphana University Lüneburg (German Research Foundation Project KFOR 1927). Its release in English gives me the opportunity to thank, first and foremost, the editors of AUP's *Recursions* series: Geoffrey Winthrop-Young, Anna Tuschling, and Jussi Parikka, who enthusiastically welcomed this work into their collection. I also thank Michael Heitz at Diaphanes for making this translation possible. My warmest thanks go to my translator Valentine Pakis, whose extraordinary skills do not just overcome the challenges inherent in the German version but also give it an additional edge. I thank the MECS directors Claus Pias and Martin Warnke for generously supporting the publication, and also my student assistants Sophie Godding and Rahel Schnitter at Leuphana University for their help with the copy-editing of the manuscript. I am further indebted to the editors of *Footprint* (Henriette Bier, Delft University, and Terry Knight, MIT), and to the editors of *Digital Culture and Society* (Mathias Fuchs, Leuphana University, and Ramòn Reichert, University of Vienna): two new sub-chapters that I have included in the translation are slightly revised and shortened versions of articles published in these respective journals. I also want to thank the team at Amsterdam University Press for the extraordinarily professional publishing process, namely Maryse Elliott, Mike Sanders, and Chantal Nicolaes. But above all, I want to give my thanks to Eva and Pepe Schauerte – my shooting stars and guiding lights.

Acknowledgements

Introduction

This book was inspired by the cold glare of a shark, which happened to meet my eyes in a train station bookstore. A full spread on the cover of a diving magazine displayed the prize-winning picture taken by the underwater photographer Doug Perrine. The image is of two copper sharks (*Carcharhinus brachyurus*) having their way with a hapless school of sardines. With sardines still stuck between their teeth, they are darting through the evasively maneuvering swarm, and the glance of one of the sharks during this feeding frenzy seems to fixate on the diver's camera. What Perrine managed to capture here so impressively is the famous sardine run – the annual migration of immense schools of sardines along the coast of South Africa. Their morphologies and dynamics number among the most fascinating phenomena of the animal kingdom that Alistair Fothergill and his team of BBC filmmakers had documented so vividly around the turn of the millennium.[1]

Not long after this encounter, I coincidentally came across this image again: Perrine's photograph happened to adorn the cover of a small brochure that, in 2005, was used to advertise an upcoming consumer trend conference in Hamburg. Given the title 'Schwarmintelligenz: Die Macht der smarten Mehrheit' ('Swarm Intelligence: The Power of the Smart Majority'), the symposium featured the keynote speaker Howard Rheingold, who had recently published his study of smart mobs, and thus shifted the focus away from sharks and schools of sardines toward the dynamics of highly concentrated network economies:

> The rapid development of information technologies has increasingly come to determine our lives, which are becoming more and more flexible, dynamic, and individual. The invention of the internet kindled a media revolution with lasting effects both on the economy and on private life. [...] Desires for community, love, and faith have found new forms of fulfillment. With the help of new technologies, autonomous individuals are able to network with one another more and more easily and inexpensively.

1 *Deep Blue*, directed by Alistair Fothergill.

Vehlken, S., *Zootechnologies. A Media History of Swarm Research.*
Taylor & Francis Group, 2019
DOI 10.5117/9789462986206_INTRO

> This has given rise to smart majorities who influence our decisions about everything from culture to consumption.[2]

The trend conference in 2005 was thus right on trend. Swarm intelligence was on everyone's lips at the time and had just lately entered the discourses of the humanities and social sciences with the publication of Rheingold's book. In the form of 'smart majorities,' swarms emerged as a metaphor for processes of coordination in the technologized present, in which the act of flexibly adjusting oneself to constantly changing parameters could be associated with a presumed potential for freedom on the part of 'autonomous individuals.' By means of ever more dynamic forms of networking – or so the metaphor of the swarm suggested – people could take advantage of an instantaneous (or at least extremely fast) infrastructure for decision-making. To achieve certain goals, people could simply coordinate themselves temporarily with others of the same mind. At the same time, it was also believed that any member of a swarm would be free to abandon the collective at any time and forge his or her own path. In highly general terms, this gave rise to an ephemeral collective figure that was ostensibly democratic (and thus welcomed with open arms by the politically correct). On the one hand, it promised to decouple political, economic, and social activity from rigid structures and organizations such as national states, political parties, corporations, and unions. On the other hand, this new form of collectivity would also revolutionize the availability of knowledge, whose traditional caretakers had been libraries and the classical (mass) media.

Institutionalizations of this sort – or so the promise went – would be absorbed on a case-by-case basis into a collaborative sphere based on fluidly and flexibly interconnected individual interests and local knowledge. Moreover, these acts of cooperation would no longer need an organizing authority but would, rather, organize themselves on the basis of the rapid interactions of their numerous participants. Swarms became a symbol for a new sort of media culture in which mobile and technical networking media converged with anti-hierarchical and distributively organized social forms. They combined a greater degree of individual freedom – in comparison with other infrastructures for networking – with a more effective logic of collective control. It is no surprise, then that they were also discussed along with new political concepts such as that of the 'multitude.'[3]

2 See the conference's homepage at http://www.trendbuero.de (accessed 18 December 2017).

3 See Hardt and Negri, *Multitude: War and Democracy in the Age of Empire*.

This observation could have been the point of departure for an analysis and critique of this collective figure's specific form of governmentality. It could have spurred an investigation into a discourse dynamic that was adopted nearly simultaneously (and this alone should cause alarm) by choreographers, subversive political groups and grass-roots networkers, military tacticians, economically-minded trend researchers, artists, and engineers – a discourse freighted with concepts ('smart mobs,' 'swarm architecture,' 'swarm energy,' etc.) that have problematized the distinction between swarms and quasi-swarms in the ubiquitous and increasingly undifferentiated use of the term.[4] It could have inspired comparative analyses with other forms of collectives or provided an occasion for differentiating between the swarm discourse and the more established discourse about networks. It could also have prompted a fundamental critique exposing putative instances of collective intelligence as mere examples of collective stupidity.[5] In what follows, however, I have abstained from taking such approaches. The aim of this book is rather to reconstruct the media-technological and theoretical conditions of possibility for such discourses and their metaphorical excesses. For – to stick with the three examples mentioned above – what is it that connects smart mobs with distributed and robotic construction processes in architecture and with the electrical grid-storage solutions of the impending post-petroleum age?

It is apparent that the discursive euphoria of recent years has been based on the idea of the bottom-up organization of swarms, a notion that has been inextricably associated with certain technical elements of mobile communication and sensor technology – regardless of whether the latter mediate between human actors, robots, or network nodes. And thus, from the outset, it is possible to voice the following suspicion: the metaphorical

4 For discussions oriented toward popular science, see Lause and Wippermann, *Leben im Schwarm*; Fisher, *The Perfect Swarm*; and Miller, *The Smart Swarm*. Regarding the fields of choreography and cultural studies, see Brandstetter et al., eds., *Swarm(E)Motion: Bewegung zwischen Affekt und Masse*. For an optimistic view with respect to politics and technology, see Rheingold, *Smart Mobs: The Next Social Revolution*. On swarms and the military, see, for instance, Arquilla and Ronfeldt, *Swarming and the Future of Conflict*. For an economic study, see, among other works, Neef and Burmeister, 'Swarm Organization: A New Paradigm for the E-Enterprise of the Future.' Notable works in the field of architecture include Kas Oosterhuis's online article 'Swarm Architecture,' http://www.oosterhuis.nl/?p=184 (accessed 19 December 2017); and his book *Hyperbodies: Towards an E-Motive Architecture*. On the motif and influence of swarms in fine art, see Miller et al., *Swarm*. Regarding the concept of 'swarm energy,' see the following website operated by the LichtBlick power company: https://www.lichtblick.de/schwarmenergie/ (accessed 19 December 2017).

5 See Seeßlen and Metz, *Blödmaschinen: Die Fabrikation der Stupidität*; and Dueck, *Schwarmdummheit: So blöd sind wir nur gemeinsam*.

force of swarms in today's narratives no longer derives, as it once did, from direct references to biological swarms. The recent discursive excitement over swarms is no longer based on reciprocal references between humans and animals, as was previously the case in the long history of comparing human and animal organisms and social forms. Over the course of this historical development, both sides variously served as models, differentiating phenomena, and inverted images for the other[6] – and various methods of comparison were academically institutionalized in fields ranging from mass psychology and sociobiology to, more recently, bionics and so-called human-animal studies.[7] Yet the central bionic question about what 'humans' might be able to learn from 'nature' has been somewhat misleading. The simple answer would be: nothing at all – at least not without certain media-technological interventions. Around the turn of the millennium, it was no longer simply animals alone and their collective behavior that were applied to the social processes of human beings. Instead, a third level was inserted between 'swarming' humans and swarming animals, a level of technical apparatuses and interfaces that first made it possible to describe 'swarm-like' interactions. This third level also enabled the technical implementation of swarm intelligence (as the phenomenon is known in the popular discourse) and engendered new dynamics in a variety of socio-economic contexts.

Like Michel Serres's figure of the parasite, it was this third level and its 'technologized' perspective that enabled humans and animals to be connected in the first place.[8] This book explores the transformation of swarms into operative collective models and how this came about by means of methods that are far from obvious. It was not simply a matter of modeling or imitating biological structures, as is often assumed. By providing a detailed media history of swarm research, I hope to show that, contrary to such assumptions, it was in fact regular *deletions* of nature that gave rise to dependable knowledge about swarms. Moreover, it was only by means of this *retreat from naturalness* that swarms could subsequently be made operative for technical applications. Swarm intelligence is based on optimizing formal relations within appropriate models, and in the case of swarms as dynamic and multidimensional multiplicities, these models are computer simulations or, to be more precise, agent-based computer

6 See Von der Heiden and Vogl, eds., *Politische Zoologie*; and Deleuze and Guattari, *A Thousand Plateaus*, 232–309.

7 See Wilson, *Sociobiology: The New Synthesis*; Kelly, *Out of Control*; and DeMello, *Animals and Society*.

8 See Serres, *The Parasite*.

simulations. Only processual, individual-based, and distributively functioning models of this sort could have given rise to a discourse dynamic that focuses on our knowledge about the particular relationality of swarms. Thus, it is no coincidence that the central concepts of this discourse – such as self-organization and collective intelligence – derive from an (information) technological context. Furthermore, this discursive euphoria that began around the year 2000 has been supported by an immense boom in the use of agent-based simulation processes in the social sciences, and these methods have allowed human social phenomena to appear swarm-like.[9] The most recent metaphorical transferences of swarms have therefore been based on a media-technological model of collective organization or *self-organization* that could in principle be applied to a variety of subject matters and was thus welcomed with open arms. For, with a few simple and local rules of interaction in place, such models of organization were able to bring to light novel, complex, and unforeseeable emergent phenomena.[10]

My focus below will thus be on the media-technological and historical conditions that made it possible, around the year 2000, for swarms to become operational as an effective model of control. How did they come to be associated with intelligence? How did it even become possible to speak about swarms as a form of *collective intelligence*? What sorts of knowledge informed the concept of the swarm at various moments in history? How long has there even been a systematic field of swarm research to generate such knowledge? Were not swarms classified for centuries as entities existing outside of order? Did they not belong to the realm of the anesthetic, in which it was impossible to assign any specific place to their incomprehensively dynamic elements?

9 See Grüne-Yanoff, 'Artificial Worlds and Agent-Based Simulation'; and Helbing, 'Agent-Based Modeling.'

10 The terms 'self-organization' and 'emergence,' which appear over and over again in scholarship devoted to the topic of labor, are not unproblematic, and each deserves a historical and philosophical study of its own. Within the framework of this work, however, they will primarily be used descriptively. In basic terms, self-organization will be understood below as a distributed organizational structure that enables adaptive, flexible, and efficient collective behavior in response to constantly changing environmental influences. A more precise definition of the term will be provided at the beginning of my fourth chapter. Following the advice of the swarm researcher Iain Couzin, I have attempted to avoid the term 'emergence' as much as possible. This is because the concept of emergence suggests far more, of course, than mere recurrence on a level of collective processes, whose appearance and whose characteristics can neither be traced back *to* nor derived *from* the features and capabilities possessed by the individual swarming elements of such nonlinear and interactive collectives. For further discussion of the concept, see Goldstein, 'Emergence as a Construct: History and Issues,' 49; Corning, 'The Re-Emergence of "Emergence"'; and Steele, 'Towards a Theory of Emergent Functionality,' 452. Regarding the meaning of the term in philosophy, see Lloyd, *Emergent Evolution*; and Stephan, 'Emergente Eigenschaften.'

Did they not belong to that class of objects which Leonardo da Vinci referred to as 'bodies without surfaces' and which, during the Renaissance, were simply regarded as being unrepresentable? Did Immanuel Kant not associate the modified but related term *Schwärmerei* (enthusiasm or fanaticism) with the distortion of reason? Even in the context of mass psychology, was not the uncanny teeming of swarms associated with social pathologies? Did swarms not evoke a fundamental epistemic fear of that which defies form? Of course, authors and natural scientists from all eras have also described schools of fish and flocks of birds with a sense of wonder and celebrated the sublimity of their collective movements. But even around the year 2000 – that is, during the age of their technological producibility – swarms continued to serve as a fitting metaphor for disseminating fear, for instance in the application of the swarm concept to new military or terrorist tactics. What had changed, however, was the reference system in which swarms could now be negotiated.

In the traditional analogy to biological swarms, teeming crowds of people were frequently described as a depraved swarming animal, acting subconsciously and thus susceptible to escalations and contagions. At the same time, however, they appeared to natural scientists as inestimable collectives that must have possessed, in order to coordinate their common maneuvers, a fascinating but uncanny (because indecipherable) common spirit – a collective soul or an inherent force that somehow controlled them. At first glance, it was still these traditional references to swarming animals that continued to appear periodically around the year 2000 in fascinating images of fish or birds pictured in movies, on television, and in a wide range of magazines and newspapers. Yet they were used to illustrate a more complex development, namely a model of control and a method for solving problems abstracted from their substantially biological origin: swarming animals, as I hope to demonstrate throughout this book, had been transformed into technically informed *zootechnologies*. Their 'intelligent' organizational potential was applicable to a great variety of subject matters, and zootechnical swarms could even be used as models for organizing human behavior. As zootechnologies, swarms began to coauthor the origin story of a particular media culture, and it is this culture that I intend to delineate here.

These developments were no longer defined by a mere sociobiological understanding of swarming animals, or by the destratified multiplicity of 'demonic animals' in the sense of Gilles Deleuze and Félix Guattari's political zoology. With an allusion to Ernst Jünger's novel *The Glass Bees*, one could rather say that, around the turn of the millennium, it was no longer animals that served as a model for humans but rather biological principles that had amalgamated with information-technological processes.[11] The neologism *zootechnologies* is meant to express that today's 'intelligent' swarms have long since combined *zoē*, the bare animal life in the swarm, with the experimental epistemology of computer simulation. Or, to put it another way, swarms of this sort make it clear that reference to animals alone is insufficient to explain what might be called complexity in humans and machines. It is rather a computer-supported perspective on animals that attributes to them an operative position from which they can, as a combination of biological knowledge and mechanical functionality, recursively produce new compatibilities between hardware, software, and wetware. Swarm intelligence is thus associated with the sort of interplay that Eugene Thacker has described as follows: "The 'bio' is transformatively mediated by the 'tech' so that the 'bio' reemerges more fully biological. [...] The biological and the digital domains are no longer rendered ontologically distinct, but instead are seen to inhere in each other; the biological 'informs' the digital, just as the digital 'corporealizes' the biological."[12] Yet unlike Thacker's concept and its connection both to *biotechnologies* and aspects of Foucaultian *biopower*, the neologism *zootechnologies* is not as strongly related to *bíos*, the concept of 'animated' life. The unanimated life of *zoē*, the 'vitality' of swarms that can only be created collectively, circumvents ontological definitions and is concerned directly with the relationalities of life within a swarm – a life that can be implemented technologically.[13] And thus swarms can serve as the object of a technically informed, cultural-theoretical history of media and knowledge that takes shape within the broader context of a theory and history of computer simulation.

By examining the treatment of swarms in the history of media and knowledge, this book traces their transformation from something existing outside of knowledge into a technically implementable form of knowledge around the year 2000. It attributes the connection between biological and technical knowledge – which allowed swarms to be reconceived as zootechnologies – to media-historical data, and it describes the genesis of swarms as productive collectives. What unfolds out of this, however, is something far more complicated than the popular narrative of swarm

11 For further reference to Jünger's *The Glass Bees*, see Bühler and Rieger, 74–75.

12 Thacker, *Biomedia*, 6.

13 On the distinction between *bíos* and *zoē*, see the informative overview by Karafyllis, 'Bios und Zoe.' For a discussion of swarm life in which the concepts of *zoē* and *bíos* have been slightly confused, see Horn, 'Das Leben ein Schwarm.'

intelligence, the dramatic arc of which always begins with fascinating natural phenomena, moves along to the dynamics of human crowds, and culminates in miniscule and blinking robotic collectives. What I present here is not simply a media history of swarm research since 1900, and certainly not the history of successive media-technological 'elucidations' of swarms, and the resulting application of transparent biological self-organizational capabilities in technical implementations. Rather, this media history crystallized in a reciprocal process in which biological phenomena, approaches, and aspects disrupted and informed technical phenomena, approaches, and aspects, and vice versa.

This book is therefore not an attempt to provide an ontological description of what swarms are or were or could be. Its aim is rather to analyze why, how, and in what manner particular dynamic collectives were understood as swarms at various points in time and in specific ways, and how these collectives were themselves able to become active in the production of this very knowledge. This sort of media history of swarm research or history of *swarm-becoming*, which investigates the respective media-technological conditions in which swarms were variously produced within specific descriptive frameworks, is thus embedded in the history of a particular form of knowledge itself. It so happens that the study of swarms has been inextricably linked to an epistemology of computer simulation.[14]

Over the course of my analysis, I have thus been less concerned with decoding the 'meaning' of swarms and their metaphorical dimension than I have been with understanding how they were (media-) historically produced as objects of knowledge at various points in time. The historical framework of my study extends from around 1900 to 2000, and thus follows an epistemic arc in which swarms shifted from being outside the realm of knowledge to being within the sphere of scientific engagement as attempts were made to address

14 This book thus formulates a unique approach to the 'object' of swarms that is based on media technology and the history of knowledge. Since I first began working on this project, a few cultural-theoretical studies have been published on the topic, most notably Eva Horn and Lucas Gisi's anthology *Schwärme – Kollektive ohne Zentrum* (2009), which analyzes the place of swarms in the history of knowledge alongside other collectives such as crowds and networks. The latter book contains an essay of my own that presents a condensed version of the arguments presented here. It also contains a German translation of Eugene Thacker's comprehensive discussion of the political dimensions of swarms in contrast to networks, and outlines their respective genealogies. For the original English work, see Eugene Thacker, 'Networks, Swarms, Multitudes,' *CTheory* (18 May 2004), http://www.ctheory.net/articles.aspx?id=423 (accessed 27 December 2017). For a work that has much to say about 'social swarms' (but does so without the perspective of the history of media and technology), see Brandstetter et al., eds., *Swarm(E) Motion: Bewegung zwischen Affekt und Masse* (2007), which I have already cited above.

them as *objects of knowledge* within the media-technological classifications of biological research. Thus, their transformation into a *figure of knowledge* was in turn reflected in their media-technologically operational applications. As a concept, an *object of knowledge* is rather precarious and can be approached with various epistemic strategies – strategies based on theories, experiments, media-technological observations, models, or computer simulations. Objects of this sort are thus themselves always subject to modulations and shifts. As objects of knowledge, swarms can thus only be produced and formed in various epistemological contexts in a specific way.[15] The term *figure of knowledge* goes back to Benjamin Bühler and Stefan Rieger, who have used it to formulate a non-traditional perspective on the relationship among humans, animals, and technology: "Animals view the human being or, to be more precise: scientists view the human being through the eyes of animals and what they see are the deficits or deficiencies not of the animal but rather of the human. [...] With the figure of knowledge of the animal, the argument is liberated from mere biologism and expanded into a figure of thought [...] whose venue is the modern order of knowledge itself."[16] What the authors regard as a sort of casuistry practiced by individual species can also serve as a way of looking at the media history of swarms. A media history in which swarms are suddenly conceived as *system animals* and are (media-) technically implemented to solve human problems must take into account a scientific and theoretical dynamic in which knowledge and technology are bound together in an intricate manner.[17] In such a way it is possible to capture the recursive connection between the biologization of computer technology and

15 It is fundamentally questionable whether swarms can be designated as 'objects' at all. In light of their ephemeral nature, their oscillation between individual interconnections and global movement, and their inherent disruptive moments, I will also associate them below with certain 'flexible' concepts that are meant to suggest this unfixability. I will thus speak of 'objects' on the same level as non-objects, non-things, or half-things (the latter was Leonardo da Vinci's term for ephemeral objects such as clouds). My use of the term *object of knowledge* owes much to Hans-Jörg Rheinberger's influential concept of the *epistemic thing*. The latter, according to Rheinberger, are entities that constitute the object of scientific inquiry; they are not necessarily objects in the strict sense but can also be structures, reactions, functions, and so on. They can be characterized as discourse objects that, by interacting with the technical things of experimental systems, describe a vague and processual 'discovery context' on the threshold of non-knowledge. See Rheinberger, *Toward a History of Epistemic Things*. In my third and fourth chapters I will discuss at greater length whether the concept of the epistemic thing is adequate for describing swarms.

16 Bühler and Rieger, *Vom Übertier: Ein Bestiarium des Wissens*, 9. All translations from works originally published in German are by Valentine A. Pakis.

17 See ibid., 10. On the concept of the *system animal*, see Von der Heiden and Vogl, 'Einleitung,' 7–14.

the computerization of biology that stands at the heart of the transformation of swarms described in this work. With differing levels of success, they were removed from a sphere of non-knowledge and transformed, as objects of knowledge resulting from their later technical applications, into figures of knowledge within an episteme of computer simulation.

This examination of swarms in terms of their place in the history of media and knowledge is thus characterized, first, by the search for adequate medial approaches to a 'body without surfaces' that demarcates, by means of its subtlety and turbulence, the boundaries of central-perspective codes. Swarms are four-dimensional collectives; they exist in a constant dynamic that unfolds in three-dimensional space as well as – and this is both their key feature and the main media-technological problem of swarm research – in an inscrutable dimension of time. On the basis of this characterization, the scope of the present project has been limited to investigating flocks of birds and, primarily, schools of fish. Other related and relevant collectives in the discourses concerned with swarm intelligence, such as social insects, will not enter my discussion because of their vastly different communication structures (e.g., pheromone traces, dance language, stigmergy in the construction of honeycombs), their different relations to topology (e.g., by referring to an architectonic and individual center – that is, by referring to a particular structure and the 'queen' as a 'reproductive organ' or by internally differentiating different 'casts' to perform various collective functions), and their consequently different orientation toward space and time.[18]

Second, swarms exhibit a mediality of their own. They can be described as relational ensembles whose relatively simple individuals possess only a limited amount of knowledge about their environment and organize themselves decentrally – that is, without any overarching authority – by interacting with their nearest neighbors. Despite this simplicity, they are capable of performing complex feats of coordination and they can adapt systematically and quickly to disruptions. Swarms exhibit emergent manners of behavior that cannot be attributed to the faculties of their individuals, and they can reorganize themselves adaptively and continuously in relation to changing environmental conditions. An adequate understanding of the mediality of swarms was not gained, however, until around 1980, when researchers first began to apply their principles to computer simulations in order to reproduce swarms' ability to self-organize by means of dynamic

18 For media-theoretical discussions of insect collectives, see, for instance, Johach, 'Andere Kanäle'; Parikka, *Insect Media*; and the studies collected in Harks and Vehlken, eds., *Neighborhood Technologies*.

models. At the same time, this conversion of quantity into new qualities also made the principles of swarms a matter of interest as the programming paradigm of so-called *computational swarm intelligence*. The latter operates with biologically-inspired software models that, unlike formalistic programming approaches, take into account potential losses of control in order to improve our understanding of contingent phenomena in the real world.

In this light, a third thesis can be formulated that is of interest to the study of media culture – that is, to the study of swarming as a cultural technique.[19] For, although descriptions of swarms have existed since antiquity, swarming in the sense of a cultural technique did not originate until the 'media-becoming' (in Joseph Vogl's terms) of swarms as 'intelligent' zootechnologies.[20] Along with this transformation, however, the concept of swarming was also fundamentally transformed – namely as a consequence of media-technological processes. Only a media-becoming could enable swarming to appear as a cultural technique. As much as possible, moreover, this media-becoming delegated the fundamental cultural techniques of image-making, writing, and calculation to automated and computerized processes, be it in the form of new object-oriented programming languages or for the sake of presenting transactional data on graphical user interfaces. Thus, within recursive chains of operation, swarm principles do not only participate in their self-description within the field of swarm research; they also coauthor processes within our culture of knowledge. They appear in economic simulations and models of financial markets, in simulations of social behavior, in simulations of crowd evacuations, and in the field of panic studies. They have become essential to epidemiology, to the optimization of logical systems, and to transportation planning. They are used to improve telecommunications and network protocols and to improve image and pattern recognition. They are a component of certain climate models and multi-robot systems, and they play a role in the field of mathematical optimization. What swarming, in its technologized and radicalized form, brings to the field of culture (or cultural techniques) is a fundamental element of culture in general: it is a dynamic structure, a topological system of inter-individual communication that has deeply affected the governmentality of the present.

19 For general introductions to the concept of cultural techniques, see Winthrop-Young, 'Cultural Techniques: Preliminary Remarks'; Maye, 'Was ist eine Kulturtechnik?'; and Siegert, *Cultural Techniques*.

20 See Vogl, 'Becoming-Media.' The term used in Vogl's original article is *Medien-Werden* 'media-becoming' (see 'Medien-Werden: Galileis Fernrohr').

All three of these aspects of a media history and knowledge history of swarms thus culminate in the foreground of a comprehensive epistemology of computer simulation. As mentioned above, the simultaneous biologization of computer science and computerization of swarm research have made it possible to think of new zootechnical connections that are not based on metaphorical transferences but rather on fundamental logics of function and control. These do not only exist *in vivo* but can also be implemented *in silico*. In terms of media history, they hinge around the question of how swarms function as 'self-organizing' multiplicities with 'emergent' features. What is of interest here, above all, are computer simulations governed by agent-based processes – which are in turn informed by biological swarms. It is only by passing through computer technology that swarms have been able to become media and operational figures of knowledge. And it was this transformation from fish into chips, so to speak, that has made it possible to speak of 'intelligent collectives.'

This book is divided into six parts. Under the chapter headings 'Deformations,' 'Formations,' 'Formats,' 'Formulas,' 'Transformations,' and 'Zootechnologies,' it brings together elements of a media-technologically informed history and epistemology of swarm intelligence. Under the guiding concept of *deformations*, the first chapter is an attempt to make the phenomenon of swarms productive for the formulation of media theory. Here swarms are treated as a materialization of the figure of the 'parasite,' as conceptualized by Michel Serres. To conduct swarm research is to study disruptive potential, and thus the field has yielded new information in the context of a comprehensive media theory of interference. This includes certain methodological insights that make the history of swarm research, as part of the winding road of 'media-becoming,' productive for concepts of media historiography that are oriented toward material cultures. The chapter closes by tracing the epistemological and cultural-technical expansion of the zone affected by swarms: the conversion of the swarm as an object of knowledge into an operative figure of knowledge was accompanied by a general shift in epistemic strategies, to the extent that self-organizational phenomena came to be applied to the study of unanalyzable problems, complex interactive processes, and inaccessible spheres of knowledge.

Concerned with *formations*, the second chapter is devoted to historical scenes in the development of behavioral biology around the beginning of the twentieth century. The latter discipline systematized knowledge about multiplicities by relying on physical instead of social models of interaction. Each of the texts discussed in this chapter was intended to formulate an explicitly non-anthropological and non-anthropocentric approach. Unlike, for instance, the discourse of mass psychology around the year 1900, behavioral biology no longer attempted to understand dynamic animal collectives from a human perspective.[21] Viewed now from a genuinely 'biological' perspective, animal collectives were disassociated from such things as 'society' and studied in terms of the 'systemic' nature of their inter-individual behavior. Techniques and media for gathering data about animal collectives thus gained a new degree of relevance, given that the human sensory apparatus could perceive little more than noise in the collective motion of swarms and that the traditional systems for recording information (diaries, hand-written observations) could not deal with the abundance of data. This period, moreover, was marked by increased self-reflection, as field researchers began to problematize their position in relation to their objects of study.

The umbrella term *formats*, which is the title of my third chapter, is meant to denote those developments which, beginning in the late 1920s, enabled swarms, as oppositional objects of knowledge, to become objects of investigation within the technically enhanced media history of biological swarm research. By reviewing behavioral-scientific publications from the field of fish-school research, this section of the book is concerned with the various attempts that were made to gain quantitative and formalizable access to the school as an object of knowledge. The goal of these studies was to describe the factors and functions behind the ability of schools to self-organize without any central authority. Over the course of these investigations, efforts were made to record schools with optical media in a variety of experimental systems, in biological research aquaria, and in the open sea. In the open sea, too, researchers tried to make schools visible by means of innovative diving techniques and sonar technology. Whether working in laboratories or in open bodies of water, researchers only began to approximate the opaque control processes of schools by retreating from nature and employing a variety of media-technological arrangements in their experiments. Within such ensembles, schools were delimited and described as specific (and always different) 'media cultures.' Again and again, however, disruptive forces came into play that interrupted the acquisition

21 See Galton, *Inquiries Into Human Faculty and Its Development*; Espinas, *Des sociétés animales*; De Tarde, *Penal Philosophy*; ibid., *The Laws of Imitation*; ibid., *L'Opinion et la foule*; Sighele, *La folla delinquente*; Bechterev, *Suggestion und ihre soziale Bedeutung*; Le Bon, *Psychology of Crowds*; Borch, *The Politics of Crowds*; Gamper and Schnyder, eds., *Kollektive Gespenster*; and Stäheli, 'Protokybernetik in der Massenpsychologie.'

of data or distorted the scientists' findings. For the technical recording media, the collectives themselves became data drifts on account of their multiple and simultaneous movements, and the environmental medium of water further concealed their control logic. Empirical research thus found itself mired in a 'technological morass.'

On the basis of this patchy empirical data, attempts were nevertheless made to construct mathematical models concerned with their geometric form or with the algorithms of the local behavior of swarm individuals. Under the term *formulas*, the fourth chapter investigates such complementary strategies for describing the dynamics and functions of biological collectives. It thereby follows traces that link biological swarm research to cybernetic ideas of 'communication' or 'information transmission.' Equipped with a new technical vocabulary, researchers began to describe swarms as 'systems' and were able to conceive of them in new ways. They were no longer regarded as an aesthetic problem but rather as *information machines* (in Serres's terms) that, operating on the basis of simple rules, could maneuver, coordinate, and adapt to external influences in a complex manner. Nevertheless, the first attempts to *simulate* swarm dynamics, in the 1970s, received little attention, a fact that was likely due to the inability of researchers to display dynamic processes visually over time.

Whereas the media-technological observational and experimental systems analyzed in my third chapter functioned above all to *suppress noise*, and the mathematical-geometrical models of the fourth chapter to a large extent *ignored irregularities* in school structure, the adequacy of computer simulation models, which is the subject of my fifth chapter, often depended on embedding and implementing moments of interference at appropriate levels of intensity and effectiveness. Concerned with the general concept of *transformations*, this chapter focuses on biological studies that, beginning around 1980, were increasingly informed by digital media – studies that experimented, for instance, with computer-supported data processing or made use of agent-based computer simulation models. In the latter, which were first employed by researchers working for Japanese fisheries, interference and noise were made operational and productive for setting the parameters and tuning the dynamic models themselves. Interference and noise, that is, acquired a *constitutive function*, and an epistemology of computer simulation enabled the opaque processes of self-organization to be addressed in a new way. It was also the case that biological knowledge about swarms made its way into the programming routines of computer science. As mentioned above, it is possible to speak of a productive chiasmus involving the simultaneous computerization of biology and biologization of computer science. Along with more recent texts from the field of biological swarm research, publications devoted to computer graphic imagery (CGI), agent-based modeling, and robotics make it clear that swarm research has come to rely heavily on digital visualization processes that productively employ precisely those disruptive functions of swarms that had baffled earlier experimental systems and methods of observation. When graphic animators in the film industry make use of swarm principles to simulate efficient dynamic collectives, they are simultaneously writing programs that provide biological swarm researchers with an entirely new and intuitive way of approaching their object of study. It is only in the computer-supported epistemology of simulation – along the epistemological third way between theory and experimentation, and especially on the related level of visual syntheticizations – that the swarm has been able to come into its own by transforming from an object of knowledge into a (computer technologically-implemented) figure of knowledge.

Thus, the media-becoming of swarms has entailed their transformation into *zootechnologies*, which have become fundamental cultural techniques for understanding and governing dynamic processes. My sixth and concluding chapter explores four decisive areas where swarm-intelligent applications have recently been deployed. First, it discusses the development of Unmanned Aerial Systems (UAS) or drone swarms. The leading hypothesis of this part is that these create a multifold 'spatial intelligence' that ranges from the dynamic morphologies of such collectives via their robust self-organization in changing environments to representations of these environments as distributed 4D-sensor systems. As is shown on the basis of some generative examples from the field of UAS, robot swarms are literally imagined to penetrate space and control it. In contrast to classical forms of surveillance, or even 'sousveillance,' this procedure could be called *perveillance*. The second part examines the dissemination of 'swarm-intelligent' applications throughout different scientific disciplines. With this focus, it highlights the importance of a variety of agent-based modeling toolkits and code libraries. The third part investigates the impact of 'swarm intelligence' on the field of architectural thinking, design, and construction. It discusses attempts to conceptually exploit swarming for architectural theory and analyzes modes of employing agent simulations for architectural design and urbanism. Finally, the fourth part turns towards the research field of crowd simulation and crowd control. Here, agent-based simulation models are used to 'calculate disaster' by modeling and thus 'pre-mediating' the dynamics of human crowds, thus turning traditional concepts of 'the mass' upside down. In all these cases, 'swarm logic' has made it possible to adapt to unclearly delineated sets of problems and clarify the operation of opaque systems. They extend the limits of what can be calculated and offer performative, synthetic, and approximate solutions in cases where analytical approaches are doomed to fail.

The history of swarm research and intelligence thus proves to be a complex interplay between epistemic, technical, aesthetic, and research-practical aspects that fluctuate beyond disciplinary boundaries, and it is the aim of this book to trace and determine their coordinates in the history of media, technology, and knowledge.

I. Deformations: A Media Theory of Swarming

Abstract

Under the concept of *deformations*, this chapter presents crucial aspects that link the phenomena of swarms to media theory. Here swarms are treated as a materialization of Serres's figure of the 'parasite.' By attending to disruptive potential, swarm research has yielded new information in the context of a comprehensive media theory of interference. This includes methodological insights that are productive for concepts of media historiography. The chapter closes by tracing the epistemological and cultural-technical expansion of the zone affected by swarms. The conversion of the swarm as an *object of knowledge* into an operative *figure of knowledge* by computer simulation signifies a general shift in epistemic strategies: self-organizational phenomena came to be applied to the study of complex interactive processes.

Keywords: media theory, parasite, noise, cultural techniques, computer simulation, epistemology

1. Theory: Noise

Amalgamations of Perplexity

"In the beginning was the noise."[1] This is not the opening sentence of Michel Serres's *The Parasite* (that would have been too prosaic). It rather concludes a paragraph in which he emphasizes the productivity of interference – the *bruit parasite*, as the French goes. Thus, instead of beginning this chapter with one hackneyed pronouncement or another about the fascinating nature of flocking birds, schooling fish, their tendency to 'hover' and 'dance,' and

1 Serres, *The Parasite*, 13.

Vehlken, S., *Zootechnologies. A Media History of Swarm Research.*
Taylor & Francis Group, 2019
DOI 10.5117/9789462986206_CH01

the apparently mystical beauty of such activity, I have decided to take the diametrically opposite approach. This is because a phenomenological or anthropological 'view' obscures the fact that, in the swarm itself, perception is always disrupted. Swarms shimmer in a field of tension between interference and organization whose discursive and historical dynamics are resistant to any subject-oriented or analytical perspective.[2] Or, in Serres's words:

> We are fascinated by the unit; only a unit seems rational to us. We scorn the senses, because their information reaches us in bursts. We scorn the groupings of the world (things like 'a flight of screaming birds,' 'a cloud of chirping crickets,' 'crowds, packs, hordes on the move') [...]. Disaggregation and aggregation, as such, and without contradiction [...] are repugnant to us. [...] We want a principle, a system, an integration, and we want elements, atoms, numbers. We want them, and we make them. A single God, and identifiable individuals.[3]

Things are entirely different with swarms: visible from afar as just a diffusely coherent dynamic of motion, when viewed in close proximity their whirring and teeming quickly exceed not only the capabilities of human perception but also the capacities of visual recording media. An occurrence that should be transmitted sensually or through media is here overtaken by the occurrence of its own transmission.[4]

In swarms it is possible to see a precedent of the medial when the act of 'swarming' comes between the observer and the swarm, and when the object of interest distrupts and impedes its own objectification. When Serres places noise at the beginning of things, then this is nothing less than, in Bernhard Siegert's words, "the beginning of media theory, of every media theory."[5] For it is only through the act of suppressing noise that mediality truly begins. Swarms address media-theoretical questions in an entirely concrete manner in that they can be understood as the *materialization of noise and disruptive moments*. The media history of swarm intelligence is thus simultaneously a contribution to both a media theory of interference as well as a disruptive theory of media.[6]

2 See Foucault, *The Order of Things*, xv.

3 Serres, *Genesis*, 2–3.

4 See Vogl, 'Gefieder, Gewölk,' esp. 147.

5 Siegert, 'Die Geburt der Literatur aus dem Rauschen der Kanäle,' 7.

6 For recent discussions of disorder and interference in media theory, see Kassung, ed., *Die Unordnung der Dinge*; and Kümmel and Schüttpelz, eds., *Signale der Störung*.

In fact, noise does indeed stand at the beginning: swarms begin with a hum, a whir, a din of multiplicity. The German word *Schwarm*, according to the Brothers Grimm, derives from "a formation found at first only in West Germanic that goes back to a root whose cognates include the verb *schwirren* 'to buzz, hum' [...] and later Norse *sverra* 'to murmur, teem, swarm.'"[7] In order to describe a swarm conceptually, a detour has to be taken through the ear. Johann Christoph Adelung defined a swarm as a "disorderly heap of living things making a muddled noise" and as "the muddled noise of a disorderly multitude."[8] Whoever is confronted with a swarm never really sees it; its visual definability is subsumed beneath its acoustics. The multiplicity of swarms is audible, and this audibility seems to be more characteristic of the conceptual definition of swarms than their confusing appearance. Etymologically, this definition is associated with a sound whose proximity to mere noise cannot be denied: "Swarm, swarming multitude. a) In its earliest attestation, the word is related to a buzzing swarm of bees [...]. b) In later uses, however, it also refers to swarming multitudes of various sorts, clearly formed on the basis of the stricter meaning of a swarm of bees."[9] A conceptual history of the swarm and of swarming thus derives genealogically from swarms of bees and the technical vocabulary of apiarists before later being applied to swarming multitudes of other types of insects, to schools of fish,[10] or even to crowds of people.

Beyond the ear (as a sound having become a word), the concept refers to a fluid and permeable boundary between the notions of multiplicity and unity. The sound of buzzing points to the impossibility of ascertaining and locating swarms both at the level of the individual and at the level of the collective. The auditory nature of swarming integrates these two levels into a *single* sound of multiplicity. As an acoustic phenomenon, this sound of multiplicity confers to swarming a necessary temporal dimension. It captures swarming's ephemeral and diffuse character without fixing it. In this most fleeting and abstract of acoustic forms, swarms can be identified without simultaneously having to be reduced to the level of individuality or collectivity. The attribution of this sound of multiplicity – this buzzing – points beyond the auditory senses to the teeming activity of the swarm and thus instantiates it as a sort of visual noise. The acoustic

7 Grimm and Grimm, *Deutsches Wörterbuch*, vol. 15, s.v. (col. 2283).

8 Adelung, *Grammatisch-kritisches Wörterbuch der hochdeutschen Mundart*, vol. 3, s.v. (cols. 1715–16).

9 Grimm and Grimm, *Deutsches Wörterbuch*, vol. 15, s.v. (col. 2284).

10 See Maaler, *Die teütsch Spraach*, 366; quoted in Grimm and Grimm, *Deutsches Wörterbuch*, vol. 15, col. 2284.

nature of swarming is reflected in the impossibility of representing it in images; here, swarms can and will only be vaguely defined somewhere between sound and noise.[11] While reading the remainder of this chapter, readers should try, at least, to keep the acoustic connotation of swarming in the back of their mind. This is because its ephemeral and indeterminate auditory nature makes it clear that *swarming* and *the swarm* should be understood in an integrative manner. Phenomena of an acoustic noise that is never completely random converge with the signature features of an identifiable swarm sound – a combination that initially seems rather difficult to comprehend and reproduce with visual strategies. This, in turn, evokes one of the central media-theoretical problems that swarms and swarming cause us to consider.

The media theorist Ramón Reichert, for instance, has asserted that *swarming* never takes place *within the swarm*. According to Reichert, swarming is external to any possible attribution of meaning: "That which *swarms* does not communicate and does not wish to be received; it does not offer itself to be read or understood, and it does not ask to be evaluated." Swarming, he goes on, has to be interpreted as an "infinitive without a subject" and should thus be kept distinct from the *swarm*. The latter is characterized by the act of integrating and then disintegrating constellations. By generating particular formations and economies of behavior, swarms are predictable "because, in their activity, the animals adhere to orders that are intelligible and valid to human beings." These orders, according to Reichert, become perceptible when one surveys a swarm as a whole and gains an overview of its unitary form and varying figures: "In the typology of its behavioral patterns, *swarming* within the swarm appears tangible as a particular form of perception [...]. In the interest of measuring it, it should be definable as an integrated *mass*." This subordination of swarming to the "rule of the number" and to certain spatio-temporal assumptions is thus also, as Reichert sees it, an illegitimate act of subordinating swarming's inherent subversion of such "spatial orientations as directions, distances, and positions" to the "wholeness" of the swarm.[12]

This analysis, however, falls short in three respects. *First*, it neglects the media technologies that make it possible to view the 'whole' of swarms in

11 In the acid-house track 'Swarm' (2007) by the Canadian DJ Richard Hawtin (Plastikman), for instance, the acoustic multiplicity tremulously and unsteadily enters the stereo space in order to fill it up only partially and for brief moments. The sounds of the song seem to be in constant motion, tugging back and forth and oscillating between the left and right audio channels.
12 Reichert, 'Huschen, Schwärmen, Verführen.'

the first place. A notion of swarms as a segmentable form can be dated in media-historical terms; the concept has changed over the years, and thus so has the understanding of what swarms are. *Second*, since at least the late 1940s – and especially since the publication of Claude Shannon and Warren Weaver's *Mathematical Theory of Communication* – it has been possible to understand medial problems as problems that do not involve the senses at all.[13] According to this interpretation, swarming is simply the source of interference whose negation enables the medial process of transmission and provides access to the swarm. And *third*, Reichert's argumentation is caught up in the dichotomy of the whole and its parts, of collectivity and individuality. Swarms, however, should rather be regarded as *heterogeneous* collectives and, beyond that, as open systems that never result in a whole entity. Swarms are never 'complete'; their size is potentially unlimited. In many respects, moreover, they maintain a relationship with their environment. It is impossible to account for swarming on the basis of a swarm's wholeness because the 'swarming' individual elements do not come together into a whole but rather constitute a relational entity with *nonlinear* interactions. Swarms are not a collection of swarming components; rather, they are an expression of a sort of 'motion intelligence' that consists of instantaneous and situation-dependent interactions on the local level of *swarming* and the feedback effects caused by the global changes of the *swarm* on these local interactions.

Swarming cannot simply be deduced from the swarm because swarms are not the cause of swarming, but rather living networks that lend a theory and praxis of complex systems to mechanistic explanatory models. As Eugene Thacker has noted, swarms exhibit recognizable global patterns, but this does not mean that the collective has primacy over the individual, or vice versa. Swarms exist on a third level where multiplicity and relations converge.[14] Both swarming and swarms – neither is conceivable without the other – negate identities and engender vagueness and opacity in regard to systems of reference and operational regulations; they make it impossible to identify, projectively, the human in the animal and they involve, as Reichert has rightly observed, "amalgamations of perplexity" because they are resistant to perceptual synthesis, mediation, and transmission.[15] Given that they had yet to be operationally differentiated at the beginning of the twentieth

13 For the earliest publication along these lines, see Shannon, 'A Mathematical Theory of Communication.'

14 Thacker, 'Networks, Swarms, Multitudes,' n.p.

15 Reichert, 'Huschen, Schwärmen, Verführen,' n.p.

century, a media theory and media history of swarms instigates a process of "sharpening the sense of possibility," in Bernhard Siegert's terms,[16] and this process ultimately leads to an epistemology of computer simulation.

Swarms act in a confounding manner. They rebuff any rational approach and yet they nevertheless operate self-reflexively. By producing signals of disruption, they cast the medial 'in-between' back onto its own mediality. What is transmitted is not messages but rather the event of transmission itself, as it focuses the ear and the eye on the presence of channels, on the locus of the medial[17] – on the space of difference that makes any definition of media so precarious.[18] And yet it is precisely these incipient amalgamations of perplexity in which, as of 1900, a knowledge of humans and swarming animals was renegotiated under media-technological conditions. These are blends of perplexity that provide hints about their medial basis in a "mixture of noise, clouds, and feathers"[19] – a mixture in which the acoustic level of swarms is combined with a visual noise.

Bodies without Surfaces

Swarms make manifest a precedent of the medial. That the two media events of interference and transmission go hand in hand is only seemingly paradoxical; it is, in Joseph Vogl's words, "a confusion in which the represented events are overtaken and intersected by events of representation, the transmitted occurrence by the occurrence of transmission."[20] The medial nexus of acoustic and optic disruptive phenomena in swarms can perhaps be made more explicit with reference to Alfred Hitchcock's *The Birds* (1963). Vogl has described how this film carries out a "program of perception" that operates both on acoustic and visible levels. Thus, at the very beginning of the movie, the birds come close together and withdraw from one another in a "wave of vibration"[21] that consists of electronic sounds (composed by Oskar Sala), which do not imitate the natural sounds of birds. This "demarcation of markedness" upends the order and locus of acoustic impressions to which listeners are accustomed.[22]

16 Siegert, 'Kakographie oder Kommunikation,' 91.

17 See Vogl, 'Gefieder, Gewölk,' 147.

18 See Mersch, *Medientheorien zur Einführung*, 9.

19 Vogl, 'Gefieder, Gewölk,' 143.

20 Ibid.

21 Truffaut, *Hitchcock/Truffaut*, 327; quoted (from the German edition) in Vogl, 'Gefieder, Gewölk,' 143.

22 Ibid., 143.

On the visual level, focus is placed on a relation between the natural optic space and its dissolution. Ever since the Renaissance, this relation has been discussed as that between central perspective and non-representable objects.[23] According to Leon Battista Alberti's work *De Pictura* (1435/36), a Renaissance painter should only attempt to depict that which he is able to see: things that he cannot see should be of no concern to him. Thus, the system space of central perspective can only accommodate those things that can be translated into its order – things that, as Hubert Damisch has noted, can be located and whose contours are defined by lines.[24] Such features do not apply to swarms. Swarms rather belong to a category of 'bodies without surfaces,' in Leonardo Da Vinci's terms. This is a category of bodies without clear forms and edges, bodies whose boundaries overlap and blend into one another: "Visible bodies are of two kinds, the first without shape or any distinct or definite edges, which though present are imperceptible. [...] The second kind of body is that whose surface defines and distinguishes the shape. The first kind is that of fluid bodies like mud and water, mist and smoke [...] whence by this intermingling their boundaries become confused and imperceptible, for which reason they are found without surface since they enter into each other's bodies."[25]

In Filippo Brunelleschi's famous mirror experiment, this problematic situation was addressed with the example of clouds. These objects, which cannot be represented according to the system of central perspective, were reflected into margins of the painted image directly from the actual sky surrounding it. They *occur* within the image. The opening made through the surface of the picture thus demarcates the boundaries of central perspective's system space: a space that can accommodate objects, planes, volumes, and contours but has no room for bodies without surfaces. As dynamic objects, the latter oppose geometricization and represent, within a chaotic realm of data, "the inability for there to be objects at all, the inability for there to be objects that can be experienced empirically."[26] To this extent, clouds, swarms, and other 'fuzzy objects' are simultaneously visible and invisible. They are perceptible only as chaotic imperceptibilities and can only be experienced as something whose substance always remains undefined under these conditions.

23 See ibid., 144.

24 Damisch, 'Die Geschichte und die Geometrie,' 11.

25 Da Vinci, *Codex Atlanticus*, 132rb; quoted, with light modifications, from Keele, *Leonardo Da Vinci's Elements of the Science of Man*, 100.

26 Vogl, 'Gefieder, Gewölk,', 145.

In later years, artists became increasingly interested in representing such forms of data drift. With the work of William Turner, as Michel Serres has demonstrated, the field of painting can be said to have entered its thermodynamic age. Turner's paintings reflect the scientific 'transformational motor' of Carnot's cycle and his formulation of thermodynamic principles:

> Turner understood and revealed the new world, the new matter. *The perception of the stochastic replaced the art of drawing the form.* Matter is no longer left in the prison of diagram. Fire dissolves it, makes it vibrate, tremble, oscillate, makes it explode into *clouds*. [...] [F]rom the fibrous network to the hazardous cloud. No one can draw the edge of a cloud, the borderline of the aleatory [...]. On these totally new edges, which geometry and the art of drawing have abandoned, a new world will soon discover dissolution, atomic and molecular dissemination.[27]

Yet the arts were not alone in being affected by data drifts; the everyday realm of perception escaped from its Newtonian confines and transformed into revolutions of energetic processes. In this regard, bodies without surfaces number among those "nervous geometries" that, as Christoph Asendorf has shown, appeared over the course of industrialization.[28] In the chaotic hustle of nineteenth-century urban society, according to Asendorf, the eye lost its overview. An event could no longer be understood as something centered on its observer. Rather, the observational system itself had shifted in various respects: observers became a part of that which they were observing. "Perception," as Asendorf noted, "was no longer oriented toward a visual axis emanating from a single point [...] but was rather organized according to the model of the field in physics. Variable amounts of stimulation, which affected various senses from all sides, were assigned to every point within a spatial area."[29] In this observational order (or better: observational *disorder*), subject and object are in a constant state of exchange. Georg Christoph Lichtenberg referred to this relation as a "storm," in which the geometrical model was replaced by an energetic model of perception.[30] New types of motion entailed new forms of perception. The act of riding on a train, for instance, opened Victor Hugo's eyes to the second dimension – to perceiving things along a line. And finally, air travel has enabled the third dimension of

27 Serres, *Hermes: Literature, Science, Philosophy*, 58 (emphasis original).

28 See Asendorf, *Ströme und Strahlen*, 119.

29 Ibid., 120.

30 Ibid.

spatial movement and opened our eyes to see things in all of the directions in which swarms orient themselves. This resulted in an irritating three-dimensional experience of space: "In these circumstances, the universe, as we have increasingly come to imagine it, is no longer Euclidean; rather, it corresponds to the geometry of Lobachevsky or Riemann. [...] If we attempt to draw a general principle from these things, we come to understand that, today, the Euclidean straight line [...] is less real than the curved lines of non-Euclidean geometry."[31]

The Birds depicts a similar excess of swarming's atomistic events, which relate back to the existence of swarms – events that, being random and unpredictable, defy the geometric and acoustic patterns that create meaning: "Before these birds tell us anything and beckon interpretations – before they can become a metaphor for family drama – they are simply there, flying and whirring, occurring and happening. Before these birds are symbols or questions, they are thus events in the film."[32] In their whirring and rustling, they lose their identifiability as individual exemplars; in their teeming and fluttering, they happen, by means of de-differentiation, "as an optical and acoustic multiplicity in which every individual element, diving down from the sky, dissipates at once into the swarm and its swarming."[33] And thus, above all, these birds defined the existence of the relations themselves; in a parasitic manner, "they make the transmission itself – not the transmitted – into an event."[34]

In such a storm of happenings, swarms moved not only into the thematic realm of 'critical' masses at the end of the nineteenth century but also into the ambit of physical models and processes of collective particle behavior.[35] This, in turn, led to the aforementioned epistemic horror of that which cannot take shape. Hitchcock's birds are events in precisely those situations in which they become the atmospheric environment for the people in the film – both on the visual and acoustic levels of sensation – and they shatter their observers' frame of perception. Thus, as pre-sensual transmission events, they stand between the observer/audience and every transmission of information or meaning. What is seen and heard is, above all, seeing and hearing itself – disrupted by the construction of a 'something,' an object of seeing and hearing. The something is that which comes between: the

31 Siegfried, *Aspekte des 20. Jahrhunderts;* quoted from Asendorf, *Ströme und Strahlen*, 121.

32 Vogl, 'Gefieder, Gewölk,' 142.

33 Ibid., 143.

34 Ibid., 147.

35 On the connections between thermodynamics, the discourses about masses, and philosophy, see Schäfer and Vogl, 'Feuer und Flamme.'

birds themselves.[36] In Michel Serres's sense of the term, the swarming of the birds behaves parasitically by establishing the priority of noise over the message and the primacy of the channel over the transmission. Here, the tripartite communicative schema of the parasite places the interferences of flocking birds in between the message of the movie and the viewing apparatus consisting of the camera, the sound, and the audience. In this regard, *The Birds* is a prototypical example of the turmoil and confusion that swarms create in two respects: first in the realm of human perception, which they disrupt, but also for technical media of observation (here the movie camera) and their perspectival and narrative spaces, which they encounter as something that defies geometry, opposes representation, and blurs the line between locations and orders.

It can be supposed that the program of perception that Hitchcock's film problematizes can also be transferred to other medial, instrumental, and technical constellations – for instance to arrangements in which swarms are meant to be studied as objects of knowledge (for instance in biology). To explain the dynamics and dynamic orders of biological swarms, it is necessary to relate the pre-empirical space that their swarming represents to a sensual approach toward their underlying manner of functioning – to what is known about their system properties. In such a scientific description of swarms, made with the aid of media-technological apparatuses, the swarming of swarms always intermits. Non-knowledge about the specific logic of swarms runs up against the reflexivity of media-technological attempts to model them in a dynamic process involving interference and its suppression. Here, a media history of swarm research becomes an analysis of various media technologies and experimental systems, and swarms become an object of knowledge in which 'fuzzy' definitions and 'irrational' behavior conflict with rational and empirical methods of putting things into focus and which involves new epistemic strategies for describing intransparent non-objects and opaque logics of control.

The Paradox of the Parasite

"The difference is part of the thing itself, and perhaps it even produces the thing."[37] This formulation indicates the problematic arena in which swarms and their swarming converge. Michel Serres did not think of deviations, noise, and interference as accidental, secondary, or supplementary processes

36 Vogl, 'Gefieder, Gewölk,' 146.

37 Serres, *The Parasite*, 13.

that are grafted onto an original and pure relationship between a transmitter and a receiver.[38] It is not first the case that there is "something 'as it really is' and then a defective image which forms the basis of our senses and understanding."[39] Serres's theory of communication does not begin with a bivalent situation in which a transmitter would like to convey a message to a receiver. Rather, as he argued as early as 1968, "to hold a dialogue is to suppose a third man and seek to exclude him; successful communication is the exclusion of the third man."[40] The original condition of communication is not, for instance, a process of unimpeded exchange but rather one defined by the parasite, and such is the case with respect to anthropology, economics, politics, and information theory.

The figure of interference precedes this relationship and is therefore its basis. The parasite – the noise – is the essence of every relation. In its familiar schematic representation, the stations S_1 and S_2 are not connected to a channel that a given parasite *P* tampers with as an equivalent to Shannon and Weaver's noise source. Rather, *P* indicates to us directly that *it* – the parasite – is the channel: "This or the third precedes the second: such is the beginning of media theory, of every media theory: 'A third exists before the second. [...] There is always a mediate, a middle, an intermediary.'"[41] This theory pushed noise and the physical materiality of the channel into the forefront of media-theoretical thinking, and the elimination of "cacography" – an act of suppressing noise – became the basis of any successful communication.[42] Serres underscores the foundational nature of noise yet again when he ventures to offer a more thorough definition of the concept of the parasite:

> Systems work because they do not work. Nonfunctioning remains essential for functioning. And that can be formalized. Given, two stations and a channel. They exchange messages. If the relation succeeds, if it is perfect, optimum, and immediate; it disappears as a relation. If it is there, if it exists, that means that it failed. It is only mediation. Relation is nonrelation. And that is what the parasite is. The channel carries the flow, but it

38 Bernhard Siegert has pointed out that Serres's concept of interference differs from that used in philosophy and the linguistic tradition. See his article 'Die Geburt der Literatur aus dem Rauschen der Kanäle,' 7.

39 Ibid.

40 Serres, *Hermès ou la a communication*, 67.

41 Siegert, 'Die Geburt der Literatur aus dem Rauschen der Kanäle,' 7–8. The quotation is from Serres, *The Parasite*, 63.

42 See Siegert, 'Kakographie oder Kommunikation.'

> cannot disappear as a channel, and it brakes (breaks) the flow, more or less. But perfect, successful, optimum communication no longer includes any mediation. And the canal disappears into immediacy. There would be no spaces of transformation anywhere. There are channels, and thus there must be noise. No canal without noise. The real is not the rational. The best relation would be no relation. By definition it does not exist; if it exists, it is not observable. This is the paradox of the parasite. It is very simple but has great import. [...] [It] is a noise of the system that can only be supplanted by noise. Thus noise – I am passing here from the human to the exact sciences; my discourse remains the same – thus noise is the fall into disorder and the beginning of an order.[43]

This collapse into disorder and the establishment of a new order provides a useful theoretical framework in which to consider swarms as objects and figures of knowledge. And this is not only the case in a study devoted to the various methods of suppressing noise but is especially so in light of the creative potential of disruptive events. The disorder of swarming and the order of swarms can thus, it is hoped, be integratively understood as explications of Michel Serres's media theory of interference. Swarms themselves should be regarded as materializations of productively applied disruptive events, and the media history of swarm research makes it clear that their dynamic and local information infrastructures are particularly well suited for them to coordinate themselves as collectives within 'disrupted conditions': "These strategies are not necessarily, or even typically, 'rational' or 'error-free.' Rather, units may make mistakes, may be prone to processing errors, and may need to rely on incomplete, possibly corrupted information. Surprisingly, simple strategies perform remarkably well in many experimental environments." This is in contrast to other interactive topologies for the collective behavior of distributed entities, which are able to function at an optimal level under 'ideal' environmental conditions but are far more susceptible to error under more 'realistic' conditions – namely in the presence of noise.[44]

For Serres (and others), the beginning of 'every' media theory is epistemologically based on Shannon and Weaver's *Mathematical Theory of Communication* and therefore in the fields of telecommunication and cryptographic communication, where the concept of noise was first used to describe interference in communication systems. In Shannon and Weaver's

43 Serres, *The Parasite*, 79.

44 Moreira et al., 'Efficient System-Wide Coordination in Noisy Environments,' 12085.

schema, interference is accorded a "systematic position," as Erhard Schüttpelz has noted. A consequence of this is that communication is (much as in Serres's work) "observable and operable as the negation *of its* negation, and otherwise not at all."[45] According to Shannon and Weaver, communication is systematically operable in that it is understood to exist within an order of interference, and thus this interference is productive. Communication now entails both the "suppression of interference" as well as the "interference of interference"; it is based, that is, on a fundamental operation of "disruptive potential."[46] Moreover, if the noise in Shannon and Weaver's model is interpreted as an additional source, then interference – as described above – can be understood both as *negation* and as a second *source* of the received signal.[47] "Interference," as Schüttpelz observed, "is now outside of and within the communication system, occupying a completely ambiguous position between being damaging and enriching – an ambiguity that has to be resolved, accepted, and reproduced in every act of communication (and in every theory of communication)."[48] Within such a systematic understanding of noise, attempts can thus be made to assign a place to apparent coincidences and individual disruptive events, and recurring patterns of interference can even be interpreted as signal sources.[49] In a relation of this sort, the line between signals and their interference becomes blurry. This priority of communication and noise makes the relations between transmitters, receivers, and observers dynamic; it lends ambiguity to their respective positions within the schema; and it causes their places to change on a continuous basis. It is this act of making relations dynamic within models of communication that Serres hoped to address and explicate with his notion of the parasite: "The guest becomes interrupter; the noise becomes

45 Schüttpelz, 'Die Frage nach der Frage, auf die das Medium eine Antwort ist,' 16.

46 Ibid.

47 In this regard, see also Kittler, 'Signal-to-Noise Ratio,' 169: "Communication (to use Shannon's language) is always 'Communication in the Presence of Noise' – and not just because real channels never do emit noise, but because messages themselves can be generated as selections or filterings of noise. Technical idealization, according to which the noise-laden output of networks counts as the function of two variables – of a signal input presumed to be noise-free and a separate source of noise – enables nothing more and nothing less than the specification of signal-to-noise ratios. In a first step, this interval indicates (on the basis of voltages, current, or power) only the quotient of medium signal amplitude and the initial degree of interference." Yet as soon as people are connected to these networks by means of technical interfaces, this ratio determines in the ear the difference between seemingly noise-free understanding ("albeit understanding that is not hermeneutical," as Kittler wrote) and unrecognizable sounds.

48 Schüttpelz, 'Die Frage nach der Frage,' 16.

49 Ibid.

interlocutor; part of the channel becomes obstacle, and vice versa. Questions: who and where is the third man? These questions have fluctuating answers, functions of noise, time, and of the new relations of equality or similarity between the terms. The same and the other change places with the third."[50]

Here, Serres's aforementioned schematic is further developed into a communication-theoretical wheel whose rotation sets in motion the position of the third – the parasite. This mobility also grants access to the place of the medium as a middle or intermediary. Under the conditions of noise and interference, the 'beginning of every media theory' can suddenly be everywhere. Moreover, this motion expands the originally triangular model into a network of dynamic relations.[51] Serres's reflections open up a number of promising ways to understand swarming and swarms. Even though his figure of the parasite is the protagonist in a theory of communication that operates, as noted above, on anthropological, economic, political, and information-theoretical levels, his central points can nevertheless be profitably applied to a media theory of interference whose field of inquiry includes swarm research. In terms of theory, this inherent operationalization of the concepts of noise and interference does much to clarify the swarm as an object of knowledge, its relationality, and its four-dimensionality. The following six points should be made to summarize my argument thus far: *First*, Michel Serres's concept of the parasite opened up a way to examine the media-material conditions of communication; *second*, this relation is only revealed in its failure – as a non-relation; *third*, disruptive events are conditions for the production of relations; *fourth*, the disorder of noise indicates the beginning of new orders; *fifth*, disruptive events have a relative character and they make rigid schemata of communication dynamic; *sixth* and finally, a theory of communication based on Shannon and Weaver is also reminiscent of the epoch-defining transition from energy to information that characterizes cybernetics.[52] A media theory of interference, which can be useful for a theoretical understanding of swarms, is thus based on the

50 Serres, *The Parasite*, 54.

51 See Serres, *Hermès ou a communication*, 11–20.

52 By applying Boltzmann and Gibb's entropy laws of statistical thermodynamics to communication technology, Shannon and Weaver reconceptualized information as a measure of the choices that can be made within a system. The greater a system's disorder, the greater its information as well. The transmission of a *particular* message consequently and likewise became a problem of probability: The calculation of transition possibilities between signals, for instance, serves to filter out interference, "which is nothing but undesired information, that is, a matter of uncertainty that is not subject to the transmitter's freedom of choice but rather to the influence of the interference's source." The quotation is from Pias, 'Zeit der Kybernetik: Zur Einführung,' 428–429.

functionality of disruptive phenomena, on the interchangeable nature of disruptive relations, on a description of rising and falling uncertainty, and on the interactive processes of dealing with disruptive events.[53]

With these aspects in mind, it is possible to oppose the initial amalgamations of perplexity with a systematic media-historical and epistemological program, at the heart of which is the investigation and description of the dynamic structures of informational exchange that exist within swarms. On account of their noise and motion, moreover, the latter require specific 'anti-disruptive' attempts to gain media-technological access to knowledge about swarm systems and the way in which they function. The issue here is not message transmission, in Shannon and Weaver's sense, but rather the relationships among observation, data transmission, and data processing. As objects of knowledge studied by biological swarm researchers, swarms *behave* in a particular manner. And the 'how?' and 'why?' of their behavior are questions to which answers have been found by means of various media-technological processes of 'decoding.' As already mentioned, the initial fascination with swarms seemed to lie in their simultaneous formlessness and coordination. In order to operationalize swarms as an object of knowledge, it was necessary to regard their swarming as a disruptive event that had to be subtracted from the equation. Yet swarming is part of the thing itself and has always been responsible, as will become clear over the course of this book, for the 'defective images' associated with swarms. Swarming designates those deviations without which swarms would be unthinkable. In addition to the phenomenon of swarming, researchers have also had to cope with the 'fundamental noise' of the natural world that swarms inhabit. Technologies for recording and modeling swarms always operate under the conditions of specific environmental media, such as the air or the sea, whose physical characteristics have to be taken into account as they attempt to focus on processes of constant interaction that are influenced by external biological factors. In these efforts to account for environmental influences, the disruptive events of swarming cause the research technologies in question to be confronted by their own materiality, by the existence of observational (and not communicative) channels, and by the attributes of modeling or simulation processes. As intransparent non-objects, swarms embody the eventful oscillation between interference and transmission, between noise and order – an oscillation that de-ontologizes our understanding of media and the medial.

53 See Kümmel and Schüttpelz, 'Medientheorie der Störung / Störungstheorie der Medien,' 10–11.

A media history of swarms is thus concerned with a dynamic 'object' that can only be understood in terms of processes. In light of the concept of media outlined above, this book examines how different media-historical and media-technological contexts (and their specific conditions) have always led to new and different understandings of swarms. Such an approach also requires, however, that the swarming of swarms be understood not only as a disruptive event in the relation between swarm collectives and the technologies used to model them but also as a reaction of swarm collectives to interference from their own environment. From this angle, the way in which swarms organize themselves is always a response to disruptive phenomena; as a global structure, they adapt according to particular rules to external and local random events in a continuous and noisy scenario of overcoming interference.

Radical Relationality

Swarms oscillate between aggregation and disaggregation, between concentration and diffusion, between order and patterns on the one hand and interference and noise on the other. A notorious problem, this oscillation between loose coupling and coordinated movement has stood at the heart of the wonder, the horror, and the study of swarms. They seem to be a hybrid of unity and multiplicity. This situation of a multitude of individual elements, out of which is formed a dynamic whole for which no single element can be held responsible, raises the question of the 'invisible hand' or 'guiding spirit' of the swarm – the question of the locus of its dynamic organization. This question further suggests that swarms, in the sum of their parts, are vaguely 'greater' with respect to their capabilities and characteristics. Schools of fish and flocks of birds, for example, are able to change directions at higher speeds than the individual animals can on their own, and social insects are able to accomplish astounding feats of collective coordination.

This relationship has been politically precarious ever since the fields of ethology and behavioral biology began to understand swarms as dynamic multiplicities: the individual elements neither serve a whole, nor are they components of a homogeneous collective body seeking to become a unit. Their collectivity is not consistent, and it is not controlled by any organizational center; rather, it is created simply on account of local interactions. Accordingly, swarms have to be regarded as genuinely time-based systems. This emphasis on collective behavior in time supersedes the static and spatially organized relationship between the individual and the whole. Swarms should therefore not be understood as homogeneous collective bodies but rather as heterogeneous collectives that constantly form and reform out of the dynamic and differentiated interactions of their local elements.

These constantly forming local neighborhoods, from which the global dynamics of swarms develop, are based on a unique principle of interaction. In flocks of birds or schools of fish, the respective movements of the nearest neighbor influence the behavior of a swarm individual – their spatio-temporal orientation converges with the 'message' that they are transmitting. The individuals in the flock or school do not leave behind any signs in their environment, such as the pheromone traces left by social insects; rather, they interact through signs of motion. Swarm individuals signal to one another with their movement.[54] In this sense, no messages are coded to be transmitted through a channel of communication that is due to the parasite. The channel converges with the motion-based message, and thus the parasites nestle into every individual of the swarm, into every nodal point of the living network: "Wherever there are channels, there is also noise" – every individual in the swarm produces noise of its own. The 'message,' which is coded in individual movements, is simultaneously interference, and this accords well with the assertion that, from the standpoint of the parasite occupying the transmission lines, there is no theoretical difference at all between a coded and a disrupted signal.[55] Every motion-based relation in swarms of this sort is thus constituted out of disruptive moments, out of a multiplicity of vaguely defined and collaterally interactive signs of motion. This relentless act of forming relationships – this ineluctable relationality – enables the coordinated and targeted motion of swarms as a whole to arise from individual, uncoordinated, and swarming movements. The swarm system works because it does not work.

Not only is the function of noise and interference crucial to the media-technological constitution of swarms in Serres's triangular model; it can also be regarded as the fundamental basis of the specific organizational infrastructure and dynamic, neighborhood-based topology of swarms. In this way, too, swarm systems are distinguished from other classical forms of networks, in which (schematically speaking) individual nodal points are connected by way of edges to particular topologies. Swarms render such structures dynamic. They confound rigid topologies because their individual 'nodal points' are always changing position, and they replace fixed edges for transmitting information with their unique movements. In network-theoretical terms,

54 For this reason, the figure of the parasite, which in Serres's work operates within a theory of signs, has to be understood more generally so as to apply to 'signs of motion' as well.

55 See Siegert, 'Die Geburt der Literatur aus dem Rauschen der Kanäle,' 10.

nodes and edges – and thus also nodes and noise – converge in them. Beyond this, the interactions that take place in swarms are not only time-*critical*, as in networks, but rather fundamentally time-*based*.

This temporal dimension plays a decisive role in determining the specific dynamic of swarms, which Michel Serres's philosophy of relations helps to bring into focus. Their collective status should be understood as a "cluster of relations" rather than as something essential.[56] As an example, Serres discusses the game known in English as 'Button, Button, Who's Got the Button?' The goal of the game is to circulate a button within a group and not to have it in one's hand when the game pauses. By constantly being passed around, according to Serres, the button assumes a precarious status: it only becomes an object when it is in someone's hands, and at that same moment the person holding it becomes a subject, because the button now designates and establishes him or her within the collective.[57]

What is constitutive for the formation of a collective is therefore an internal moment of movement, a time-based state of continuous oscillation, a sort of undecidability: "What must be thought about, in order to calculate the 'we' is, in fact, the passing of the [button]," in Serres's words.[58] The condition of collectivity is the incessant reconnection of connections, the constant interaction between individuals and their nearest neighbors. A *quasi-object* builds this relation, but it only does so by being absent. It is defined by being passed around, by its transmission. This understanding yields a redefinition of the collective as an ensemble of accelerated transmission events – not as something composed of individual elements but rather as "the transubstantiation of being into relation."[59] Within a swarm, the individuals change their positions constantly in an effort to evade attention from the outside (from predators, for instance); their aim is to be absent within the totality of the swarm collective.

In this sense, swarms embody a radical relationality in which any essential definition dissolves. Questions concerning something like the 'ontology of the swarm' thus always fall short. Their existence can only be thought of in terms of their *relational being*, and this changes from one historical moment to the next. In two respects, swarms thus have to be regarded as

56 Serres, *The Parasite*, 225.

57 See ibid. In his chapter titled 'Theory of the Quasi-Object' (ibid., 224–234), Serres discusses a nexus of human relations and therefore uses the terms 'I' and 'we.' In order to avoid historical and terminological complications, the terms 'swarm individual' and 'swarm collective' will be used throughout this work.

58 Ibid., 227.

59 Ibid., 228.

transformational phenomena: constant transmission events, incessant movement in space and time, and ongoing reformation define their constitution as a perpetual and discontinuous *becoming* – including the parasites dwelling within them – and further define them as a fall into disorder and the beginning of new (media-technological) orders. Serres's process of 'essential' dissolution marks a historical shift after which the scientific study of swarms began to focus on the 'how?' of distributed organization and to lay to rest notions of their miraculous essence and higher order.

This defining relationality also contains a third level, which, as Eugene Thacker has noted, is not that of the individual or the collective but rather the level "where multiplicity and relation intersect."[60] What is 'essential' to swarm collectives – and this is what the field of swarm research has been attempting to define, from various perspectives and with various technological arrangements, since the 1930s – is thus the act of existing within ensembles of relations. Local interactions and global patterns – swarming and the swarm – can therefore be understood in integrative terms as mutually necessary conditions of this biomorphic, dynamic figure. Clusters of relations within a shifting constellation: the concept of the parasite ties a media-theoretical classification of swarm collectives back to the beginning of this chapter – to the singular noise of multiplicity and to the acoustics of the swarm, which makes this unity and multiplicity seem compatible and yet irreducible to one another. In a media history of swarm research, this relationality is accorded a systematic position. The oscillation between individual interactions and collective movements disrupts media-technological attempts to record and model swarms and thus initiates research dynamics in which the rules and dynamics of *becoming a swarm* are constantly reconceived in light of changing media-historical information. The question of this relationality, moreover, is accompanied by a *media-becoming* of swarms in which their disruptive relations can ultimately be transformed into productive and operational systems.

2. Historiography: Recursion

Media-Becoming

That which is true of swarms likewise applies to media: there are no media in a substantial and historically stable sense. The question of what media

60 Thacker, 'Networks, Swarms, Multitudes,' n.p.

can be at all should therefore be placed, on a case-by-case basis, within medial historiographies that take into account the fact that the status of media as scientific and systematizable objects always results in them being "that which they store and mediate under conditions that they themselves create and are."[61] Accordingly, media-theoretical analyses are not simply concerned with devices or codes, and they are in no way restricted to particular techniques, symbols, or forms of representation. Such analyses are rather concerned with the dual eventfulness of media: *First*, they communicate this themselves as specific events along with the events that they communicate. *Second*, this happens with the "tendency to extinguish themselves and their constitutive participation in these sensualities and thus to become, at the same time, imperceptible and anaesthetic."[62] The dichotomy of swarms between noise and self-organization, which results from their ability both to form and deform, makes it difficult to place them in a proper media-historical perspective. Swarms are a marginal case of media-theoretical underexposure, and doubly so. A media history of swarms is not only concerned with the attempts that have been made to record and model, by means of various media, living networks, which for their part operate medially. For, at the same time, the media technologies used to analyze swarms have each transformed their object of investigation in different ways. Various observational and experimental systems have each created their own specific knowledge about swarms. In doing so, they have had to account for swarms as media of interference that repeatedly undermine such attempts at analysis and emphatically underscore their own mediality. As media events themselves, swarms come between the dual media events that are the focus of medial historiographies. This reciprocal relationship can be described quite adequately with Joseph Vogl's concept of *media-becoming*.[63] For, ever since the beginnings of systematic and scientific swarm research, around 1930, the processes of interaction and control between swarm individuals have been the central matter of interest. The main issue, in other words, has been the internal mediality of swarms. Under a range of historical conditions, various discourses have formulated various concepts of swarms; concepts such as *information exchange* and *self-organization*, for instance, could of course not appear until the development of information-theoretical and cybernetic principles. The idea of 'digital swarms' first gained significance around 1990 in the context of a scientific

61 Engell and Vogl, 'Vorwort,' 10.

62 Ibid.

63 Vogl, 'Becoming-Media,' *passim*.

boom concerned with 'complex' and 'emergent' phenomena, a situation that in turn owed much to material innovations related to rapid technological advances in the field of computing speeds and to changing programming paradigms. Nevertheless, all of the disparate media-technological arrangements discussed in this book go hand in hand with a media-becoming that ultimately materialized, on various occasions, in the transformation of swarms from objects of knowledge to figures of knowledge. The concept of media-becoming makes it possible to describe certain constituents that are relevant to the transformation of swarms – as agents of the uncanny, of non-knowledge, and of interference – into *systems* that, as of the 1990s, have been ascribed something like *swarm intelligence*. The *first* constituent of this transformation could be called the *de-naturalization of the senses*. The study of swarms required the development of media-technological perspectives that were not simply extensions of human vision or human senses more generally. The observation of swarms depends above all on engaging with media-technological materialities that enable them to become recognizable, measurable, and systematically describable structures. It is only through this engagement that they are able to appear as objects of knowledge. Rather than being mere extensions of the senses, such media techniques "create the senses anew: they define the meaning of vision and sensory perception, turning any and all visible facts into constructed and calculated data. Ultimately, all the phenomena and 'messages' they produce bear the mark of theory. The sensory evidence transmitted by these messages is conveyed alongside the procedure by which that evidence was established."[64] This becomes explicit, for instance, in various aquarium designs, cameras, or visualization and acoustic detection systems.

Second, this media-becoming is characterized by the *production of self-referentiality*. For, in addition to locating its object (or non-object), the application of technical media in swarm research also locates its observer, the researcher. Among other things, this unity of observation and self-observation has led underwater researchers and filmmakers to learn particular swimming techniques and to develop new breathing apparatuses in order to be able to work 'like a fish among fish.' Or they have resulted in ornithologists attempting to camouflage themselves in certain environments in order to study the behavior of flocking birds. In short, the observers observe themselves in these situations in order to remove themselves as much as possible from the observational relation. It is the case, moreover, that this process of self-referentiality can also lead to a

64 Ibid., 17 (the quotation has been slightly modified).

multiplication of observational arrangements and thus to a multiplicity of swarms (depending on the media technique being used). Throughout the media history of swarm research, swarms have therefore repeatedly been reconceptualized and understood in new ways, for, as Vogl has remarked, the "correct observation can only be expressed in the conditional."[65]

Third and finally, the media-becoming of swarms is also characterized by the *generation of an anesthetic field*, and thus it is associated with a productive disruptive function: "The critical point of the historical analysis of media is not to be found in what a medium makes visible, tangible, audible, readable, or perceptible; it is not so much located in the aesthetic of the data and information provided by a medium but rather in the anesthetic side of a media process."[66] The technical observation of swarms therefore also documents the relationship or difference between the visible and the invisible. And this is the point of application at which, in the 1990s, a fundamentally different 'view' of swarms was taken by means of agent-based computer simulation processes and their unique way of operating. Within such simulation systems, swarms themselves have become media endowed with a specific sort of *swarm intelligence* for solving complex problems.

Repetition and Variation

However, this reciprocity between the computerization of biology and the biologization of computer science, which mutually inform one another, is itself problematic. To specify, one could call this complication a historiographical *figure of recursion*. In terms of computer science, recursion is defined as the reapplication of a processing instruction to a variable that is itself the output of this instruction: "The value of the variable changes every time it runs through the loop, and the effect of repetition is not the production of identity but rather an example of predefined variation. [...] Recursion combines repetition and variation with the aim of creating something new."[67] My understanding of the term here is somewhat different from that of the cultural theorists Ana Ofak and Philipp von Hilgers, who introduced the idea of recursion as a historiographical concept.[68] In the recursive process between biological swarm principles and

65 Ibid., 18.

66 Ibid., 20.

67 Ernst, 'Der Appell der Medien,' 185. Here Ernst provides a reference to Winkler, 'Rekursion,' 235.

68 See, among their other works, Ofak and Von Hilgers, eds., *Rekursionen: Von Faltungen des Wissens*.

computer-generated simulation environments, two areas involving opaque processes of self-regulation (each with only partially known or vaguely defined parameters) come closer together. If, in the media history of swarm research before 1930, it is possible to find 'data' and case studies in which researchers attempted to approximate swarms without generating any real *data* about their object (or non-object) of study, then this can be said to characterize the earliest stage of the media-becoming of dynamic swarm formations. This was associated with a shift toward media-technological recording processes. Following this media-historical transition, knowledge about swarms was further developed by means of the data processes of technical media that attempted to address the data problem of earlier field researchers. Yet this simply caused the data problem to shift. The technical media used in swarm research involved additional data drifts. In this case, too, the methods of observation and recording were only able to yield results by confronting various moments of interference. The output of technically supported swarm research, in turn, consisted of such a vast mess of data that this *transcript* of swarm activity could only be managed by means of automated analysis.

The third step in this genealogy would be that of recursive functionality. Under the condition of possibility created by agent-based computer simulations, which were first developed in the 1990s, swarms came to be applied to themselves. Software models, simulation models, and visualization models that had been inspired by biological swarm research were reimported into the field of biological swarm research itself for the purpose of studying, by means of computer experiments and simulations, the behavior of four-dimensional dynamic collectives *in silico*. It was only when swarms' intransparent structures of control and order were no longer described or recorded but were rather implemented as writing processes themselves that it became possible to readdress the data problem yet again. Beyond coding – beyond the composition of 'digital swarms' and agent-based program and simulation environments – there also occurred a scenario-based approximation of the *description* of biological swarms. Over the course of this development, data became reversible in competing scenarios and could always be rewritten in new and different ways (and could record themselves as writing processes). As media, swarms provide the intermediary that is needed for their own description. Only when they became *writing processes* did they become *describable*. The non-object of the swarm thus reflects in an exemplary manner the epistemic interrelation between the historiography of media and the media of historiography, an interrelation that defines the research program of medial historiographies.[69]

3. Epistemology: Computer Simulation

Mindsets of Messiness

If, in their media-becoming, swarms first became sufficiently describable in a recursive operation between biology and computer technology, this development was not exclusively related to the formulation and technical implementation of a specific figure of knowledge. In addition, it is fruitful to discuss swarms in light of the more comprehensive *historical epistemology* of computer simulation. In this case, swarms can be regarded as a telling example of the fact that scientific objects – as thinkers such as Gaston Bachelard, Georges Canguilhem, and Hans-Jörg Rheinberger have pointed out – are not empirically or historically given and that "objects of analysis do not exist in a pre-discursive order of natural things that can be experienced by means of a theory of representation."[70] Swarms make it possible to clarify the extent to which an epistemology of computer simulation contributes new aspects to these reciprocal effects. At the same time, the question of studying swarms with the assistance of agent-based computer simulation models raises further questions involving the epistemological dimension of dynamic visualizations and image-generating processes, which are features of the models themselves. The latter no longer represent behavior that is natural or somehow observable; rather, they produce and *present* conjectural scenarios that generate an approximate knowledge of swarm behavior and system behavior by means of a differential way of knowing – that is, by means of the trial-and-error science of computer simulation (and thus with all the abstraction and openness that such models entail). In this context, swarms function as 'space-generating' processes: because they can hardly be described with reference to Euclidian spaces (in the sense of their *location*), they themselves can become a topological space, given that their global behavior is governed by the implementation of internal parameters and that the interaction of their elements 'occurs' in the real time of multi-agent systems and occupies its own dynamic swarm space. Ultimately, knowledge about swarms is

69 See Engell and Vogl, 'Einleitung.'

70 Barberi, 'Editorial: Historische Epistemologie und Diskursanalyse,' 6.

thus created by a form of "synthetic history."[71] Swarm researchers have thus created scenarios and image sequences of a 'world of the swarm' and of the context in which a swarm's inter-individual interactions can take place with the global ability to move as a collective. At the same time, however, these biological swarm studies have also informed the development of computer simulation models. A reevaluation of the swarm as an intransparent object of knowledge was thus accompanied by a general reevaluation of epistemic strategies. Even knowledge cultures can become media cultures of intransparency when they turn their attention to unanalyzable problems, complex processes of interaction, or other sets of problems that are vaguely defined: as Kevin Kelly remarked as early as 1994, "Science has done all the easy tasks – the clean simple signals. Now all it can face is the noise; it must stare the messiness of life in the eye."[72] Self-organizing multi-agent systems with distributed information infrastructures thus extend the 'world of the swarm' to a variety of scientific disciplines in which their application plays a role. Agent-based computer simulations create, as Eric Bonabeau has pointed out,[73] a demonstrably swarm-induced 'mindset' – and thus, too, a specific media culture in which 'swarm-intelligent' tools and applications become operational. Recently, this interwoven media history of swarms and computer technology has made it explicit that swarms have been reassessed as a productive – instead of disruptive – form of non-knowledge. Swarms are simultaneously the object as well as the principle of agent-based simulation models and, over the course of this simulation-based research, methods of graphically visualizing simulation results have played a decisive role in the process of refining basic parameters. Moreover, the agent-based description of swarms has been characterized to a considerable extent by applications from the field of graphic design, whose 'distributed behavioral models' were in fact developed to produce special effects in films. And thus swarm intelligence was, at least in part, made in Hollywood.

The base function of this knowledge is the act of "seeing in time."[74] In its state of temporal 'thrownness' (*Zeitgeworfenheit*) – or, better, in its state of having been designed in time (*Zeitentworfenheit*) – computer simulations are able to animate mathematical models, that is, endow them with life in 'real time.' In this way, it does not exhaust itself into a mere expansion of existing epistemological

71 See, for instance, Pias, 'Synthetic History,' 176.

72 Kelly, *Out of Control*, 25.

73 Bonabeau, 'Agent-Based Modeling.'

74 Wilson, *The Cell in Development and Heredity*, 77.

strategies. Computer simulations represent more than simply an improvement of numerical calculation methods by means of the processing speed of computers. It can rather be attributed an entirely unique epistemological status of theoretical experimentation: it is here that pragmatic operationality has supplanted the need for precise theoretical foundations. It is here that categorical truth-claims are replaced by provisional knowledge. Here, in other words, "the performance on the computer is more important than the model's derivation and its accuracy of calculation."[75] Unlike the case of theories, computer science is less concerned with what is true or false than it is with pragmatic utility.[76] The hypothetical character of knowledge in this field is underscored by the different and competing models of swarm simulation; instead of confirming one another's findings and producing certainties, they have rather generated a spectrum of opinions and viewpoints. Where computer science focuses its attention is on the *relations* that exist within systems. At this point, swarming as an object of knowledge encounters the epistemology of simulation. The relational being of swarms, with its intersections of the microscopic and macroscopic, can only be adequately captured by a technology that itself bisects the distinction between the epistemic and the technological thing – that is, by a technology that focuses on *knowledge relations*. The knowledge of swarms and that of computer simulation go hand in hand: that which cannot be addressed adequately *in vivo* and *in vitro* can be recorded *in silico*.

The Governmental Constitution of the Present

The recursive coupling of swarm-inspired agent-based modeling and swarm research, however, entails an even graver consideration. Agent-based models were first implemented by means of object-oriented programming. Both agent-based modeling and object-oriented programming can thus be assigned to the same paradigm, one that Frederick Brooks subsumed under the concept of "growing" (in its dual sense of 'increasing' and 'cultivating').[77] To a certain extent, control and 'intelligence' are here delegated to a self-regulating system.[78] And within the paradigm of growing, which inclines toward self-organization and procedurality, swarms appear as a digital cultural technique *par excellence*, one that enriches the study of cultural techniques with a zootechnological dimension.

75 Küppers and Lenhard, 'The Controversial Status of Simulations,' 271.

76 Sismondo, 'Models, Simulations, and Their Objects,' 247.

77 Brooks, 'No Silver Bullet: Essence and Accidents of Software Engineering.'

78 See Parikka, *Insect Media*.

The concept of cultural techniques was originally developed in 2000 at the Humboldt University in Berlin and has been further refined since then, especially at the Bauhaus University in Weimar. With technical media likewise occupying a central position in its theoretical formulations, it bypasses having to posit a medium *a priori* by reexamining the context of media and culture: taking into account the *technē* dimension of 'culture,' which has etymological roots in agriculture, cultural-technique research allows for an integrative analysis of the *operational chains* (in Bruno Latour's terms) that precede the formation of media before the latter and then place, in a 'reality-forming' way, everything that they store, process, and transmit under the conditions that they themselves create and are. Everything that we know, we know through media – but the media of this knowledge are based on elementary cultural techniques. Counting thus existed long before the concept of the number; the latter existed long before sophisticated mathematics; and the latter, in turn, is much older than the applied mathematics that is reflected in so many apparatuses and techniques, which have themselves initiated new operational chains. Moreover, counting with ten fingers, for instance, means something different from counting with an abacus, and the latter practice is significantly different from counting done by a computer. Proceeding from basic 'body techniques,' which were first conceived by Marcel Mauss,[79] cultural-technique research therefore opens up a new way of analyzing the extent to which human actors have always been decentered toward particular media and technical objects. It reconstructs the practices into which media are integrated and which, in turn, the media themselves create. These practices are also problematized in regard to methods of generating and representing *objective* data in the sciences,[80] and they are further examined in light of the political, administrative, and biological 'versions of human beings.'[81] That said, researchers have so far given little attention to the practices of scientific computer simulation together with the elements mentioned above and others (the ludic and theatrical, for instance).[82]

Casey Alt has taken an even more radical approach, for he has identified object-oriented programming to be the material foundation of our entire understanding of computers as media. Alt conceptualizes this medial

79 Mauss, 'Techniques of the Body.'

80 See the inaugural issue of *Zeitschrift für Medien- und Kulturforschung* (2010), which is devoted to cultural techniques.

81 The phrase derives from Seitter, *Menschenfassungen: Studien zur Erkenntnispolitikwissenschaft.*

82 An exception in this regard is Pias, *Computer Game Worlds.*

relation as a 'society of objects' within a computer, the communication of which takes place both among the objects themselves, at the program level, as well as with human users by means of interfaces. Thus, the user is likewise conceived of as a programming process, and object-oriented programming begins to structure, more than just metaphorically, our daily lives: "Object orientation increasingly mediates how we work, play, fight and love"[83] – from video game communities to social networks to the flow of information at modern businesses. To this list, agent-based modeling contributes the realm of knowledge and science. For, from the media-historical threshold where the epistemic conflation of biological and technical findings and models yielded an extensive and novel understanding of the principles of regulation and self-organization that govern swarms, these principles became operable as figures of knowledge in various fields of implementation and for various technological applications. From the first thought experiments on robot collectives composed of simply designed individuals at the end of the 1980s to the physically implemented versions of today, the researchers operated according to the following motto: "[U]sing swarms is the same as 'getting a bunch of small cheap dumb things to do the same job as an expensive smart thing'."[84] The logic of swarms introduced a new type of economy to technological processes, an economy based on the flexibility of model environments, on a distributed mechanism of control and regulation, on the independent creation of unpredictable solutions, and on high levels of fault tolerance and reliability. Swarms integrated themselves as components of the evolutionary software designs with which mathematical optimizations could be executed – in the form, for instance, of particle swarm optimization.[85] The latter designs were in turn implemented for problems of multi-objective optimization, that is, for processes involving multitudes of reciprocal and mutually constraining variables. Their field of application has extended from industrial production processes to logistics planning to the optimization of network protocols.[86] Moreover, the interactional intelligence of swarms can play a role wherever there are time-sensitive problems of coordination and transference between numerous particles; such problems present themselves, for instance, in traffic simulations, social simulations, panic simulations, consumer simulations, epidemic simulations, simulations of animal collectives, in the behavior of aerosol in climate models, and even

83 Alt, 'Objects of Our Affection,' 298.

84 Corner and Lamont, 'Parallel Simulation of UAV Swarm Scenarios,' 355.

85 See Kennedy and Eberhart, 'Particle Swarm Optimization.'

86 See Engelbrecht, *Fundamentals of Computational Swarm Intelligence*.

in the case of organizing building materials. Swarms create information by means of formation.

Swarms and the algorithmics of their relational being can be called 'intelligent' whenever a matter concerns the (independent) government and planning of interactions in space and time. Their applicability to agent-based computer modeling and to distributed technological collectives is indicative of their effectiveness as a novel cultural technique. As such, swarming is characterized by the fact that it was produced in the area of tension between biology and computer science. Originally regarded as mere disruptive phenomena, swarms emerged as operational media technologies. As an addressee of this cultural technique, humans were at first only an unintentional part of the equation. Strictly speaking, swarming did not exist as a cultural technique before its media-technological manifestation, that is, before it became applicable in the field of computer science as a novel epistemic process and as a solution configuration for a multitude of complex problems.[87] Moreover, the influence of the cultural technique expanded even further when the 'crowd logic' of its behavior came to be employed as imitable particles in social simulations. Around 2000, at the latest, swarm intelligence and agent-based modeling emerged as a powerful and irreversible element of the current media culture. It is as *zootechnologies* that they have developed into a relevant cultural technique, and as such they have enabled and initiated novel engagements with opaque areas of knowledge, with disruptive phenomena, and with technological and systemic correlations that otherwise would have been difficult to ascertain.

At the same time, they produce and even demand – like the paradigm of object-oriented programming – a *Zeitgeist* and worldview in which cultural processes are characterized more and more by the multiple and dynamic interactions of autonomous and self-optimizing 'actors.' Once aware of the lasting effects of swarming as a cultural technique on our current media and knowledge cultures, at least as described here, one should be quick to distrust the highly touted potential of social swarming and the grass-roots-democratic 'nature' of human techno-collectives. This holds true even despite the elevation of the discourse, in the past few years, to sophisticated media-theoretical levels (see in this regard the work of Tiziana

87 As a term used in mass psychology, or as an obsolete element of military tactics, the concept of swarming was chiefly employed to signify the *dissolution* of order, that is, the act of 'swarming all over.' It was not then conceived of as representing the relational, procedural, and structural intermediary domain between the individual and the collective, namely the very domain that, according to Eugene Thacker, defines the dynamics of swarms.

Terranova, Luciana Parisi, Olga Gurionova, Howard Slater, or the issue of *Limn* devoted to 'crowds and clouds'). Ultimately, whoever belittles recent revolutions with the journalistic banalities of swarm logic – 'Facebook revolution,' 'Twitter revolution,' and so on – deliberately overlooks the extent to which the cultural technique of swarming has come to define our situation. Swarms should no longer be understood simply as advanced manifestations of older forms of collective behavior. It is much rather the case that they have gained relevance as structures of organization and coordination. These structures have become effective against a backdrop of an opaque culture – one defined by the permanent flexibility of various domains of life – and they have become effective namely as optimization strategies and zootechnological solutions *within* these very domains. At the heart of swarming, as a cultural technique, is thus the governmental constitution of the present itself, in which operationalized and optimized multitudes have arisen from the uncontrollable data drift of dynamic collectives. There is no going back from this.

II. Formations

Abstract
Concerned with *formations*, the second chapter is devoted to historical scenes in the development of behavioral biology around 1900. The latter discipline systematized knowledge about swarms by relying on physical instead of then popular social models of interaction, e.g. in mass psychology. It developed a genuinely 'biological gaze' that was determined to study animal collectives in terms of the 'systemic' nature of their inter-individual behavior. Techniques and media for gathering data thus gained a new degree of relevance, replacing the human sensory apparatus, which perceived little more than noise, and traditional systems for recording information (diaries, hand-written observations), which could not deal with the abundance of data.

Keywords: social instinct, animal psychology, behavioral biology, thought transference, marine biology laboratories, observation versus experimentation

> [Konrad] Lorenz was once visited in Seewiesen by an English colleague, and when the latter inquired about the *lavatory*, Lorenz understood *laboratory* and replied: "Oh, we don't have one; we are doing everything outside."[1]

The development of a genuinely biological and ethological perspective on animal collectives by no means represented a clean break from the related fields of animal psychology and mass psychology, which had been heavily influenced by the well-known works of authors such as Auguste Forel, Francis Galton, Alfred Espinas, Gustave Le Bon, Gabriel de Tarde, and Scipio Sighele.[2] Beginning around 1990, a methodological arsenal was

1 Wuketits, *Die Entdeckung des Verhaltens*, 104 (my emphasis). It was in Seewiesen, which is near Munich, where, in 1958, Konrad Lorenz and Erich von Holst founded the Max Planck Institute for Behavioral Physiology. In 1997, the institute was closed because of budget cuts.
2 See Galton, *Inquiries into Human Faculty and Its Development*; Espinas, *Des sociétés animales*; Forel, *Les fourmis de la Suisse*; Maeterlinck, *The Life of the Bee*; De Tarde, *Penal Philosophy*;

Vehlken, S., *Zootechnologies. A Media History of Swarm Research.*
Taylor & Francis Group, 2019
DOI 10.5117/9789462986206_CH02

gradually developed that made it possible to investigate the behavior of social insects, flocks of birds, and schools of fish without having to rely on mere anecdotal overviews or to focus on interactions between humans and animals. Systematic approaches and experimental methods found their way into zoological research practices, and only then did the data produced by zoologists become comparable and 'scientific.' The protagonists of the early stages of ethology were especially concerned with formulating a natural-scientific basis for their research that could serve as a counter-position to the vague psychological attributions – the poorly defined and anthropomorphizing concepts of instinct, emotion, and intelligence – that had previously characterized descriptions of animal collectives. Of course, these older approaches were not ousted altogether. Theories about the quasi-metaphysical levels of understanding possessed by collectives and theories about their miraculous collective instincts and 'souls' persisted even within the newly institutionalized field of behavioral biology. The dismissal of such concepts, however, allowed a form of natural-scientific swarm research to develop that, in its conception and approach, would later make it possible to describe swarm formations as productive collectives.

For many years, there was nothing that could be called an overarching ethological research plan, let alone a lasting institutionalization of progressive research approaches. Rather, contributions to ethology between 1900 and 1930 came both from university contexts as well as from the pioneering achievements of amateur researchers, who were essentially carrying on the tradition of nineteenth-century natural scientists. Furthermore, new institutions were established, such as research aquaria and research facilities at fisheries, where the new form of studying animal collectives could be practiced and where a productive academic discourse was developed. As Lynn K. Nyhart has shown, a new research landscape was gradually cultivated in which different discourses, methods, theoretical constructs, and institutional parameters overlapped and were in conversation with one another.[3]

In this evolving and ever-changing research landscape with a growing number of methodological approaches, technical apparatuses, and theoretical considerations – in short, with a uniquely developing *zoo-logic* – animal

idem, *The Laws of Imitation*, 59–88; idem, *L'Opinion et la foule*; idem, *Les crimes des foules*; Sighele, *La fola delinquente*; Bechterev, *Suggestion und ihre soziale Bedeutung*; Le Bon, *The Crowd*; and Freud, *Group Psychology and the Analysis of the Ego*. More recent studies include Johach, 'Schwarm-Logiken'; Stäheli, 'Protokybernetik in der Massenpsychologie'; and Werber, *Ameisengesellschaften: Eine Faszinationsgeschichte*.

3 Nyhart, 'Natural History and the "New" Biology.'

collectives were not understood and written about as something naturally given. On the contrary, they were made into objects of study in a specific way and with a particular biological vocabulary. This development was characterized by a gradual transition toward experimental methods, toward quantifiable and empirically supported physiological descriptions, and not least toward the implementation of media-technological instruments in response to the difficulties entailed by the study of dynamic multiplicities. Friedrich Kittler's observation about the effects of writing instruments, storage devices, and data processing systems on the field of anthropology applies just as well to the study animal collectives: "The hard science of physiology did away with the psychological conception that guaranteed humans that they could find their souls [...] It became obsolete as soon as body and soul advanced to become objects of scientific experiments. The unity of apperception disintegrated into a large number of subroutines."[4]

This successive *formation* of a type of swarm research with a natural-scientific self-perception and on the basis of various 'subroutines' can be described as a theoretical and discursive shift away from psychology – as a sort of 'de-psychologization.' Setting aside the idea of collective instincts in their explanation of swarm behavior, the new ethological approaches followed a course that led away from (animal) psychological concepts and models toward those involving the physics of motion. An additional zoological component can be identified as well: whereas earlier approaches had associated 'fuzzy' concepts with the fuzzy phenomenon of swarms, the aim now was to make the discourse more precise and scientific, that is, to give sharper definition to the object of research (namely by means of scientific experimentation). Anthropomorphic characterizations of collectives – references to their excitable, contagious, or escalatory mass behavior – gave way to experimental arrangements and manners of description in which agglomerations did not fuse into a single mass but rather oscillated, as interactive quantities, between unity and multiplicity. This difference-logical principle, according to which both sides – the individual and the collective – are constitutive for the dynamics of the system as a whole, paved the way for a systematic investigation of the *operations* that take place between its elements. In this case, a sort of relational knowledge was cultivated without any recourse to human psychological concepts, and this knowledge situated unity and multiplicity into a new and physics-based reference system.[5]

4 Kittler, *Gramophone, Film, Typewriter*, 188.

5 For a more detailed analysis, see Vehlken, 'After Affects.'

In this chapter I will trace various lines in the cultivation of ethological research between 1900 and 1930 that formulated decisive figures of thought and methodological approaches for the development of biological swarm research. This will require a cursory overview of the early stages of behavioral biology and the discipline's different approaches to swarms. The chapter will look first at a *dispositif* of observation that gained popularity as more and more behavioral scientists began to conduct work in the field. There – or better: *above* the fields – it was possible not only to observe the dynamics of large flocks of birds and analyze their basic manner of functioning. It was also possible to observe the reactions of animal collectives to external influences and see how their structures changed in response to environmental factors. The relation of animals to their environment was likewise discussed by researchers studying schools of fish in aquaria and laboratory tanks, and these early efforts would lead the way to the experimentalization and mediatization of swarm research – to the media-technological *formatting* of swarms – that became common after 1930. This chapter is thus devoted to the epistemic features that enabled swarms, as an object of research, to be transferred into the context of replicable experimental arrangements and reduced complexity. The goal, of course, was to conduct more precise studies of their behavior, but even these early efforts faced certain conceptual and technical problems that would have to be dealt with more intensively in the subsequent attempts to *format* swarms.

In sum, from around 1900 to 1930 swarms were freed from the anthropomorphizing tendencies of earlier researchers and became a matter of interest to a type of behavioral biology that applied its own language and perspective to animal collectives. These biologists understood them as dynamic systems whose decisive functions did not depend on the characteristics of the collective or the condition of the individuals but rather on the manner and constitution of the *relations* between the individuals. In this discourse, the psychic ability of swarms to be excitable or contagious was no longer an issue; the focus was rather on the measurable and verifiable *sensory-physiological potential* of the individuals and the way in which *information* was transferred between them.

1. Odd Birds

Suddenly, as at a signal, they all launch themselves toward the center of the field; the hundred companies unite in one immense flock, and presto! the drill is on. The birds are no longer individuals, but a single-minded myriad,

> which wheels or veers with such precision that the flash of their thousand wings when they turn is like the flicker of a signal glass in the sun.[6]

On May 24, 1984, the journal *Nature* published a short article by the biologist Wayne K. Potts of Utah State University.[7] The article is concerned with how flocks of birds are able to change directions in such a coordinated way. Potts's analysis of this phenomenon is based on his own footage of flocks of dunlin (*Calidris alpina*), and he comes to the conclusion that the so-called 'chorus-line hypothesis' might be applicable to swarm research. Potts demonstrated the existence of a control impulse that can influence the members of a flock as a group and even individually: as soon as an individual moves from the margin toward the relative center of the flock, it stimulates its neighbors to move as well, and this motion spreads like a concentric wave throughout the group.

Such 'maneuver waves,' according to Potts, govern the dynamic collective movement of large flocks in the sky. What is unique about them, however, is their speed. At the beginning of the process, it is slower than the reaction time of any individual bird of the species (as established in laboratories), but then it soon accelerates to as much as three times this speed. Whereas the 'maneuver wave' in the flock ultimately takes just 14.6 milliseconds to propagate from one bird to its neighbor, a dunlin in a laboratory requires 38.3 milliseconds on average to react to a neighboring bird startled by a flash of light. Instead of exhibiting this standard delay of 38.3 milliseconds, in other words, the birds in a collective can somehow transfer information much more quickly. Potts argued that the birds anticipate the arrival of the wave in their given area of the flock, as though they were taking part in a human chorus line: "Films taken of human chorus lines indicate that rehearsed manoeuvers, initiated without warning, propagate from person to person approximately twice as fast [...] as the 194-ms human visual reaction time."[8]

Potts's article was rightfully criticized by his colleagues in the field, especially because it does little to explain how the obstructed view of individuals in the middle of a flock, where they are densely surrounded by neighbors on all sides, allows them to adapt to a maneuver wave that is approaching from some distance away. Such details aside, however, Potts's hypothesis can serve as a fitting example of how long biologists have failed to explain the precise functions and interactions within dynamic, four-dimensional

6 Long, *How Animals Talk*, 113.

7 Potts, 'The Chorus-Line Hypothesis,' 345.

8 Ibid., 345.

swarms. Admittedly, more recent studies have since identified pressure-sensitive sensors in the wingtips of birds, which, in addition to their sharp vision, enable their bodies to react very quickly to changes in air pressure (much like the lateral line in fish, which is sensitive to water pressure).[9] Nevertheless, it remains the case even today that many aspects of the collective behavior of avian flocks do not have a satisfactory explanation. Starlings, for instance, leave their nests in intervals of approximately three minutes. This creates a pattern that generates, on radar screens, a distinctive signature that resembles concentric waves intermittently emanating from the nest. Large flocks of starlings can produce up to twenty of such waves upon their departure, and no one has been able to explain the timing and regulation of this behavior, not to mention its basic function.[10]

From today's perspective regarding the transmission of information within swarms, the chorus-line hypothesis formulated in the 1980s already seems like a rather peculiar way to explain the informational architecture of flocking birds. During the early scientific stages of observing flocks, however, this area of ornithology was not only a hotbed of truly odd theories (involving such things as telepathy, collective instincts, or emotional contagion); it also attracted researchers who were themselves some really odd birds. One man in particular combined both of these aspects: the British ornithologist Edmund Selous, whom Potts cited in his article and whose theory of 'thought waves' was given a retrospective in a 2004 issue of *New Scientist*. With seemingly boundless passion for his area of research, Selous set new standards for ethological field work. As an amateur scientist, however, he was susceptible to making rather speculative assumptions about the swarm behavior of birds. Whereas his methodological innovations have earned him a good deal of posthumous recognition, a number of his ideas, including those at the heart of his theory of flocks, were a matter of contentious debate during his own lifetime.

Sportsmen without Swarm Spirit

Bearing in mind Winston Churchill's legendary response to a reporter who asked him how he had managed to reach such an old age – "No sports!" – a learner of the English language might be somewhat less irritated by the fact that the British term *sportsmen* does not refer exclusively to people who actually play sports. It rather refers to people who simply *attend* sporting events – to

9 See, for instance, Brown and Fedde, 'Airflow Sensors in the Avian Wing.'
10 See Vines, 'Psychic Birds (Or What?).'

place a bet on their favorite team or horse, for instance – and thereby (one can assume) increase their chances of having a heart attack on account of rising blood pressure. Around the beginning of the twentieth century, however, the term had a wider semantic sphere, for it could also refer to big game hunters. In this sense, *sportsmanship* played a large role in Edmund Selous's family history: his brother Frederick was one of the most famous big game hunters of his day. He went on safaris with Theodore Roosevelt, wrote thrilling accounts of his hunting adventures, and was mentioned in a 1894 issue of *Vanity Fair Album* in an ironic but rather revealing way: "[E]lephant, rhinoceros, lion, hippopotamus, giraffe, zebra, quagga, hyena, koodoo, hartebeest, duiker, oribi, klipspringer, tsessbe, and antelope of all kinds; many of which animals are now all but extinct, having been killed off by railways, by civilization and by Selous."[11] Edmund Selous had also been fond of hunting, though in a passage of his book *Bird Watching*, which was published in 1901, he described himself as being a bit of a blunderer and a poor shot: "For myself, I must confess that I once belonged to this great, poor army of killers, though happily, a bad shot, a most fatigable collector, and a poor half-hearted bungler, generally."[12] Unlike like his brother, however, Selous became a passionate convert from the art of hunting to the art of detailed observation:

> But now that I have watched birds closely, the killing of them seems to me as something monstrous and horrible; and, for every one that I have shot, or even only shot at and missed, I hate myself with an increasing hatred. [...] [F]or the pleasure that belongs to observation and inference is, really, far greater than that which attends any kind of skill or dexterity [...]. Let anyone who has an eye and a brain (but especially the latter), lay down the gun and take up the glasses for a week, a day, even for an hour, if he is lucky, and he will never wish to change back again. He will soon come to regard the killing of birds as not only brutal, but dreadfully silly, and his gun and cartridges, once so dear, will be to him, hereafter, as the toys of childhood are to the grown man.[13]

Unlike the situation in the United States, where researchers such as Charles O. Whitman and William M. Wheeler had begun around 1900 to establish

11 Anonymous, 'Men of the Day, No. 585: Mr Frederic Courtney Selous.' Quoted from Burckhardt, *Patterns of Behavior*, 77. Today – ironically enough – the largest nature preserve in Africa bears his name.

12 Selous, *Bird Watching*, 335.

13 Ibid., 335–36.

zoological institutes at universities and systematically study the behavior of animals in the wild, in the United Kingdom such work was still being done by amateurs, who nevertheless made some of the most creative and prominent contributions at the time to the study of animal behavior. Their main difference from earlier natural scientists was that, instead of simply collecting specimens of animals for the sake of making taxonomic classifications, they attempted to describe the *undisturbed* behavior of living animals as precisely as possible.[14] And while sportsmen observed this behavior with the goal of bringing it to an end (by shooting at it), the naturalist disdained such pleasures and collected new knowledge instead of trophies. Natural scientists, according to Selous, are more intelligent hunters: their version of hunting transformed into a scientific interest in life, an interest that still required an 'adventurous' approach but still lacked, at the time, any established scientific instrumentation. On the British Isles, at least, ornithology was the natural discipline for field work of this sort.[15]

Selous spent hours, days, and weeks making the most minute observations of the way various types of birds behaved in the wild. While doing so, he recorded everything he observed (and the time he did so) into what he called an "observational diary of habits."[16] This practice posed, as Selous himself remarked, a few problems of its own: "One has [...] often to scribble very fast to keep up with the birds, and so must leave a few things to be added."[17] As one contemporary witness recalled: "Back at his lodgings, he would copy out his notes and elaborate upon them. Later he might add something else if it remained fresh in his memory. He prided himself on recording *all* that he saw."[18] Moreover, he also described his ever-improving tactics of observation: the mimetic methods he used to camouflage himself and become one with the environment of the birds under observation, coupled with his insistence on remaining in his hiding place for extremely long periods of time. Only in such a way was it possible to maintain a nearby view of his object of study over an extended time frame.

14 See Burckhardt, *Patterns of Behavior*, 69. Aside from Edmund Selous, other noteworthy researchers of this type include Henry E. Howard, Frederick B. Kirkman, and Edward Armstrong. The zoologist Julian Huxley and the ethologist C. Lloyd Morgan, whose influence extended beyond the field of ethology in Great Britain, were exceptional in this regard because they held academic positions.

15 See Selous, *Bird Life Glimpses*, v–vi.

16 See, for instance, Selous, 'An Observational Diary of the Habits – Mostly Domestic – of the Great Crested Grebe (*Podicipes cristatus*).'

17 Ibid., 173.

18 See Burckhardt, *Patterns of Behavior*, 82.

In addition to using this specific ornithological recording system – which consisted of his eyes and a pair of binoculars, paper and a pencil, meticulous attention to detail, patience, and often his own self-made "turf huts" on Scottish moors – Selous was also concerned with linking his observations to Darwin's theory of evolution. He was especially interested in the connections between the evolution of behavior and natural and sexual selection. Particular patterns of movement and elaborate mating rituals might not derive, he thought, from the ad hoc psychological instincts of animals, which were both much maligned and yet frequently postulated at the time, or from some vaguely defined form of animal intelligence but rather from a long process involving various selective impulses and physiological factors: "[M]any actions of birds which seem now altogether intelligent and purposive (and, no doubt, are so to a very large extent) will be found to betray traces of a nervous and non-purposive origin."[19] And yet despite all of this, Selous's texts are characterized by a convoluted prose style that did little to win them acceptance within the zoological community (his constant verbal attacks against the "armchair ornithologists" and "thanatologists" working at natural history museums probably did not help his case either).

His two-hundred-page treatise on the swarm behavior of various birds, for instance, is organized so haphazardly that it is only possible to find any systematic information in it thanks to the index. In this work, Selous provides detailed examples of synchronic and collective flight behavior, but he combines these descriptions with truly (and necessarily) speculative reflections about their organization. For, as Selous notes, flocks are and move "more and faster than the eye can take in" – they could not be fully comprehended with his way of observation. In any case, he recognized in his observations that flocks lack a leader or sentinels that control their swarm dynamics and that function, as in the hierarchies of social insects, as organizational authorities: "The whole group acts thus as though it were a single bird. If a fishing-net, stretched on the ground, were to go up and float away, the one has to imagine every knot of every mesh to be a bird, and everything between the knots invisible, to have a perfect simile of what has just taken place."[20]

In contrast, Selous could only account for a flock's rapid changes in movement by referring to a fast form of communication and, for lack of a better explanation, he wondered whether this might not take place by means of

19 Selous, 'An Observational Diary,' 173.
20 Selous, Thought Transference (Or What?) in Birds, 94.

some sort of thought transference among the birds – hence the title of his book *Thought Transference (or What?) in Birds*:

> What, with us, is rational intercourse, with conversation, which probably weakens emotion, may be with birds in numbers a general transfusion of thought in relation to one another, on the plane of bird mentality – such thought corresponding more to our feeling than to what we call such, for it is out of feeling, surely, and not *vice versa*, that thought has evolved. This, then, may be the great bond between individuals in a species, probably acting through a sensation of well-being in one another's society which, when well developed, leads to gregariousness in rising degree.[21]

Selous thus postulated something like communication on the level of emotional affect, that is, preconscious processes that, in evolutionary terms, must have come before that which we call thought and that take place without any interference from this rational level. He was not the only person to offer this hypothesis. As early as 1919, the American natural scientist William J. Long had published an anthology titled *How Animals Talk* in which he discussed the allegedly 'telepathic' capabilities of animals.[22] In his chapter titled 'The Swarm Spirit,' Long also described at length his observations of the swarm dynamics of starlings and plovers – observations whose "emotional excitement" far surpassed anything that could be experienced while hunting:

> That you may visualize our problem before I venture an explanation, here is what you may see if you can forget your gun to observe nature with a deeper interest: [...] Your 'stand' is a hole in the earth, hidden by a few berry-bushes [...]. As the day breaks you see against the east a motion as of wings [...]. Those are plover, certainly; no other birds have that perfect unity of movement; and now, since they are looking for the source of the call they have just heard, you throw your cap in the air or wave a

21 Ibid., 115. The idea of such "thought waves" in birds, as Gail Vines has noted, must be considered not only in the context of the contemporary pseudo-scientific theories about thought transference between people but also in terms of the new media technology at the time – wireless radio and radar, for instance – and the hotly debated issue of wave theories in physics. See Vines, 'Psychic Birds,' 48.

22 See Long, *How Animals Talk*, 102–125. Drawing on the ideas of (unspecified) African tribes, Long asserted that the interplay of all biological senses gave rise to a sort of super-sense, which he referred to as *chumfo*: '[E]very atom of him, or every cell, as a biologist might insist, is of itself sentient and has the faculty of perception. Not till you understand that first principle of *chumfo* will your natural history be more than a dry husk, a thing of books or museums or stuffed skins or Latin names, from which all living interest has departed' (ibid., 33).

> handkerchief to attract their attention. There is an answering flash of white from the underside of their wings as the plover catch your signal and turn all at once to meet it. Here they come, driving in at terrific speed straight at you! [...] On they come, hundreds of quivering lines, which are the thin edges of wings, moving as one to a definite goal. [...] Suddenly, and so instantaneous that it makes you blink, there is a change of some kind in every quivering pair of wings. [...] [E]very bird in the flock has whirled, as if at command, and now is heading straight away.[23]

Unlike Selous, Long was not a passive observer. Rather, he used acoustic and visual signals to evoke certain reactions from his objects of study and then marveled at the baffling synchronicity of their movements. In contrast to the authors of other 'bird books' at the time, who proceeded from the assumption that the birds within a flock were not governed by individual decisions but rather by means of a collective impulse or instinct that simultaneously affected all of them and thus often drew comparisons to the 'hive mind' of swarming bees, Long firmly rejected the idea of such an external authority: "Indeed, I doubt that it ever holds true, or that there is in nature any such mysterious thing as a swarm or flock or herd impulse. [...] In other words, the swarm instinct has logically no abiding-place and no reality; it is a castle in the air with no solid foundation to rest on."[24] He rather believed that the origin of synchronous reactions had to be sought in and between the swarm individuals themselves – for instance on account of his observation that warning signals from one individual on the edge of a flock would 'silently' spread and lead to a collective reaction. By 'silently,' he meant that the warning was not produced by means of any acoustic signal but was rather transmitted in a different, immediate way – in the sense, for instance, of the sort of 'blind understanding' that exists between trusted friends. This impulse could bring about this sort of instantaneous synchronization because, he thought, it was a learned and practiced form of communication that functioned affectively and did not involve the time-consuming interpretation of acoustic signals:

> I conclude therefore, naturally, and reasonably, that [...] my incoming plover changed their flight because one of their number detected danger and sent forth a warning impulse, which the others obeyed promptly because they were accustomed to such communications. There was

23 Ibid., 106–108.

24 Ibid., 109–112. On the idea of an external 'All Mind,' see Newland, *What Is Instinct?*

> nothing unnatural or mysterious or even new in the experience. So far as I can see or judge, there is no place or need for a collective herd or flock impulse, and the birds [...] have no training or experience by which to interpret such an impulse if it fell upon them out of heaven.[25]

Rather, Long believed that it was certain emotions, such as fear, whose transmission was responsible for the coordinated behavior of flocks. This was a transmission of 'oscillations,' so to speak, which (like light waves in the unfamiliar medium of the ether) could move around in a likewise unfamiliar medium and bring about a flock's rapid waves of movement.[26]

Whereas Selous and Long thus denied the existence of vital forces such as 'swarm spirits,' which had been the stuff of poets like Maurice Maeterlinck,[27] the hypotheses of these two nature lovers, which were based on detailed observations recorded into their diaries, nevertheless existed in a poetological context that involved the widespread theories of mass psychology at the time as much as did certain para-scientific influences that were not terribly different from spiritism. Although they argued against the postulate that 'spiritual' influences were being exerted on a flock by an external agent, they assumed that some sort of 'mental' level of communication existed between its individuals. Because concepts such as *information* and *information transmission* were still lacking at the time, Selous and Long relied instead on the idea of *thought transference* – on a form of communication below the threshold of consciousness that functioned in an unclear manner but was not necessarily specific to animal psychology. Rather, they believed that the swarm behavior of animals exhibited a form of organization whose affective communication could, in certain situations, be observed in every life form, and therefore also in humans. At issue was therefore not a new form of animal psychology but rather a level of communication that represented a fundamental condition of possibility for every sort of social life and relativized the special position of human beings in the field of biology.[28] In their studies, para-psychological concepts can be regarded as images intended to express communication-theoretical circumstances for which a suitable vocabulary had yet to be developed.

The way in which Long, and especially Selous, attempted to observe flocks leads to two important conclusions. *First*, the textual form of their

25 Long, *How Animals Talk*, 116–117.

26 Ibid., 117.

27 Maeterlinck's poem *The Life of the Bee* was originally published in French in 1901.

28 See Long, *How Animals Talk*, 121–125.

observations reveals how the flocks of birds – the objects of their research – repeatedly defy being fixed in writing:

> And now, more and faster than the eye can take it in, band grows upon band, the air is heavy with the ceaseless sweep of pinions, till, glinting and gleaming, their weary wayfaring turned to swiftest arrows of triumphant flight – toil become ecstasy, prose an epic song – with rush and roar of wings, with a mighty commotion, all sweep, together, into one enormous cloud. And still they circle; now dense like a polished roof, now disseminated like the meshes of some vast all-heaven-sweeping net, now darkening, now flashing out a million rays of light, wheeling, rending, tearing, darting, crossing, and piercing one another – a madness in the sky.[29]

Despite a growing accumulation of new concepts, the object of observation was not understood with any more clarity. Rather, it remained undefined because of this very surplus, vanishing behind a net of concepts that, instead of capturing the object of description, in fact expanded it with further associations. The succession of verbs, adjectives, and nouns in the text may be a decent prosaic attempt to describe the many parallel events that were ongoing in the flock, but it in no way provides any material that could be subjected to scientific analysis. Swarms, in short, are not a proper object for paper tools.

Second, the works of these amateur researchers demonstrate how observing the dynamics of flocks always entails an act of self-observation. In order to develop a particular perspective on the research object, they either had to eliminate themselves as observers from the observational system by means of camouflage and 'blending in with the environment' or (in Long's case) to do so in a specific way, for instance by appearing as a 'decoy.' Despite such efforts, an anonymous reviewer for the journal *Nature* disqualified Selous's 1932 book about thought transference as untenable – a critique that refers directly to the insufficiencies of observation: "The crux lies, of course, in the interpretation, and the reader may doubt whether the human eye is not deceived by an appearance of simultaneity that is in fact an extreme rapidity of imitative action: the author, indeed, seems to give part of his case away when he describes instances in which the movement could be seen spreading through the flock."[30] The fundamental issue here is thus the spatial and

29 Selous, *Bird Life Glimpses*, 141.

30 Anonymous, '*Thought-Transference (Or What?) in Birds*, by Edmund Selous: Short Review.'

temporal resolving power of the human observational apparatus and its inability to distinguish such things as simultaneity and rapid processes of synchronization. The reviewer doubted, in other words, whether Selous's ornithological recording system had done justice to its object of knowledge.

As much as a decade earlier, the zoologist Robert C. Miller had refuted William Long's ideas about the existence of a telepathic level of communication in flocks of birds. Miller focused instead on what he called a "spread of impulse," which was disseminated through the known sensory organs of the birds and enabled them to coordinate as a collective. In his article "The Mind of the Flock," which appeared in 1921, he repudiated the theories of thought transference, hypnosis, and swarm spirit in one fell swoop. Interestingly enough, the starting point for his critique was Gustave Le Bon, even though the latter had made many assumptions about 'suggestibility' in his work on crowd psychology. In any case, Miller drew a more explicit theoretical connection to popular psychological scholarship than did Selous and Long.[31] As is well known, Le Bon had looked for analogies in chemistry to describe the fact that the collective ability to coordinate is not simply a summation of individual capabilities; rather, something new is at play – a game of relationality: only this can explain how, in chemical processes, elements combine to form new substances whose properties are entirely different from those of the constituent elements themselves.[32] Miller was not convinced by this comparison: "But this analogy, admirably as it states the case, hardly helps us towards an explanation of it, since the origin of the new properties insisted upon is quite as obscure in the one instance as in the other."[33] In Le Bon's view, the coordination of collectives was based on the three factors of suggestibility, contagion, and a sort of collective consciousness, but because Le Bon himself had described contagion as a function of suggestibility, Miller saw no reason at all to regard these three factors as being distinct. All three areas would therefore have to be integrated, Miller thought, in order to provide an adequate description of the 'group mind' that enables flocks to coordinate. What, then, could the incipient field of behavioral studies contribute to such an integration?

Even Miller, in his article, expressed his distaste for anthropomorphizing hypotheses that attempted to explain the organization of bird flocks with reference to leading individuals who were even thought to use 'vocal commands': "Unfortunately I was unable to profit by this information, as

31 Miller, 'The Mind of the Flock.'
32 See Le Bon, *The Crowd*, 15–16.
33 Miller, 'The Mind of the Flock,' 183.

the crows of my acquaintance apparently spoke a different dialect."[34] Such ideas, he thought, had no place in critical studies, because they were little more than mystic speculations and because birds should not be regarded as "diminutive human beings with wings and feathers." Whether it was speculations about the cosmogony of an 'All Mind,' which supposedly bound all life forms together and enabled swarm individuals to organize as elements of this collective intelligence, or whether it was the telepathic capabilities postulated by Long, a 'super-sense' that enabled impulses to be transferred between the individuals – Miller dismissed such explanatory models because their speculative level could not hold up to empirical scrutiny and because they contributed nothing to the scientifically verified facts that had already been discovered. Interestingly, Miller began his argument by pointing out the *imperfect* manner in which flocks coordinate their behavior:

> Unfortunately for such views, the group-mind is not at all the perfect instrument that they assume. It often stumbles in a manner unworthy of an All Mind, and hesitates in a fashion inconsistent with the idea of a perfectly functioning natural telepathy. Furthermore, we are able to trace among gregarious forms a progression from a simple to a complex type of organization; in the case of the more loosely organized groups we are able to explain behavior in terms of known facts of psychology, and it is logical to suppose that greater complexity is a difference, not of kind, but of degree only.[35]

The fact that flocks 'stumble' led Miller to conclude that their collective movements did not take place on a hypothetical level of synthesis but were rather enabled by the familiar sensory organs of the individual birds. It was not, in his opinion, a mysterious form of mediation that governed the simultaneous reactions of a flock's collective motion; they were rather synchronized by means of known sensory organs according to a process. Over the course of this process, moreover, there were moments of disruption brought about by the sense-based transmissions within the collective as well as by external influences upon these transmissions: "When the [...] [birds] behave all as a unit, it is by the method that I have termed the 'spread of impulse'. [...] [T]he impulse spreads, not telepathically, but through the

34 Ibid. For a hypothesis about leadership roles in flocks of birds, see Kessel, 'Flocking Habits of the California Valley Quail.'

35 Miller, 'The Mind of the Flock,' 184.

ordinary channels of sight and hearing, and the flock follows suit."[36] Miller believed that the spread of these impulses could be observed best in loosely organized collectives, where the process takes place more slowly, but also that the same phenomenon (being a quantitative matter, as he thought) must also apply to the transmission of information in denser swarms, in which the impulses would happen too quickly to be observed and yet had to function according to the same principle. They could be stimulated, he argued, by such things as hunger or a perceived threat, and thus they perhaps indicated a basic function of swarming itself. If, for instance, a predator happened to be noticed by just a few individuals, the spread of impulses enabled by the proximity of swarming could serve to warn the collective as a whole:

> If an enemy appears, it is sighted perhaps by only one or a few in the flock; from them the impulse spreads, almost instantaneously in this case, but through the medium of sound, to the others, so that those birds who may not have seen the enemy unite in the 'confusion chorus.' There is nothing in their behavior to suggest telepathy, or any other mysterious type of psychic communication.[37]

Life forms that gather in swarms are especially sensitive to the impulses of their neighbors, which they pass along almost instantaneously, thus enabling the spread of impulses to take place extremely quickly. Dismissing the notion of a 'super-sense,' Miller instead argued that the transmission of signals in flocks of birds was analogous to the conduction of stimuli in the nervous systems of certain invertebrates: "In a medusa, for example, or a sea-urchin, the part of the body immediately stimulated first responds; coordination of action takes place slowly, spreading from part to part, until at least the whole organism is in motion. No part controls the rest. No reactions are controlled by the central nervous system."[38] Thus it could be said that the coordination of avian flocks was now at last being studied from a genuinely biological perspective. This perspective revealed a particular form of animal organization that unfolded as a process of neighboring activities, and it also took into account the reciprocal effects between swarm systems and the environments in which they move. Regarding the latter, Miller discussed the 'circumstances' to which flocks are exposed, and he did so with reference

36 Ibid.
37 Ibid.
38 Ibid., 185.

to the work of the German biologist Jakob von Uexküll: "Von Uexküll has called the sea-urchin a 'republic of reflexes', and remarks ingeniously that 'the legs (spines) move the animal', as contrasted with higher animals, where 'the animal moves the legs'. Whichever part takes the lead depends upon circumstances, and the rest of the body gradually cooperates. [...] The flock behaves as a sort of *primitive organism*."[39]

Here we finally have a perspective on the organization of swarm collectives that does not rely on hypnosis (as in Le Bon's work), telepathy (as in the work of Long and Selous), or on vitalistic collective spirits and social instincts (as in the works of authors like Maurice Maeterlinck and Piotr Kropotkin). Rather, it treats this type of organization as something analogous to nervous systems that exchange signals without any regulatory center. The focus was no longer on a mental level of animal intelligence or a special form of animal psychology but rather on the mere exchange of signals and their representation in movements. As Selous, Long, and Miller agreed, efforts to decipher the precise way in which flocking birds communicate depend on close and detailed observation, and yet the collective dynamics of large and dense swarms push our observational powers to their limits.

Wave Events

In a certain sense, these early efforts to observe flocks of birds can be related to a theory that the Austrian psychologist Fritz Heider developed during the 1920s and first published in 1926 under the title 'Ding und Medium' (Thing and Medium).[40] Heider was concerned above all with the *physical* conditions that underlie (human) perception, and thus he expanded a discourse that had previously been limited to physiological, psychological, and cognitive factors. For him, perception was always linked to a form of mediation, and his text investigates the material constituents that make it possible to connect the function of mediation, on the one hand, with the mediated realities on the other. Heider called the former *media* and the latter *things*.

A central aspect of his theory is the concept of 'wave events' within media, as in the transmission of sound waves in the atmosphere. In the latter example, he wrote, it is not the sound waves that should be regarded as media but rather the physically describable material context that serves

39 Ibid.
40 Heider, 'Ding und Medium,' reprinted as *Ding und Medium* (2005). My quotations are from the reprint.

as the condition of possibility for their creation and dissemination: the air. The wave event that spreads within the medium of the air (for instance) then crashes against things and is transformed in that, by crashing into things, the wave event becomes perceptible. Heider thus invites us, as Dirk Baecker notes in his preface to the essay, "no longer simply to distinguish between things and the 'nothingness' surrounding them; rather, he introduces to the observation of this nothingness the idea of the loose coupling of elements."[41] Heider thus describes mediality in the physical sense as a particular capability of material forms. The latter do not necessarily connect oscillatory events with other oscillatory events in a *causal* manner, but they are susceptible "to being stimulated into oscillatory events from the outside."[42] Or, to quote one of Heider's own examples: "Within things, it is possible to ascertain a stronger and weaker emergence of media characteristics. We can also experience this with our senses. We can feel through a soft body to a firm one. The material is a medium to us just as much as the air, through which we hear and see."[43] Heider thus situates the distinction between thing and medium in their physical density, in the variously tight connections between their elements, with *things* understood as tightly connected multiplicities and *media* as (relatively) loosely connected. In this case, the distinction itself remains differential, for, depending upon the context, a thing can also function as a medium. A wall, for instance, can be an object of perception that reflects sound, but it can also serve as a medium of perception by, for instance, transmitting the sound waves produced by someone knocking on it.[44]

Mediation and mediated are here differentiated into variously concentrated connections between elements, which are internally (as in the case of 'solid' things) or externally conditioned and which can also vary in terms of the way that they exert influence (intermittently or continuously).[45] And in Heider's work, too, the vague figure of the swarm has a place within this differentiation:

> A chair consists of a multitude of elements. How is it that this multitude is nevertheless a unit? It is not only a matter of subjectivity that I unify this multitude into a unit by comprehending it in a certain way. It is not

41 Baecker, 'Vorwort,' 15–16.
42 Engell, 'Zur Einführung,' 303.
43 Heider, *Ding und Medium*, 43.
44 See Engell, 'Zur Einführung,' 303.
45 See Heider, *Ding und Medium*, 36–40.

> a matter of arbitrarily combining a few elements of the chair and a few elements of the air into a unit. That would result in a meaningless unit. [...] [T]he chair is something that distinguishes itself quite well from adjacent things, and also in a purely physical manner. [...] Between the elements of the chair there exists a level of dependency that does not exist between the elements of the air. [...] In short, solid objects do not easily yield; their elements do not shift against one another. [...] The elements of the air separate easily; there is not always the same group in the same arrangement. A swarm of gnats changes this arrangement completely.[46]

The tight connections of solid bodies distinguish them from the medium that surrounds them. A chair, however, would not only cut through the loose connections of air particles; it would also cut through a swarm of gnats. And yet not only can the latter alter the 'even looser' connections of air particle; on account of the internal condition of its dynamic structure, it can also rearrange itself as an ephemeral unity-in-multiplicity.

In order to recognize and perceive such loosely connected arrangements within media, it is necessary, according to Heider, to make the *wave events* occurring within them 'physically effective.' The order that is only latently contained in media has to become explicit. And this requires an appropriate set of analytic tools:

> The individual, which can be extracted from the flurry of events taking place within the medium, must correspond to an objective [*dinglich*] unit. [...] The task of analyzing the multiplicity of events is now to be undertaken by a number of mechanical apparatuses used in conjunction with our sensory organs [...]. All of these apparatuses are similar to resonators, to the extent that they filter out events and suppress anything that distorts or conceals them.[47]

However, the 'primitive organisms' mentioned above (which was Miller's designation for swarms) made it difficult to identify the 'objective unit' in question. The event filters of human perception are simply too crude to function as 'resonators' and produce meaningful results, and behavioral biologists had not yet developed the necessary mechanical apparatuses to serve as more precise filters in the study of swarms.

46 Ibid., 51–52.
47 Ibid., 98–99.

Nevertheless, Heider's work makes it possible to situate the early stages of swarm research in media-theoretical terms, and his concept of tight and loose connections is reflected in the recurring problem that swarm researchers faced in their attempts to differentiate between *unities* and *multiplicities*. This concept also aligns with the idea, which I proposed in my introduction, that swarms became operational not only as objects of knowledge – that is, as 'things' with specific properties – but also as *media* and thus as figures of knowledge and epistemological techniques. In this *media-becoming*, it is swarms that, in a wave event of their own, transmit and make perceptible external impulses and therefore inform the world of things that surrounds them. They mediate between the world of things and the world of media, and this is why they cannot be classified as one or the other. Swarms problematize the space of difference between categories; they are problematic *in themselves*. They are diffuse figures of difference that, as connective *processes*, provide a temporal aspect to the punctual differentiations between *thing* and *medium*.

At least four points can thus be made about the early biological engagement with swarms. *First*, Heider pointed out that organisms can be understood as systems of connections. *Second* – and this is a way of thinking that can be found in Jakob von Uexküll's work as well – organisms are things in the world that, by means of a medial level of senses and motor skills, perceive the world in which they live and produce knowledge by means of their own activity.[48] *Third*, Heider's theory addresses the problem of extracting underlying and latent orders from media – orders that have to objectify themselves and that become perceptible when their wave event is interrupted with an apparatus. *Fourth* and finally, Heider argued that organisms have an interactive relationship with their environment that is physical – a physicality that is not mechanically invested but at least energetically so (though not necessarily with any clear causality) and that, beyond the concept of connection, could also be applied in later discourses to the concept of information.[49] This focus on the physical also defines the approaches adopted by the other significant biologists and natural scientists who will be discussed in this chapter and who, in efforts to establish a natural-scientific basis for their research, likewise distanced themselves from psychic and psychological approaches.

48 See Baecker, 'Vorwort,' 19.

49 This is neither here nor there, but it somehow seems necessary to point out that it was Niklas Luhmann who had rediscovered Heider's 'Thing and Medium' in the 1970s and incorporated the ideas of that essay into his own distinction between form and medium. See, for instance, Luhmann, *Die Politik der Gesellschaft*, 30–31.

The Psychology of the Fish School

> Man in his thirst for knowledge / ponders and studies his whole life long / only to be resigned to see: / At bottom, he can understand nothing.[50]

Although the hypotheses of Long and Selous were not of lasting influence, the way in which they studied animal behavior became paradigmatic for later ethology. Their recording systems may not have provided the most accurate information about the activity of swarm collectives, but they contributed a good deal to the establishment of a specific behavioral-biological research perspective within the field of biology. Their methodological innovations more than made up for the flimsiness of their theories. That said, it remains questionable how, from the elegiac descriptions of "all that one can see" (in Selous's words), it might be possible to distill any data that could be useful for scientific analysis. How could the experiential knowledge of observation and description ever yield any truly scientific results beyond the mere production of a "taxonomy of living beings"? Such questions were not addressed by the field researchers at the time; in fact, they could not even be raised until the next generation, when certain improvements had been made in the ability to measure what is visible. And although their hypotheses tended in part to be somewhat mystical, they were nevertheless able to identify the organizational authority of avian flocks – much like researchers such as William Morton Wheeler, who at that time was conducting field studies with social insects[51] – in the interactions that took place among the swarm individuals: in the relationality between swarms and their underlying processes of exchanging signals.

One answer to these questions was provided by the zoologist Karl von Frisch. In 1938, Frisch, who is best known today for his work on the perception and communication of honey bees (for which he was awarded a Nobel Prize in 1973), was likewise studying the behavior of swarms. His focus at the time was not on birds or bees but rather on fish or, to be more precise, on Eurasian minnows (*Phoxinus phoxinus*). These are small fresh-water fish that can be found throughout Europe. His study illustrates the great extent to which 'animal psychology' transformed during the first decades of the

50 Frisch, *Erinnerungen eines Biologen*, 183: "Der Mensch in seinem Wissensdrang / Sinniert und forscht sein Leben lang, / Um dann verzichtend einzusehn: / Im Grunde kann er nichts verstehn."

51 See, for instance, Wheeler, *Social Life Among the Insects*; idem, 'The Ant-Colony as an Organism'; and idem, 'On the Founding of Colonies by Queen Ants.'

twentieth century and the wide range of research practices and contexts in which the concept was applied. It also illustrates the operations and strategies of scientification that were undertaken to comprehend swarms and other animal collectives as worthy objects of knowledge and to fill previously vague psychological concepts with meaningful content. While observing a school of minnows in the open water over the course of several weeks, Frisch made the following 'chance observation':

> After a month, our minnows were so trusting that they allowed themselves to be touched without any shyness. One could splash around them in the water, and not even this would chase them away. Then one of them got caught on the sharp edge of the metal feeding tube. The others watched their floundering comrade until he was freed by me and swam away. At first it was like Job's news disseminating throughout the entire school. A growing sense of unease began to spread, and after a while – perhaps half a minute had passed – all of them fled away and the enticing food was left untouched.[52]

Frisch was fascinated by this reaction, and so he designed special experiments to explain the fish's behavior. At various places in the pond and at varying intervals, he would remove a single minnow from its school and then return it to its fellows a few minutes later in order to see whether the victim was somehow able to communicate this negative experience to the others. Yet the school never reacted. Next, he tried to determine whether there were certain movements that an injured fish might enact to warn the rest, but his observations provided no evidence of any warning signals:

> When I scratched the tail of a minnow with my forceps, the entire school once again fled away in fear. When we discovered that minnows have an outstanding sense of hearing and that they can also produce sounds under certain circumstances, our first thought was that an injured fish could incite its comrades to flee by means of a warning signal. This, however, was not the case. I removed another minnow from the school and crushed its head with the forceps so that it was instantly dead and thus also, of course, entirely mute. Then I punctured its bladder (because otherwise it would not sink) and let it fall underwater to the feeding station. […] [The others] nibbled at it, and it often took several seconds – at most half a minute to a minute – before they noticed something. Then it would look as though

52 Frisch, 'Zur Psychologie des Fischschwarms,' 601.

> something abominable had dawned on them. They would withdraw from the prey, some of them seemingly terrified; a bout of aimless scurrying would follow, and often the entire school would then form into a dense mass. After that, even the most minor event would cause the school to rush away and vanish into deeper water.[53]

Could it be that the mere *sight* of an injured fish might arouse fear in its fellows? In order to test this hypothesis, Frisch mixed a chopped-up minnow into the food that fish enjoyed (mashed earthworms) and tossed the "unrecognizable clumps" back into the water. In this case, too, a short period of time would pass before the school exhibited the same frightened behavior, even though "the sharpest eye [...] could not have recognized the form of the fish." The same thing happened when he poured "minnow extract" (that is, water in which chopped minnow had been placed for a given period of time) near their food source, which pointed to the conclusion that the school's reaction must have something to do with a chemical connection.[54]

Frisch now began to search for the location of this chemical material. A series of experiments revealed that the reaction could be induced by the ground-up skin of the minnows. And in order to learn whether this reaction was species-specific, Frisch conducted a number of experiments with the skin of other types of fish: Would the minnows be scared off by these as well? He wanted the latter tests to be unaffected by the moods of schools swimming in the wild, which, because they could be frightened by any number of factors and because they would swim away and never return, would be "unusable" for further experiments. Frisch thus recreated a lake "on a smaller scale in the laboratory."[55] Placed in an 80 × 50 × 60 cm aquarium, the minnows exhibited – much to Frisch's satisfaction – the same behavior as in the open water. Here, however, "the individuals were much easier to see. [...] The good view over all the fish participating in the experiment also enabled me to carry out an objective protocol [...] with a stopwatch in my hand."[56] He was thus able to create detailed tabular protocols in the sense of the behavioral *ethograms* that would become essential to the methodological inventory of natural-scientifically driven ethology.[57]

53 Ibid., 602.
54 See ibid., 602–603.
55 Ibid., 603.
56 Ibid.
57 See Jahn et al., eds., *Geschichte der Biologie*, 583.

Frisch took his experiment systematically further by making detailed lists of all the fish species whose skin did not cause any fear (all outside of the family to which the minnows belonged) and he determined, by cutting off the corresponding nerve fibers, that it was the sense of smell and not the sense of taste that allowed the fish to perceive and react to the skin of their fellow minnows. Finally, he posed the question of the biological utility of this phenomenon: Did it protect the school from suffering additional losses, in the event that a predator had captured and injured one of their own? Then why would schools often distance themselves just a few meters from the place they were fleeing and then soon rejoin there as though nothing had happened? According to Frisch, this question, and that of whether such behavior was also a feature of other schooling fish, would require further experiments – his article was only meant to be a point of departure and not the final word.[58]

Frisch's article makes it clear how far, in the field of ethology, the concept of animal psychology had come to differ from previous approaches, which literally sought to assign anthropomorphic and sociomorphic characteristics to the behavior of animals and animal collectives. Frisch's approach was entirely different. Instead of presenting an anecdotal series of accounts about the behavior of animals in certain situations, he developed a systematic method that gradually excluded environmental factors and various senses that might have been involved in the minnows' reaction. His experiment was thus sensitive to a critique of any 'psychology of animals' that had been voiced by Nikolas Tinbergen (among others): that which cannot be objectively observed can also not, in a strict scientific sense, be the alleged cause of certain behavior. Rather, such causes would have to be ascertainable by means of natural-scientific methods. For ethology, in short, 'psychological' ascriptions such as fear could only be a conglomerate of external and internal conditions for processes within an organism or collective, and such conditions manifest themselves in regular and measurable symptoms: heart rates, dilated pupils, trembling, frightened reactions, becoming still, or making violent movements. These symptoms, in turn, are associated with sensory organs, nervous systems, hormones, and muscles that all work together in particular ways.[59] From this perspective, things like fear or terror can be broken down into a wide variety of micro-relations within animal organisms and collectives and between such organisms and their environment.

58 See Frisch, 'Psychologie des Fischschwarms,' 606.

59 See Tembrock, *Angst: Naturgeschichte einer psychobiologischen Phänomen*, 17–18.

In the example at hand, the psychology of a school of fish was broken down into basic sensory functions, and various experimental processes of elimination revealed that this psychology was species-specific and that it was based on the fish's sense of smell. In conducting these experiments, Frisch relied on the controlled environment of an aquarium; here a medial, observational *dispositif* was a fixed element of his ethological research. As much as possible, influencing factors were isolated and systematically tested, and the behavior of the school of minnows was analyzed statistically. Instead of using vague terms such as 'sociability' or 'altruistic instinct,' which Piotr Kropotkin (for instance) employed in his work on cooperative behavior,[60] Frisch endeavored to conduct 'empirical social research.' His studies involved making comparisons with other types of fish in order to identify typical features and reactions. His text, moreover, lacks any anthropomorphic terminology whatsoever; instead, Frisch made use of straightforward concepts to describe the execution and observation of his experiments, and he quantified the results in ethograms. In a footnote, he even made a point to mention that any textual description of the observed phenomena would be insufficient: "When lecturing, I have shown a film to demonstrate the behavior of minnows in the aquarium while eating normal food, while eating food mixed with perch skin, and their frightened reaction to eating food mixed with minnow skin. Words are unfortunately an insufficient substitute for the immediate impression."[61] This reads precisely like a commentary on researchers such as Selous. His footnote underscores the importance of moving images to describing the movements and dynamics of swarms.

The development of the concept of animal psychology into a chemical-physiological approach, as in Frisch's work, or into kinetic-physical models did not win universal approval in field of ethology. In his book *On Aggression*, which was originally published in German in 1963, Konrad Lorenz, who was likewise a Nobel Prize winner and a proponent of empirical and natural-scientific ethology, described the formation of fish schools in anthropomorphic terms as something that is prohibitive of democratic decision-making. Like Frisch, Lorenz was also interested in the frightened reactions of minnows. However, he described them in a far more undifferentiated manner as an amorphous collection of identical elements and spoke simply of an 'infection' through which perceived dangers were circulated. He also introduced the well-known but rather vaguely defined

60 See Kropotkin, *Mutual Aid: A Factor of Evolution.*

61 Frisch, 'Psychologie des Fischschwarms,' 605 (footnote).

concept of the 'herd instinct,' which, he believed, had certain political implications:

> Inside the shoal there is no structure of any kind, there is no leader and there are no led, but just a huge collection of like elements. Of course these influence each other mutually, and there are certain very simple forms of 'communication' between the individuals of the shoal. When one of them senses danger and flees it infects with its mood all the others which have perceived its fear. How far the panic in a big shoal spreads, and whether it is able to make the whole shoal turn and flee is purely a quantitative question, the answer to which depends on how many individuals become frightened and flee and how intensively they do so. [...] The purely quantitative and, in a sense, democratic action of this process called 'social induction' by sociologists means that a school of fish is the less resolute the more individuals it contains and the stronger its herd-instinct is. [...] Again and again, a small current of enterprising single fish pushes its way forward like the pseudopodium of an amoeba. The longer such pseudopods become the thinner they grow [...]. Generally the whole advance ends in precipitate flight back to the heart of the school. Watching these indecisive actions one almost begins to lose faith in democracy and to see the advantage of authoritarian politics.[62]

In this context, Lorenz also summarized the results of an experiment that had been conducted with minnows by Erich von Holst. The latter involved locating the precise part of a minnow's brain that governs its ability to participate in a school. Having allegedly done so, Holst removed the part in question from a few specimens and returned them to their schools. The 'pithed' minnows swam around fearlessly without any regard for the school's cohesion, so much so that they became the leaders of their respective groups. This is because they demonstrated, according to Lorenz, a sort of decisiveness that was lacking in the normal individuals. He criticized the lack of autonomy exhibited by ordinary minnows (and ordinary men as well), though in this case, ironically, the 'outstanding' individuals were those with brain damage.

2. On the Edge

As we have seen in the previous sections, the early research efforts in the budding fields of ethology and biological behavioral science lay outside of

62 Lorenz, *On Aggression*, 139–140.

mainstream research interests, and the first impulses behind biological swarm research derived from methods used in non-biological disciplines. As a field of study, swarm research operated on the margins of biology as the latter was developing into a professional scientific discipline. It is perhaps no surprise, then, that the scientific study of biological collectives did not take place at large university institutes and the laboratories associated with them but rather on the periphery, namely where smooth and striated space converge, where charted knowledge encounters the dynamics of constant motion, and where the land meets the sea: in marine-biological research stations. What is particularly interesting, however, is that the swarms being studied in these places evoked additional 'marginal phenomena.' These methodological and operative marginal phenomena will be the focus of my discussion below.

Of course, even the marginal zones of biology had their research centers. For several decades, one of these was the Marine Biological Laboratory (MBL), which is located in the small town of Woods Hole on Cape Cod. It was founded on a dictum pronounced by the Swiss-American zoologist Louis Agassiz: "Study nature, not books."[63] During the summer months, Woods Hole became a "seaside magnet" for the brightest biologists in the United States, and Charles Otis Whitman, the pioneer of American behavioral biology, not only gave his groundbreaking lectures on animal behavior there but also served as the laboratory's founding director from 1888 to 1908.[64] At Woods Hole, the study of such questions could be approached through a combination of field work and laboratory experiments. Thus it was possible, on the one hand, to subject natural settings to experimental analysis and, on the other hand, to verify approaches from the laboratory by means of the wide evolutionary panorama of marine life. These studies would also form the basis for Warder C. Allee's work on animal aggregations, which I will discuss later in this section.

Woods Hole can thus be regarded as a 'lighthouse project' within a particular historical and historiographical constellation in which classical natural-historical research branched into new biological-scientific areas of study. Ethology and long-term 'life-history studies' of living organisms gradually found homes through the establishment of a large number of new

63 For the history of the Marine Biological Laboratory, see its homepage at http://www.mbl.edu/ (accessed 17 February 2018). As early as the 1870s, Agassiz had founded, in a town near Woods Hole, a sort of precursor to the MBL that was likewise devoted to the study of marine life.

64 See Burkhardt, *Patterns of Behavior*, 19. Here Burkhardt refers in particular to Whitman, 'Animal Behavior.'

research laboratories for zoological studies.[65] And if the aim was to study highly developed life forms, then this often involved marine research stations that had been founded with certain economic incentives in mind. After all, the study of how organisms interact with their environment, which today we would call 'ecological' approaches, and the discovery of knowledge about the migratory and breeding behavior of economically valuable (schooling) fish such as herring or mackerel often received enormous financial support from governments. In such cases, as in the German research station founded on Helgoland in 1892 (and at similar institutions founded by other European countries), research projects driven by the interests of the fishing industry were often on equal footing with foundational biological research: "In Germany," according to Lynn K. Nyhart, "a surprising number of zoologists ended up working for the fisheries industry in the decades around the turn of the century."[66] As at Woods Hole, the possibility of combining and contrasting observations 'from open nature' and experiments in laboratory aquaria represented a methodological and epistemic leap forward that should not be underestimated, especially in comparison to the efforts of ornithologists to observe flocks of birds. Despite the numerous challenges and problems that they presented, which I will discuss in the next chapter, schools of fish could at least be studied under more or less controlled conditions and in closed and monitorable aquaria. This was not a luxury enjoyed by those researching avian flocks.

Seeing Fish: Between Observation and Experimentation

Against the backdrop of the developing landscape of biological studies and research directions outlined above, the beginning of the detailed study of the organization and interactions of fish schools can be dated to 1930. Yet even before that, within the context of the aforementioned establishment of marine research stations around the end of the nineteenth century, studies were conducted – by the British biologist William Bateson, for instance – that examined not only the optimized use of bait but also the 'shoaling' behavior of mullets.[67] I have singled out his essay 'The Sense-Organs and Perceptions

65 See Nyhart, 'Natural History and the "New" Biology,' 436–438.

66 Ibid., 436.

67 See Bateson, 'The Sense-Organs and Perceptions of Fishes.' Bateson also made essential contributions to the rediscovery and popularization of Gregor Mendel's theory of inheritance and even coined, in 1906, the term *genetics*. Incidentally, the word *shoal* designates a loosely organized collective of fish swimming around in the same area, while *school* designates a group of fish swimming together in the same direction.

of Fishes' as an example because it was also groundbreaking in its reflections about methods and in its awareness of the problems involved with generating knowledge from 'blurry' phenomena. This explicit engagement with the complex system of fish schools can be regarded as a sort of precursor to cybernetic thinking (as William's son Gregory Bateson later asserted, his father "was certainly ready in 1894 to receive the cybernetic ideas").[68] The reciprocal effects between swarm individuals and their sensory basis, which Gregory Bateson would discuss at the Macy Conferences sixty years later, had already captured the interest of William Bateson by the end of the nineteenth century.

Already here, as in later generations of texts and experiments related to swarm research, one sees the same epistemic divide that had characterized the early observations of ornithologists and would not be closed for decades. Regardless of the fact that the sensory organs of aquatic life forms are hardly comparable to those of terrestrial animals, simply because they operate in different 'living conditions,'[69] William Bateson stated the fundamental problem involved with studying schools of fish: "From the nature of the case, moreover, satisfactory evidence as to their conduct in the wild state is scarcely to be had, so that it is necessary to depend largely on observations made upon them while living in tanks."[70] Therefore, it must be kept in mind that, strictly speaking, his findings applied only to the behavior that fish exhibited under artificial laboratory conditions – even though great effort was made, of course, to recreate their natural living conditions as accurately as possible.[71]

He conducted his experiments at the Marine Biology Association in Plymouth, which had been founded, like the institute at Woods Hole, in 1888. Here, in designing the aquaria and physiological laboratory, scientists applied the most advanced knowledge of fishkeeping at the time to approximate the natural living conditions of the fish being studied.[72] The prominent zoologist Anton Dorn, who was then working at a research aquarium in Naples, was brought to the site several times to serve as an adviser.[73] For his own observations, Bateson used the longest of three rectangular tanks

68 Bateson, *Steps to an Ecology of the Mind*, xxii.

69 See Bateson, 'The Sense-Organs and Perceptions of Fishes,' 225: "To interpret their behaviour by comparison with our own is even more clearly an inadequate treatment than it is in the case of the other lower animals."

70 Ibid.

71 See ibid., 226.

72 See Allen and Harvey, 'The Laboratory of the Marine Biological Association at Plymouth.'

73 See Southward and Roberts, 'The Marine Biological Association 1884–1984,' 162.

in the basement of the facility, though it seems as though he did not use any technical observational media in his research. Instead, he simply described in great detail what he had been able to *see*:

> By day the whole shoal of about fifty little ones stays together more or less. [...] At night they lie *on the surface* of the water, and seem not to swim about as a body, nor are their heads all pointing one way as they generally are by day. The shoal seems at no time to have any leader, but will sometimes follow the front fish until one of those that are behind makes a dart elsewhere, when the whole shoal turns round and follows. They certainly have no tendency to follow the largest fish in the shoal, or indeed any fish in particular.[74]

Bateson made three main observations. *First*, he stressed his impression that fish schools should be regarded as a single *body* that sets itself apart from its immediate environment with its marginal edge. *Second*, this body functions without a leader; the schooling individuals follow neither the largest nor the strongest fish among them – there is no 'pack leader' in a school. And *third*, the schooling body is defined by the directionality of the individuals that are swimming in parallel. However, Bateson made all these observations by daylight; in the case of mullets, for instance, their behavior thus seemed to be strongly dependent on their sense of sight.[75] At night, his mullets allegedly (and strangely enough) preferred to sleep comfortably on the surface of the water instead of shoaling or schooling. He did not, however, extend this conclusion to other types of fish but rather supposed that species like herring maintain their schools during the night as well.[76]

Bateson's notes on the shoaling behavior of mullets are a collection of the observations that he made through the glass walls of a large aquarium, and he hypothesized that the sense of sight was highly relevant to the successful formation of a shoal. At the same time, his approach is indicative of a second sense of sight, namely a specific 'biological gaze.' Evelyn Fox Keller has pointed out the great importance of observation in the history of biology, which the Canadian biologist N. J. Berill once called an "eminently and inherently visible science."[77] The entire nineteenth century was pervaded

74 Bateson, 'The Sense-Organs and Perceptions of Fishes,' 249–250 (emphasis original).

75 Ibid., 250.

76 Ibid.

77 Keller, *Making Sense of Life*, 211. Here Keller refers to Berill, 'The Pearls of Wisdom: An Exposition,' 4.

by the question of whether observation or experimentation was the more adequate method for biological research.[78] Throughout, (media) technologies played a decisive role to the extent that they expanded the realm of the observable and always provided new manners of observation. This was especially so in the field of animal physiology, in which the primacy of observation over experimental methods endured far longer than in fields such as botany, where experimental approaches were used as early as the eighteenth century.

Toward the end of the century, when Bateson was conducting his experiments in Plymouth, this dispute over methods had still not run its course, at least not in Germany, even though prominent physiologists such as Emil Du Bois-Reymond or Hermann Helmholtz had of course been working with experimental methods as early as the 1860s. For instance, in his 1897 book *Mechanik und Biologie*, which was directed against Wilhelm Roux's work on developmental mechanics (*Entwicklungsmechanik*), Oskar Hertwig still cited Johannes Müller's comparison of observation and experimentation. In a lecture delivered in 1824, the latter had remarked: "Engaging with living nature takes place by means of observation and experiments. Whereas observation is simple, assiduous, diligent, honest, and without any preconceived opinions, experiments are artificial, impatient, busy, flighty, passionate, and unreliable."[79] For Roux, on the contrary, "descriptive research methods are incapable of providing any sure evidence for causal connections"[80] – a statement that owes much to the French physician Claude Bernard, who in 1865 had defined biology along the lines of a 'hard' natural science: "The necessary conditions of every phenomenon are absolutely determined. [...] To have determinism for phenomena, in biological as in physico-chemical sciences, we must reduce the phenomena to experimental conditions as definite and simple as possible."[81]

Bateson's studies took place at the intersection of these two methodological approaches. At the beginning of his text, he remarks that he had been

78 See Querner, 'Die Methodenfrage in der Biologie des 19. Jahrhunderts.'

79 The lecture – 'Von dem Bedürfnis der Physiologie nach einer philosophischen Naturbetrachtung' – was published as the first chapter of Müller's *Zur vergleichenden Physiologie des Gesichtssinnes des Menschen und der Thiere nebst einem Versuch über den menschlichen Blick* (Leipzig: C. Cnobloch, 1826), 1–38. Müller's text is quoted here from Querner, 'Die Methodenfrage in der Biologie,' 420.

80 Roux, 'Aufgaben der Entwicklungsmechanik der Organismen,' 75. Quoted from Querner, 'Die Methodenfrage in der Biologie,' 420.

81 Bernard, *An Introduction to the Study of Experimental Medicine*, 67, 71. The original French edition of this book appeared in 1865.

appointed "to make observations on the perceptions of fishes," and yet a few lines later he also writes: "In addition to this I have also made some experiments."[82] Moreover, his problematization of the artificial laboratory situation, which I discussed above, brings his observations awfully close to Johannes Müller's 'artificial experiments.' In general, it is difficult to make a conceptual distinction between the two methods, given that observations are always a part of experimental processes. Classical empiricists such as John Stuart Mill or even Claude Bernard already regarded experiments as a special form of observing natural processes – *induced* observations, as Bernard called them. In experiments, possibilities for observing facts are *created* by varying the parameters that operate differently in nature.[83] This strategy of *discovering* a putative hidden reality can be radicalized further by treating all 'facts' as social constructs (as Bruno Latour does) or by regarding them as purely technical effects of experimental arrangements – and thus as the *invention* of reality.[84] Experimental observation does not proceed in a purely descriptive manner but rather interrogates natural phenomena within the framework of particular ideas or theories: "Experiments rub up against the reality that they are asking about," as Michel Serres and Nayla Farouki maintain.[85] A few additional criteria apply as well: the experimental conditions have to be controlled so that an experiment can be reproduced and tested. The philosopher Allan Franklin has defined the experiment as follows:

> One of its important roles is to test theories and to provide the basis for scientific knowledge. It can also call for a new theory, either by showing that an accepted theory is incorrect, or by exhibiting a new phenomenon that is in need of explanation. Experiment can provide hints toward the structure or mathematical form of a theory and it can provide evidence for the existence of the entities involved in our theories. Finally, it may also have a life of its own, independent of theory. Scientists may investigate a phenomenon just because it looks interesting. Such experiments may provide evidence for a future theory to explain.[86]

82 Bateson, 'The Sense-Organs and Perceptions of Fishes,' 225.

83 See McLaughlin, 'Der neue Experimentalismus in der Wissenschaftstheorie,' 211–212. For an in-depth discussion of the relationship between theory, observation, and experiments, see Hacking, *Representing and Intervening*, 149–275.

84 See McLaughlin, 'Der neue Experimentalismus,' 214.

85 Serres and Farouki, eds., *Thesaurus der exakten Wissenschaften*, 252.

86 Franklin, 'Experiment in Physics.'

In his monograph *Unter Beobachtung* (Under Observation), Christoph Hoffmann rehabilitated the importance of the observer, which has been underestimated in the history of science. Scholars like Ian Hacking, for instance, have thus downplayed efforts (like those of William Bateson) to observe and report things, claiming that they have been less important than getting "some bit of equipment to exhibit phenomena in a reliable way"[87] – less important, that is, than constructing a functional experimental system with reproducible results. As sources of knowledge, the objects being studied are no longer the primary focus of observation. For, as Hoffmann notes, "not only has the task of observation been largely externalized; under these conditions, observing can no longer be the primary way to register and identify a phenomenon with the senses. Rather, observing is defined [...] as viewing, inspecting, and evaluating a technically stored occurrence that is being displayed by an apparatus."[88] What has taken the place of perception is information, which is mediated by technical apparatuses, and the act of 'being attentive' to the latter. As an activity, observing has become a skill.[89]

This development was inevitable in the field of fish-school research, given that the observation of schooling fish will always overtax the capacities of human perception (as noted above, ornithologists face the same problem). Formally, such sense-defying noise calls for an 'informatization' and requires a careful analysis of technical images (in this case). The same limitations of the senses, as Hoffmann discovered, were discussed long ago in Jean Senebier's work *L'art de observer*: "The senses are too limited to be aware of everything and too imprecise to gauge everything properly; they need just as many aids to discover microscopic creatures as they do to see the moons of Saturn."[90] Technical apparatuses of observation extend the capabilities of the human senses; they become, in Hoffmann's words, "second-order tools that can be understood as building upon the first-order tools of the senses. In the practice of observation, the logical subordination of tools to the senses is reversed into a subordination of the senses to the tools."[91]

In this two-sided reversal of order, observation blends together with technical apparatuses and various experimental approaches into an epistemological amalgamation that never fully hardens. The goal of *experimental systems* – which, though an ahistorical concept in this case, are nevertheless

87 Hacking, *Representing and Intervening*, 167. Quoted from Hoffmann, *Unter Beobachtung*, 23.

88 Ibid., 25.

89 Ibid., 26.

90 Senebier, *L'art de observer*. Quoted from Hoffmann, *Unter Beobachtung*, 43.

91 Hoffmann, *Unter Beobachtung*, 43.

quite applicable to fish-school research – is not simply to confirm theoretical hypotheses empirically, as Karl Popper claimed.[92] Drawing on the work of Ludwik Fleck, Hans-Jörg Rheinberger has maintained that such systems are rather distinguished by their irreducible fuzziness. For, "[i]f a research experiment were well defined, it would be altogether unnecessary to perform it. For the experimental arrangements to be well defined, the outcome must be known in advance; otherwise the procedure cannot be limited and purposeful."[93] It is only as something vague that experimental systems have the potential to provide previously unknown answers, and to do so to questions that the researchers involved might not have even known to ask. Only in such a way do they become a "machine for making the future,"[94] which is somewhat akin to writing 'synthetic histories.'

Yet even in light of the vague orientation of experimental systems, the unclear status of swarms poses a fundamental problem. They are not just an aggregation of individuals, which would mean that they could be reduced to their elementary components. They do not function according to the rules of multipliable inter-individual mechanics, even if their behavior is seemingly stereotypical and reducible to a few rules. At the moment when a swarm is formed, such a quantification transforms into the new qualities of the total system, qualities that cannot be causally derived from the level of the individuals. As a total system, the swarm is neither 'dividual' nor a new form of individual; rather, it is both at the same time, *in* time.

Already in Bateson's work and in later systematic biological research on schools of fish, the creation of the simplest possible experimental conditions (in Claude Bernard's sense of the term) was an impetus in the production of knowledge. In order to cope with the complexity of the total system, researchers presupposed the existence of a causally describable mechanics that would enable them to break down the interactive conditions of single swarm individuals into basic, isolatable processes. In doing so, the scientists examining swarms conflated the questions of simple experimental conditions and good observational conditions; around 1890, observation and experimentation thus seemed to enter a productive relationship. At the beginning of the twentieth century, this relationship was further fostered by the increasing demand for 'objective' research methods, a trend which

92 See Popper, *The Logic of Scientific Discovery*, 89. Hans-Jörg Rheinberger has located the origin of the concept of "experimental systems" in biochemistry and molecular biology. Here I am using the term in Rheinberger's sense, namely as the totality of epistemic and technical things. See Rheinberger, *Toward a History of Epistemic Things*, 24–37.

93 Fleck, *Genesis and Development of a Scientific Fact*, 86.

94 Jacob, *The Statue Within: An Autobiography*, 9.

had already begun, as I discussed above, to transform the field of animal psychology into comparative behavioral research or ethology. This was a shift that, in the case of studying schools of fish, essentially depended on laboratory media and observational media that defined the existence of the ever-changing appearance of the school in question.

The field of tension between observation and experimentation would later be replicated in the similarly problematic relationship between experiments and (mathematical) models, which themselves depend on visual evidence: "In fact, its particular reliance on visual evidence may shed some light on the troubled history of mathematics in biological science, [...] [which] is rather a tension between imagining and seeing – that is, an opposition between what may be imagined with the help of mathematical and mechanical models and what can actually be seen with one's own eyes."[95] These tensions erupt in the methodological distinction between that which can been seen with one's own eyes and that which can only be imagined with the help of experimental methods and mechanical or mathematical models: a tension between 'imagining and seeing' in which – to borrow Bernard's terms once again – the imagination is indispensable to studying natural processes. Bold, free, intuitive, and willing to contradict existing theories, an imagination of this sort is endowed with the 'experimental spirit.'[96]

The Psychomechanics of the Periphery

William Bateson relied on his 'biological' sense of sight in order to approximate this 'seeing-in-time.' More than thirty years passed before his research would influence another "working hypothesis,"[97] which examined schools of fish "in the light of cold reason"[98] from a similar point of departure but was, in epistemic terms, ultimately more oriented toward a strategy of modeling. Its author, the Norwegian marine biologist Albert Eide Parr, likewise developed his theory on the basis of observing fish in an aquarium:

> While watching the movements, especially the milling of a small school of chub mackerel [...] in captivity in the tanks of the New York Aquarium, the author perceived the possibility of a comparatively very simple set of

95 Keller, *Making Sense of Life*, 211.

96 See Bernard, *An Introduction to the Study of Experimental Medicine*, 206. Quoted in Querner, 'Die Methodenfrage in der Biologie,' 427.

97 Parr, 'A Contribution to the Theoretical Analysis of the Schooling Behavior of Fishes,' 1.

98 Spooner, 'Some Observations on Schooling in Fish,' 422.

> reactions, which would explain the apparently complicated and mysterious behavior of the fishes in question [...], in the hope that it may arouse the interest of those who by good fortune or occupation will be able to gather further information concerning the very interesting phenomenon of schooling among fishes.[99]

From his 'watching,' Parr formulated a principle of 'chaotic' or self-organizing systems *avant la lettre* to account for the creation of complex structures or behavioral patterns that are based on just a few simple rules – rules which, in the physiological tradition, he called 'reactions.' Parr, too, regretted some of the factors that hindered the study of schooling fish:

> It is most unfortunate that the species showing the schooling performances most clearly, as for instance herrings, sprats and mackerels, usually are of such delicate nature that it is practically impossible to keep them alive for any great length of time or in any numbers in captivity, and the opportunities to make observations in the field, though not infrequent, are too dependent upon chance to be especially pursued by a single student.[100]

In a laboratory setting, as I have already mentioned in my previous chapters, it can even be difficult to study those species of fish that school in a stable manner – in the sense of there being a mutual attraction between the schooling individuals over a long period of time and regardless of environmental influences – and that maintain a constant state of motion.[101] Yet it was this very type of school that interested Parr and is also the media-theoretical focus of the present book: swarms that become the environment itself, so to speak, and whose 'independent character' with respect to the external factors that might influence their formation suggests that their existence can only be a *relational existence*, and one that depends on the internal factors between the swarm individuals and the swarm as a whole.[102] Parr's suspicion that the mysterious formation of fish schools could be broken down into simple mechanisms of communication between animals went

99 Parr, 'A Contribution to the Theoretical Analysis of the Schooling Behavior of Fishes,' 1.

100 Ibid.

101 See ibid., and Spooner, 'Some Observations on Schooling in Fish,' 422. Today, for instance, the Leibniz Institute of Marine Sciences at the University of Kiel has one of the few aquaria in which herring are kept in captivity. See https://www.aquarium-geomar.de/ (accessed 22 February 2018).

102 See Parr, 'A Contribution to the Theoretical Analysis of the Schooling Behavior of Fishes,' 2.

against the grain of the tendency, which was still popular at the time, to psychologize and anthropomorphize animal behavior and thus to explain the formation of swarms as an effect of certain 'social instincts.' To this idea, Parr replied: "The internal factor keeping together a herd of animals of any type is generally referred to as a social instinct [...]. The term, however, seems void of any logical definition or analytical description, and the meaning it conveys is therefore very vague."[103]

Parr did not believe that such a 'conscious' act among the individuals to form into a school for the sake of their own protection, for instance, could explain the 'exquisite harmony' with which the fish moved together in terms of their speed, direction, and maneuverability. There is not enough time or space to make decisions of this sort. Moreover, Parr also observed that, "under certain circumstances" (namely when having to make a 180-degree turn, which, in the case of observing a school of fish on land, was usually caused by the wall of an aquarium), the fish would form into a toroidal structure, which also spoke against the idea of the 'conscious' formation of the collective.[104] In this case, the schooling individuals would constantly swim in a circle and were seemingly unable to break from this circular movement without the influence of sufficiently strong external stimuli – without, for instance, the factor of 'fear' being injected into the system. This stereotypical behavior seemed to confirm Parr's suspicion that the mystery behind the formation of schools had nothing to do with social behavior and that the apparent 'society' of fish was rather the *result* of mechanically integrated and automatically occurring reactions.[105]

Given that Parr relied to some extent on the psychological concept of 'fear,' which he left insufficiently defined, it must be said that his work contains some of the very features that ethnologists such as Willian Morton Wheeler and, later, Konrad Lorenz or Nikolaus Tinbergen would dismiss as unscientific. That said, Parr's approach nevertheless shifted the understanding of external influences: he was not interested in how such a factor could be defined in detail but rather in how it became effective and could modulate the spatial structure of the school. 'Fear' became visible and legible in characteristic collective movements. Studies of schooling fish thus revealed particular behavioral programs that would come into play in response to the visible threat of a presumed predator. Schools would visualize these

103 Ibid., 2–3.

104 In the next chapter I will discuss in greater detail the problems that rectangular aquaria pose for the continuous motion of fish schools.

105 Parr, 'A Contribution to the Theoretical Analysis of the Schooling Behavior of Fishes,' 3.

microdynamics by dynamically modulating their global structure – by enacting a behavioral program that regulated the disruptions and 'fear' caused by predators in a productive way.[106]

The basic unit in Parr's studies was thus any pair of fish belonging to a "species which habitually live in schools," and in such pairs he identified three main types of behavior: immediate attraction between the individuals upon eye contact, parallel orientation, and the maintenance of a specific distance from one another. This basic "psycho-mechanical equilibrium" between forces of attraction and repulsion could be scaled up into congregations with many schooling individuals, which would be distributed with equal density within a given area by exerting forces of attraction (and repulsion) upon their neighbors on all sides.[107] Parr thus focused here on the significance of neighboring, *local* psychodynamics to the dynamics of the school as a whole. He also remarked, however, that a dense and differentiable school of fish could not arise from such a structure alone. Rather, the schooling individuals would simply fill up the entire space of the aquarium. In order to explain the formation of a dense congregation, he pointed to the central role played by the congregation's periphery:

> In any number of specimens, however, some will always have to be at the side of the columns. These peripheral specimens certainly are under constant stimulation from one side only, i.e. from the next specimen towards the centre of the school, as they have no companions on the other side. In the peripheral files on the two sides of a school one should therefore expect to find a constant tendency to seek towards the centre. [...] [T]he reactions caused by this tendency may serve to explain the condensation of the school as a whole and the subsequent maintenance of a constant density of the individuals in space.[108]

It may seem odd to speak of 'two sides' of a school of fish, but this can be attributed to an act of reductionism that was necessary because Parr had

106 For a more detailed discussion of swarms and the psychological concept of fear, see Vehlken, 'Angsthasen.'

107 The term 'congregation' denotes an *active* aggregation of individuals in which the aggregation itself is the source of mutual attraction. Such is the case in swarms, and it is for this reason that I use the term here instead of the less precise term 'aggregation.' The latter designates a collection of individuals that can come together either passively (e.g., by means of currents) or actively (e.g., by being drawn to a food source) and that do not typically remain cohesive for a long time. See Parrish et al., 'Introduction – From Individuals to Aggregations,' 4–6.

108 Parr, 'A Contribution to the Theoretical Analysis of the Schooling Behavior of Fishes,' 5–6.

converted his observations into a simple, two-dimensional 'theoretical case' on paper, which was meant to demonstrate the plausibility that the tendency of school to condense might be an effect of its peripheral fish. In his opinion, the one-sided attraction of the peripheral individuals toward the center would ultimately lead, by means of a sort of chain reaction, to a new psycho-mechanical equilibrium with reduced space between the schooling individuals. Parr called this factor "an automatically transmitted tendency to turn inwards,"[109] which, for the sake of simplicity, he modeled according to the orientation of each individual to just *one* neighbor on the left and right. The periphery acted like a wall that was difficult to break through for the individuals in the middle of the school, and this was also because any temporary tangential motion away from the center would cause – "for purely geometric reasons" – the outer 'columns' of fish to swim at a greater deflection angle, which would in turn strengthen the stimulus for the fish to orient themselves in parallel to their neighbors. Here Parr cited the mathematical principles behind his psychomechanics, as supported by his observations.[110]

Parr's idealized model thus visualized a possible factor that might explain how the specific and evenly dense structure of fish schools comes about, though he did so by breaking the process down into a series of discrete steps which, as he himself acknowledged, actually take place simultaneously and continuously. This factor of 'automatic attraction' made any assumptions about diffuse 'social' or 'altruistic' instincts superfluous.[111] According to Parr's understanding, fish schools are Cartesian animal-machines – automata in which information is disseminated on a local basis and converted into an all-encompassing form of mechanics. At the beginning of the 1950s, the biologist Charles M. Breder drew upon Parr's concepts and formalized them mathematically. He did not, however, base his formula on empirical data but rather altered its values at random until it yielded a family of curves that, as he postulated, could be used to describe various forms of schooling aggregations.[112]

Even within Parr's 'object' of observation – his machine-like school of fish – the visual played a decisive role, for he believed that the mechanical coupling of individual elements depended on the sense of sight. In another

109 Ibid., 16.

110 Ibid., 20–22.

111 Ibid., 9. The concept of an "altruistic instinct" can be found in David Starr Jordan's book *Fishes*, 41.

112 See Radakov, *Schooling in the Ecology of Fish*, 26, and Breder, 'Equations Descriptive of Fish Schools.'

section of his text, he described a series of experiments with which he had hoped to test this connection:

> Some specimens were taken out of the tank and, after the eyes had been dried with clean cotton and then covered with a layer of vaseline mixed with lamp-black, they were again returned into the tank wherein the rest of the school was constantly milling. [...] 3 specimens lost the vaseline-cover within 2 minutes and joined the mill, but in 4 other specimens the cover adhered for a long time and these gave very convincing results. These individuals did not at any moment join the milling activities of the school, but kept moving separately around in all direction over the entire tank, striking its walls, even in spite of the fact that they would very often pass directly through the mill, sometimes even colliding with the milling specimens but always without showing traces of a tendency to join them. [...] As a control, 10 other specimens were taken out of the water, were submitted to the same treatment, except the covering of the eyes and were then returned. [...] 9 of the ten had joined the mill within 1 minute, while the tenth had in some way become hurt and quickly died.[113]

As a result of such experiments, Parr asserted that visual stimuli were responsible for *controlling* the interactive behavior of the individual schooling fish. Even the function of the eyes in the ability of the fish to adopt an equal distance from one another could be explained, he thought, for there could be a particular range of focus that might determine the 'correct' distance that had to be maintained. Yet Parr also mentioned an additional possibility, which had been proposed earlier by George Howard Parker and J. T. Cunningham: the sense of sight could be decisive for long-distance communication, whereas a fish's lateral-line organ would come into play for short distances. His experiments with blinding mackerel, however, had no effect whatsoever on their orientation, even when swimming through the school.[114]

Just two years later, Guy Malcolm Spooner built upon Parr's work and broadened the experimental side of fish-school research. Experimenting with various partitions and mirrors in small laboratory tanks, he too demonstrated the significance of the sense of sight to the formation of schools, and he further noted the extent to which the behavior of the fish under investigation (perch, in his case) was affected by their container. Whereas

113 Parr, 'A Contribution to the Theoretical Analysis of the Schooling Behavior of Fishes,' 22–23.
114 Ibid., 24. Here Parr refers to Parker, 'The Function of the Lateral-Line Organ in Fishes.'

they exhibited "considerable activity" in the large aquarium at Plymouth, where they swam around throughout the entire tank and were attracted to activity taking place outside of the water, in the smaller tanks (c. 1.5 × 0.8 × 0.35 meters in water volume, with only the front side made of glass so as to minimize environmental influences) they displayed a fearful sort of behavior, they reacted to disruptive influences more than they reacted to each other, and they sought protection in the corners of the aquarium.[115]

Spooner tested whether the fish could communicate with one another through glass and whether they would orient themselves in front of a mirror as they would to their neighbors. Both of these questions were affirmed, which further suggested that the behavior of a school depended to a great extent on the sense of sight. Yet Spooner's series of experiments, too, yielded just an approximate determination of the relevant factors in the formation of schools. What is more, he shied away from suggesting that there might be any strict determinism in the behavior of the schooling individuals (an idea that was still explicit in Parr's article), and he stressed the conjectural and inconclusive nature of his findings: "Unfortunately, the reactions of the fish to each other and to a mirror are not sufficiently cut-and-dried to provide a basis on which accurate comparisons can be drawn. [...] For any given fish it is impossible to predict definitely how it will behave, but it is possible to say how it will most probably behave [...]. But it is not possible to measure this probability [...] accurately."[116] In the end, Spooner was unable to determine any unambiguous correlations between the reactions of the fish and the experimental system as regards, for instance, whether poor lighting conditions might affect the intensity of the fish's reactions.

Because of the unclear definition of the relevant factors in the formation of schools, Spooner could do little but point to certain probabilities, and he had to refrain from making any predetermined or linear assumptions about cause and effect. He encountered problems not only in making accurate observations but also in designing his experiment and analyzing his data. In the early research concerned with the behavior of schooling fish, at any rate, Wilhelm Roux's "causal research methods" were made to seem ambiguous and poorly defined. The tension between visual indeterminacy and statistical determinacy – the problem of making calculations on the basis of non-knowledge – drew a constitutive epistemological line through fish-school research and would remain an issue for years to come.

115 Spooner, 'Some Observations on Schooling in Fish,' 426–428.

116 Ibid., 444.

Whereas Karl von Frisch, in his 1938 article on the psychology of fish schools, praised the "visible" advantages of laboratory experiments conducted with aquaria and described the observable behavioral patterns of minnows (some of which he observed in the wild) with detailed charts of empirical data, Spooner's studies revealed problems that essentially invalidated any efforts to quantify this very sort of behavior. Frisch, it seems, was somewhat rash to characterize research aquaria as institutes for applying objective methods to fish-school research, and this is also true in light of the media-technological problems which will be the focus of my next chapter.

Animal Aggregations

While the first systematic studies of fish schools were taking place at the New York Aquarium and at the laboratories in Plymouth, Warder C. Allee was conducting research on the phenomenon of 'animal aggregations.' Beginning with an article from 1923 and continued with a series of experiments under the header 'Studies in Animals Aggregations,'[117] he developed a research program devoted to understanding the general and unconscious need for individual animals of the same species to be near one another and its evolutionary implications:

> [T]here is, in effect, a deleterious effect from *under-crowding* as well as the more familiar one of *over-crowding*. The phenomenon of a better group survival, as contrasted with individual survival, was tested against [...] artificial environmental factors. More important was the demonstration of the reality of an unconscious cooperation, which he referred to as *proto-cooperation*, in a wide diversity of animal forms.[118]

Allee's program for researching the physiology of animal aggregations thus developed from one that was at first based on the physical conditions in a given environment to one that examined the physiological and chemical 'connections' between individuals by controlling the parameters of interindividual behavior. "In the end," according to his biographer Karl P. Schmidt, "he was obviously more interested in principles than in practice."[119] For

117 See Allee, 'Animal Aggregations: A Request for Information'; idem, 'Studies in Animal Aggregations' (a series of essays published in a variety of scientific journals from 1923 to 1933); idem, 'Cooperation Among Animals'; and idem, 'Animal Aggregations.'
118 Schmidt, 'Biographical Memoir of Warder Clyde Allee, 1885–1965,' 10.
119 Ibid., 16.

Allee, "proto-cooperation" meant a form of positive influence that a group exerted on its individuals – an influence that, in the case of more simple forms of life, should not be investigated on the basis of psychological factors (as in the case of human mass psychology; here Allee cites Gabriel Tarde's *Laws of Imitation*) but rather on the basis of collective physiological factors whose functions could be described in physical-chemical terms and whose functionality could be described in quantitative terms:[120]

> In contrast to the more-or-less automatic aggregation in response to odors, light or shade, moisture, favorable niches, and other environmental factors, there are the much more definitely social situations in which animals collect as a result of positive reaction to the presence of others like themselves. The aggregation of male midges 'dancing' in the quiet atmosphere, or the formation of schools of fishes or flocks of birds illustrates this widespread phenomenon.[121]

In his series of essays, Allee investigated various forms of marine life, but he never concerned himself with any 'social' organisms (of which schooling fish represent a special case). He did, however, refer to certain principles of inter-individual connectivity such as William M. Wheeler's and Theodore Schneirla's descriptions of ants interacting by means of trophallaxis.[122] What motivated him as a young ecologist was rather questions about *why* such interaction occurred and about the evolutionary benefits that might underlie an aggregation of organisms. He was less interested in *how* the specific inter-individual interactions took place between the organisms of an aggregating species. In the spirit of Kropotkin, Allee focused on cooperation instead of on a post-Darwinian notion of competition related to the individually advantageous 'struggle for existence.'[123] However, he integrated his examination of cooperation with certain quantitative considerations. For him, the issue was not conscious or unconscious altruism or some sort of

120 See Allee, 'Animal Aggregations,' 395–397, 410.

121 Ibid., 394.

122 See ibid., 410.

123 With his essay 'The Struggle for Existence and Its Bearing Upon Man,' the biologist Thomas Henry Huxley (aka 'Darwin's Bulldog') propagated a rather extreme version of Darwin's ideas that focused on the "struggle against the environment": "From the point of view of the moralist, the animal world is on about the same level as a gladiator's show. The creatures are fairly well treated, and set to fight; whereby the strongest, the swiftest and the cunningest live to fight another day." See Huxley, 'The Struggle for Existence and its Bearing Upon Man.' Quoted from Dugatkin, *Cooperation Among Animals*, 6.

social instinct to ensure the survival of other individuals; rather, he believed that a combination of factors made individual survival more likely on the basis of specific group processes – a perspective that he shared with the physiologist Ralph W. Gerard.[124]

Two of Allee's principles are especially interesting here. First, he maintained that there is a geometrical relation between the mass and the surface of a body, and certain surface-to-mass ratios are more advantageous than others in that they provide better protection against external influences. This principle, moreover, applies not only to the body of an individual organism but also to the 'body' of a (relatively dense) aggregation of living beings. Thus, as is well known, if the surface of sphere is increased by a factor of two, its volume increased by a factor of three. In Allee's work, too, aggregations are thus problems involving margins or the periphery: "Each cell in a temporary or permanent aggregation or in a multicellular organism presents less surface to the outside world than does one that leads an independent existence. As a result, the danger of harmful exposure to environmental effects is decreased, and, on the other hand, the difficulty of respiration, of individual food getting, and of receiving external stimuli is increased."[125] According to Allee, aggregations achieve a sort of 'pre-social equilibrium' between the two poles of under-crowding and over-crowding.

Second, Allee was concerned with problems of scaling, and his experiments demonstrated that a particular lower limit of individuals was necessary for an aggregation to have beneficial features. In the case of simple organisms (like sponges, for instance) this could be a certain number of cells that enable damaged areas to regenerate. If this number is not reached, regeneration will not occur.[126] Allee also alluded, however, to group effects that higher forms of life exhibit while looking for food. Thus, he compared the fishing behavior of small and large groups of cormorants:

> The food-procuring behavior of many different kinds of animals changes, depending on the number present. The group fishing of the double-crested cormorants near San Francisco gives an example of elaborate and flexible group cooperation. These cormorants may fish singly, in small coordinated flocks of from ten to twelve, or in larger flocks that may contain as many

124 See Schmidt, 'Biographical Memoir of Warder Clyde Allee,' 24. A few years later, Ralph W. Gerard would speak at the Macy Conferences about the relevance of studying 'communication between animals' to cybernetics. Gerard's work will reenter my discussions at various points below.

125 Allee, 'Animal Aggregations,' 397.

126 See ibid., 397.

> as 2000 birds. Fishing usually begins before the larger flocks are fully formed. The basic pattern in small flocks consists of a circle with all birds facing the same direction. This pattern changes with the large flocks; then, a long, narrow, well-packed line moves forward, fishing as it goes. Some cormorants swim at the surface, others dive and swim at the same rate; those left behind by the rapid advance take to the air and fly forward again to become members of the line of fishers. The large flocks swim decidedly faster than do small fishing groups – an example of another kind of social facilitation; they also pursue a given school of fish until the hunger of the cormorants is satiated, or until the school escapes. Thus the persistence of a large flock is greater than that of a small one.[127]

These observations on the phenomenon of nonlinear scalability, which were necessarily made in the wild and are therefore somewhat vague and imprecise, are worth considering for at least two reasons. First, they pose an epistemological question that gets to the heart of the dispute among biologists over the methods of observation and experimentation. For, under these conditions, to what extent are laboratory experiments with small groups or even with individuals of an aggregating species applicable to the behavior of the same individuals in their natural environments, where they are typically part of much larger collectives? Second, they further compound the difficulty of accurately defining the constituents of such multiplicities and the connections between their units, given that any increase of an aggregation's quantity entails qualitative changes.

In several respects, the foregoing pages have been concerned with marginal phenomena. On the margins of university-based zoological research during the first decades of the twentieth century and in newly founded marine research institutes, margins became a problem and an operative function that arose from investigations of swarm phenomena, aggregations, and congregations. With a novel combination of observational field research and experimental methods, attempts were made to furnish swarms, in their oscillation between unity and multiplicity, with new margins and limits. Swarms became an object of research within the walls (margins) of marine research aquaria, and this involved questionable acts of reductionism as regards the effects that such an environment might have on various levels of the aggregation. At the same time, however, some of the research

127 Ibid., 411. Here Allee refers to Bartholomew, 'The Fishing Activities of Double-Crested Cormorants.'

approaches discussed above revealed the constitutive function of the margin or edge for the formation of swarms and other collectives. It was shown that swarms could only achieve a psychomechanical equilibrium through the tendency of their marginal individuals to orient themselves toward their fellow species on the inside – and, in principle, the reduction of the size of a congregation's margin in relation to its mass could be regarded as a protective factor and thus as an evolutionary development of animal collectives. Yet these early marine studies also characterized swarms and other multiplicities as fundamentally dynamic and time-based objects that could not be comprehended with static methods. Already at this stage, swarms appeared as an object of knowledge for which Rheinberger's concept of the *epistemic thing* is not quite applicable. A more fitting term, I think, would be *epistemic aggregations*.

Engaging with the margins has also pointed to attempts to master the defining movement whose development has been traced throughout the course of this chapter: the *formation* of a biological perspective on the dynamics of collectives that, on the basis of connective principles that could be explained with scientific hard facts, operationalized the latter in the sense of systematic approaches. Thus was formed a specific point of view from which it became a matter of interest to engage with swarming and to study swarms as *objects*. Its formation was outlined above with reference to forms of *relationality* modeled on the physics of motion and processes of information transmission. This new zoo-logical perspective on swarms and other multiplicities established a way of thinking in terms of sense-based, physiological connections – in terms of information processes that lead to the self-regulation and self-organization of a dynamic equilibrium. Hence, it can also be said that swarms exemplify the threshold – described by Michel Serres – between two epistemes: that of the *transformational engine* of thermodynamics and that of the *informational engine* of a field of knowledge that would later be known as cybernetics.[128] For it was this perspective, which was still largely based on observation and only used in technologically simple experiments, that paved the way for profound new media-technological *formats*.

128 See Serres, Hermès IV: La distribution, 43–58.

III. Formats

Abstract

With the umbrella term *formats*, this chapter explores the history of technical enhancements of swarm research between 1930 and 1980. It is concerned with the various attempts that were made to gain quantitative and formalizable access to the swarm by *suppressing noise*. Efforts were made to record swarms with optical media in a variety of experimental systems, and in the open sea researchers additionally tried to make swarms visible by means of innovative diving techniques and sonar technology. Again and again, however, disruptive forces like the internal movements of the collectives or the distortive effects of the environmental medium of water interrupted the acquisition of data. Empirical research thus found itself mired in a 'technological morass.'

Keywords: fish school, epistemic things, smooth and striated space, Jean Painlevé, history of sonar, oriented particles

1. Fishy Business: Media Technologies of Observation and Experimentation

> What is a fish? A fish is a back-boned animal which lives in the water and cannot ever live very long anywhere else. Its ancestors have always dwelt in the water, and most likely its descendants will forever follow their example.[1]

Schools of fish have not been studied as complex systems for very long. It was not until the end of the 1920s and the beginning of the 1930s that researchers first engaged with questions concerning the possible parameters and observational media that would be necessary to gain any knowledge about them. In his 1931 article on schooling behavior, Guy Malcolm Spooner remarked: "The phenomenon of schooling has received surprisingly little

1 Jordan, *Fishes*, 1.

Vehlken, S., *Zootechnologies. A Media History of Swarm Research.*
Taylor & Francis Group, 2019
DOI 10.5117/9789462986206_CH03

attention either from fishery investigators or from those studying animal behaviour."[2] The Russian biologist Dimitri Radakov similarly commented: "The phenomenon of schooling has undoubtedly been known since ancient times, at any rate since our ancestors began to catch fish. But it was not until comparatively recently that special investigations of fish were launched: at the end of the 1920s."[3]

At the beginning, investigations of the form and structure of fish schools were not as strongly motivated by the sense of fascination that can be found in literature or, as we have seen, in the work of early swarm researchers such as Edmund Selous. For, beyond this fascination and awe before "nature's higher plan" (as Jakob von Uexküll called it),[4] there were economic interests that incentivized the study of fish schools, and such interests were often explicitly mentioned as the driving factor behind certain research projects. In short, knowledge about fish schools was primarily thought to be useful for the commercial fishing industry. Thus, as I noted in the previous chapter, the laboratories and aquaria used for experiments were typically located in institutes that were devoted above all to optimizing fishing methods or were founded with the express goal of creating a productive link between science and industry.[5] Albert Parr, for instance, justified his research as follows: "The problems involved are of special interest to human society because several of the most typically schooling species are also among the economically most important ones, partly gaining their economic importance through the very schooling habit itself, which is the necessary basis for most of the fishing methods adopted for the exploitation of the species in question."[6] At this stage, before the gradual blending of biological research and computer-supported media techniques, the simple if somewhat dubious aim of fish-school research was to increase the profits of the fishing industry.

For several reasons, fish schools are especially well-suited to problematize certain issues that lie at the heart of the present book. *First*, and pragmatically, there have been a number of scientific (mostly biological) studies of

2 Spooner, 'Some Observations on Schooling in Fish,' 422.

3 Radakov, *Schooling in the Ecology of Fish*, 8.

4 See Von Uexküll, 'Die Bedeutung der Planmäßigkeit für die Fragestellung in der Biologie.'

5 Marine research institutes of this sort included the Laboratory of the Marine Biology Association of the UK in Plymouth, the Marine Biological Laboratory in Woods Hole, USA (both discussed in the last chapter), a variety of centers for Soviet research described by Dmitri Radakov, and the research aquarium housed at the Department of Agriculture and Fisheries for Scotland in Aberdeen.

6 Parr, 'A Contribution to the Theoretical Analysis of the Schooling Behavior of Fishes,' 1.

the phenomenon of swarm formation in fish, many of which I already discussed at some length in the previous chapter. *Second*, many species of fish form highly stable congregations or 'schools' (the German term is *Fischschwarm*, 'fish swarm'). The latter multiplicities move in a highly regular and synchronized manner through three dimensions and, viewed from a distance, this activity makes it appear as though they are moving as a *single* body. At the same time, however, they are also capable of making extremely dynamic maneuvers. In certain situations, for instance when under attack by a predator, the whirring movements of the schooling individuals can confuse their attacker in an elusive flurry. Rule-bound behavior and the individual's random deviations and reactions integrate schools into heterogeneous totalities in which 'the school' (as a unifying concept) converges with 'schooling' (as a free-floating form of movement that defies representation). This dichotomy elicits special forms of neighbor-based organization and communication, as in the so-called 'waves of agitation' that emerge from individual interactions. *Third*, fish schools are an example of *placeless* swarms. They are formed not only as a reaction to environmental factors (such as the need to find food or partners), which would mean that they only take place in relation to particular places (as is the case with certain coral reef fish). And they do not operate in relation to a center like a beehive, anthill, or termite hill (together with their respective queens). It would also be conceivable to write a media history of swarm research by looking primarily at studies devoted to flocks of birds. Yet, even more so than research on schools of fish, the latter studies are dependent on advanced media-technological processes and apparatuses. After all, avian flocks can hardly be investigated in a laboratory context, and their dynamics in the wild are at least as difficult to follow as those of schooling fish. Systematic studies of this sort did not really begin until the 1970s, despite the early fieldwork that I discussed in my section 'Odd Birds' above. What is decisive here is that the fundamental models of interaction for explaining the rules governing the 'traffic' of swarms (in Julia Parrish's terms) are applicable to both flocks of birds as well as schools of fish. For this reason, a later chapter will examine the topic of tracking avian flocks with computers and discuss this topic in conjunction with more recent methods for analyzing fish schools.

Fourth, the development of fish-school research is characterized by shifting epistemic strategies that, until the end of the 1920s, were still essentially based on observations made with the naked eye. More and more experimental approaches gradually began to enter the picture, some of which involved mathematical models whose spatial structures were described with geometrical relations and whose dynamics were represented according to physical laws. Here it is possible to identify a movement "from measurement to models"[7] – a trend that raises a *fifth* question, namely that of the visibility and visualizability of fish schools, and this in two respects. One the one hand, there were experiments based on visual scanning and grids, and these ranged from samples monitored by the human eye to those recorded by various optical and acoustic media (and, finally, by computer-supported methods). These experiments were attempts to allow for 'seeing-in time,' which was meant to 'capture' the movements of schools in time and space in such a way that the data material for describing schooling dynamics could be produced from analyzing *images*. On the other hand, this question of visibility is related to a 'realistic' representation (and later, too, to a literal *animation*) of mathematical fish-school models. The goals were to make the (relatively simple) basic rules of schooling comprehensible as dynamic processes and interactive relations, to account for the automatic organization of collectives, and thus later to make the synthetic sequence of images produced by simulation models compatible with images produced by technical media of observation. The approaches to fish-school research discussed in this chapter are thus essentially based on reducing noise as much as possible, on excluding disruptive moments from the medial arrangements of observational and experimental systems or models. Here, fish-school research seems to be concerned above all with suppressing noise and interference. It could be said that these media-technological arrangements stripped schools of their 'naturalness' in order to treat them as objects of knowledge.

Through such processes, fish schools are meant to be broken down and atomized as finely as possible in order to become analyzable and definable as a quantity – as a composite mass that encompasses the non-representable act of 'schooling' in its global structures. In both cases, however, schools of fish pose a fundamental medial problem: first as a problem of *recognition* and second as a problem of *representing* an 'object' that repeatedly defies definition and objectivization in the dichotomy of global, dynamic unity and local, chaotic multiplicity. As 'objects' of this sort, fish schools point back to the classical problems that accompanied the development of perspectival representation, that is, to problems relating to 'non-things or half-things' (in Leonardo da Vinci's terms). This shift in the geometric horizon – this expansion of the realm of perception to include movements in three dimensions and the possibility of the observer becoming part of the observed event – also opened up entirely new epistemic horizons.

7 Parrish et al., 'Introduction – From Individuals to Aggregations,' 9.

The problems associated with making perspective grids for fish schools – again, bodies without surfaces – became relevant to biologists whose research was concerned with the complexity of a given school's essence. Attempted at first with naked-eye observations, then by means of analog processes (such as chronophotography, film, and sonar), then by means of mathematical models, and finally by means of computer simulation processes – a transition described by Peter Galison as one from "computers as tools" to "computers as nature"[8] – each of the media technologies used to this end brought with it a specific set of problems and new ways of approaching fish schools as an object of knowledge that engaged in various ways with their multidimensionality, fuzziness, and intransparency.

By means of a media-theoretically informed historical epistemology of fish-school research that takes into account media-technological materialities, observational arrangements, mathematical models, and computer simulation systems, it is thus possible to delineate a field of research that reflects the program of a historiography of biology that François Jacob proposed in his book *The Logic of Life*. In a review in the newspaper *Le Monde*, none other than Michel Foucault referred to Jacob's work as "the most notable history of biology that has ever been written," and he noted how the book "invites us to reconfigure our thinking in a fundamental way."[9] It was Jacob's aim to do away with the notion there is a continuous history of sequential ideas that can be followed from the present through the past. In his own words:

> The alternative approach to the history of biology involves the attempt to discover how objects become accessible to investigation, thus permitting new fields of science to be developed. It requires analysis of the nature of these objects, and of the attitude of the investigators, their methods of observation, and the obstacles raised by their cultural background. The importance of a concept is defined operationally in terms of its role in directing observation and experience. There is no longer a more or less linear sequence of ideas, each produced from its predecessor, but instead a domain which thought strives to explore, where it seeks to establish order and attempts to construct a world of abstract relationships in harmony not only with observations and techniques, but also with current practices, values, and interpretations.[10]

8 Galison, *Image and Logic: A Material Culture of Microphysics*, 692.
9 Foucault's review is quoted here from Rheinberger, 'Nachwort,' 345.
10 Jacob, *The Logic of Life*, 11.

This new form of historiography no longer focused on broadly conceived and teleologically oriented scientific developments but rather on conducting cursory searches for those 'stages of knowledge' which Foucault would later refine, theoretically, with his concept of the *dispositif*, which requires precise descriptions of the transformations and conditions that make certain objects of knowledge and analytical perspectives more probable (and thus, in a sense, 'truer') than others.[11]

"It is," as Claude Bernard once wrote, "the vague, the unknown that moves the world."[12] In his book *Les sciences de l'imprécis*, Abraham Moles similarly offered a systematic analysis of the functionality of vague phenomena in the sciences. More recently, drawing upon the work of these authors, the historian of science Hans-Jörg Rheinberger has employed his concept of the *epistemic thing* to describe those discourse objects that take shape in the interplay between the technical components of an experimental system and the ongoing (media-) archaeological rewritings and overwritings of (layered and interwoven) histories and their respective strategies of investigation. Epistemic things, as Rheinberger defines them more precisely, "are material entities or processes – physical structures, chemical reactions, biological functions – that constitute the object of inquiry. As epistemic objects, they present themselves in a characteristic, irreducible vagueness."[13] By means of this concept, Rheinberger directs the attention of a historically reflective epistemology onto the "contexts of discovery" of nascent scientific experiences.[14] For the latter, conceptual indeterminacy and vagueness are not something disadvantageous but are rather productive and guiding.

By characterizing science as being fundamentally provisional – epistemic things embody, according to Rheinberger, that which is not yet known – they are indicative of the fact that time is at play in the history of science and that science does not simply unfold in time. To repeat the words of François Jacob, every experimental system could thus be called "a machine for making the future."[15] Michel Serres and Bruno Latour, as Rheinberger notes, likewise advocated for this perspective of historical epistemology. Whereas Serres stressed that research is not knowledge but rather a steady act of

11 See ibid., 12.

12 Bernard, *Philosophie: Manuscrit inédit*, 26. Quoted from Rheinberger, *Toward a History of Epistemic Things*, 11.

13 Rheinberger, *Toward a History of Epistemic Things*, 28.

14 Ibid. The term "context of discovery" stems from Hans Reichenbach's *Experience and Prediction: An Analysis of the Foundations and the Structure of Knowledge*, 6.

15 Jacob, *The Statue Within*, 9. Quoted from Rheinberger, *Toward a History of Epistemic Things*, 28.

tinkering and faltering, Latour has underscored the mutability of objects of science. The latter exist only as 'lists' of activities and properties, so that with every new item added to the list – that is, with the growth of science beyond the object in question – the object is redefined and acquires a new shape. Thus, the question of epistemic things is always associated with a historical index, and it takes place in the even-sided interaction between human and non-human actors.[16] According to Rheinberger, new objects of science are never simply discovered or unveiled and 'brought into the light.' Rather, they are gradually produced and formed within technical arrangements. For Latour, moreover, research processes are manifested through the elimination of their predecessors' unambiguously definable and medial processes of representation. Objects of knowledge can only be produced in a series of referential acts such as ordering, distinguishing, recording, transferring, marking, or filtering.

The efforts of swarm researchers are aimed at acquiring knowledge about the structures and functions that enable a qualitatively different entity with new features and operational potential to be created out of a mere aggregation. Hence, at first glance, swarms perhaps seem to be epistemic things *par excellence* in their hybridity between local relations and global structures. As both material 'objects' and objects of science, they inhabit a realm of optic vagueness and structural intransparency, and they are processed with methods and techniques of clarification and obfuscation: the dividing line between the two is defined by the media-technological processes that are being implemented. Simply put, the *creation* of vague phenomena is likely when such phenomena are being *processed* by analog media. In the case of digital media, on the contrary, a different form of vagueness is consciously implemented in order to increase the precision of calculations. Taking into account the statistical variance of a phenomenon, for instance, leads to the reduction of such variance in the technical image, and filter algorithms smooth over the visualizations of distorted or distorting objects. The question that needs to be asked at the beginning of this chapter and the next, however, is whether swarms do not in fact go beyond Rheinberger's definition of epistemic things and ultimately allow the latter to be critiqued. The question, in other words, is whether we are here dealing with dynamic systems that can only be investigated by means of an epistemology that is able to combine entirely new epistemic and technical approaches: an epistemology of biologically inspired computer simulation.

16 See Serres, 'Introduction'; and Latour, *Science in Action*, 87–88.

In the case of fish schools, however, the literal 'vagueness' of the object of knowledge is always in relation to a type of organization that holds the entity together as a global structure that is flexible yet coherent and dynamically stable. Schools of fish create a visual and perspectival paradox. If one imagines a large school of sardines in crystal-blue water, it is easy to observe from a distance the fascinating coordination of the collective as it changes directions or evades predators. Just as John Steinbeck and Jules Michelet reported long ago, the maneuvers of such schools are so smooth that it seems as though they possess their own form of life or vitality.[17] Yet the closer one comes to this non-object, the greater one's view of it is distorted and its begins to seem more like a flurry, like image noise, like a chaotic dervish, like something beyond the visible. The reciprocal relationship between the local behavior of the schooling individuals and the global behavior of the collective is thus of central interest: at what point and under what conditions does a mere quantity of similar elements transform into qualitatively different and new properties that cannot simply be derived from the characteristics of the individual elements themselves? Are these global structures and patters mere epiphenomena of biological aggregations and congregations, or do they have adaptive functions – that is, are they somehow linked to external phenomena from the environment? Is it possible to speak of forms of self-organization, and is this always an effort to uncover their purpose – as Eugene Thacker has remarked, is the issue of swarms one of "pattern or purpose"?[18]

All of these questions are significant when a decision has to be made about what sort of perspective ought to be adopted when examining the non-object of swarms. Is it best to approach them from the micro-perspective in order to comprehend the behavior of the swarming individuals? What sort of statements can be made about the morphology and structure of swarms by investigating them from the macro-perspective? The question of perspective is thus always a question of its own conditions, which might be co-determined by such things as media technologies or environmental factors. In the case of aquarium experiments, for instance, only a limited number of schooling individuals can be involved, whereas, for many years, the resolution of acoustic scanning technologies in the open sea would only allow for a crude overview of activity. Following Michel Foucault, one could say that different diagrammatic operations lie at the basis of every media-based method of analysis – from the act of producing spatial and

17 See Steinbeck, *The Log from the Sea of Cortez*; and Michelet, *The Sea*.
18 Thacker, 'Networks, Swarms, Multitudes,' n.p.

temporal grids of visual processes to the stochastic definitions and statistical determinacy involved with digitally visualizing acoustic raw data.[19]

The level of visualizing measurements, experimentally generated data, and models is thus essential to any analysis of fish schools. The vagueness and intransparency of 'perceiving' them with the senses or with sensory technology and the static nature of mathematical and geometric models ultimately encounter one another in the strategies of computer-supported simulations and visualizations. As objects of knowledge, fish schools – like the *placeless* collectives at the heart of this chapter – appear in various dynamically changing *epistemic aggregations* that, as mentioned above, seem to expand upon Rheinberger's concept of epistemic things. The media history at hand is certainly not a teleology, advancing through 'better' and 'better' technologies that gradually seem to reveal the secrets and rules of school formation and continually refine our perspective on their structures. My concern is rather to show how schools of fish, as objects of knowledge, have always appeared to be something different and new from the perspectives of various media technologies and experimental systems. The aim is to demonstrate how, as objects, they have repeatedly been redefined and given a new form (in Bruno Latour's sense of the terms).[20] And this is a question that cuts to the core of what one might call *media historiography.*

Over the course of this redefinition, empirical observational data and experimental findings are repeatedly compared with one another and treated as a sort of feedback system. Here, the medial ensembles and *dispositifs* of observation that existed *before* the advent of computer-simulation processes will be productively examined in light of the more recent concept of 'material culture,' as it has been understood above all by Latour and Rheinberger. Both of these authors think about natural-scientific knowledge in terms of processes; their focus has been on historically contingent combinations of materialities such as apparatuses, architectures, recording systems, and storage devices as well as on forces of resistance and conditions of possibility in the production of knowledge. Thus the question of swarms as figures of knowledge also presents itself as a question about their relation to various forms of observation and representation:

> From an epistemological perspective, a crucial factor is whether any distinction is made in such experimental systems between the object of investigation and the means by which it is represented. To overstate

19 See Deleuze, *Foucault*, 34–35.

20 Latour, *Science in Action*, 131–132.

> matters somewhat: Was it not inside a botanical garden that a plant first became a taxonomic category? Was it not within the gravitational field of an ultracentrifuge that a cell organelle first became a manageable entity and thus something relevant to science? Thus, in a fundamental sense, that which Bachelard called 'scientific reality' is a matter of debate. Historically, this raises the question of how the graphisms were created that enabled this reality to be represented. They imply a technological semantics that does not at all merge into that which is typically called a materialization.[21]

To this extent, it can be supposed that swarms have always been transformed into something different by means of various experimental systems, technical media, and mathematical models. That which comes into view subsequently guides the view itself. From a media-historical perspective, the 'scientific reality' of swarms is contingent upon various *formats* and is *transformed* as a result: from a figure of the chaotic or wondrous – from a metaphor for irrational thinking and behavior – into a problematic object of scientific research. In what follows, I hope to demonstrate how, by means of a media-technologically induced *retreat from naturalness* and in light of more recent concepts, the supposedly 'wondrous' aspects of swarms and the older tendencies to anthropomorphize and sociomorphize them have been given a new perspective. In short, I intend to show that it has become possible to describe processes of self-organization.[22] Such descriptions derive the complexity of a swarm's overall behavior from the nonlinear connections between the swarming individuals, which are equipped with just a few basic attributes and capabilities. Following my discussion at the end of the previous chapter of the earliest fieldwork and laboratory studies concerned with swarms, the present chapter will now turn to the systematic biological experiments on swarms that were conducted between 1930 and 1980, that is, *before* the era characterized by the widespread use of digital media. That said, I will also examine why and in what way, as of the late 1940s, swarm research became a matter of interest to the newly established field of cybernetics. For it was in this context that biological swarm research, now equipped with the concept of *information transfer*, first began to engage with information-theoretical and computer-technical considerations. Geometric and mathematically formalizable models were devised in order to reconstruct the functional

21 Rheinberger and Hagner, 'Experimentalsysteme,' 20–21.

22 Although the concept of self-organization was first established during the 1960s and in the context of cybernetics, the dynamic organizational structure of swarms had already become an object of systematic biological research as early as the late 1920s.

rules of swarms. Moreover, this happened well before such models were put to productive use, which was not until the 1980s.

One of the main theses of this book unfolds along with this media history, namely that schools of fish first came into their own through the implementation of computer-supported simulations and only thereafter, as changed figures of knowledge, were they able to be open to new and productive discourses. Before that turning point, the media history of fish-school research has to be told – for long stretches – as a history of failure: the failure of various media-technological attempts to turn swarms into objects of knowledge. In the introduction to their 1997 book *Animal Groups in Three Dimension*, for instance, Julia Parrish and her colleagues could not help but remark that the "[m]ethod sections in several fish schooling papers from the 1960s and 1970s are full of agonizing descriptions of the number of frames analyzed."[23] In these experiments, the 'truth' about the organizational principles of swarms was mired in a "technological morass" of observational media and experimental arrangements.[24] In what follows, I will trace back many of these attempts to format fish schools and outline the reasons for their failure.

This chapter is thus organized around three research perspectives – each with its own specific set of questions – and the ways in which they solved (or partially solved) a variety of problems. These perspectives provide a sketch of how it was attempted to *format* schools of fish in a number of divergent ways and with various types of media. Two of these epistemic strategies were devoted to optically analyzing schooling into the particularity of single schooling individuals – the first in research aquaria and the second in open water. Another approach made use of acoustic processes in order to ascertain schools by inverting this perspective, namely from the outside to the inside. What these three approaches had in common was the environmental medium of the sea or water, and this situation posed at least three noteworthy sets of problems. The *first* pertained to the question of viewpoint: divers and scientists submerged themselves beneath the surface of the sea in order to observe life under water like a 'fish among fish.' The *second* issue was that of seeing clearly through the rigid glass boundaries of research aquaria. And a *third* problem was that of revealing visual 'invisibilities' by implementing entirely different – namely, acoustic – eyes. In all three of the experiments in question, the technical recording systems and the manners of gathering, processing, and archiving data were matters of debate and of course became part of the experiments themselves.

23 Parrish et al., 'Introduction – From Individuals to Aggregations,' 10.
24 Ibid.

2. Plunging into the Deep

As Marshall McLuhan knew all too well, media theory is occasionally a fishy business. It is no coincidence that McLuhan, having grown up in the "backwaters" of Canada (in the words of his biographer Philip Marchant), sent a 'message to the fish' in 1968 that begins with the following consequential observation: "One thing about which fish know exactly nothing is water."[25] In this text, McLuhan's message is about the medium. Like fish in relation to their natural environmental medium, human beings tend to know nothing at all about environments that have been created by new technologies. The decisive difference, however, is that fish always move around like proverbial fish in water, whereas nothing similar could be said about us: "[T]he fish has an essential built-in potential which eliminates all problems from its universe. It is always a fish and always manages to continue to be a fish while it exists at all. Such is not, by any means, the case with man."[26] Under (media-) technical conditions, that is, our relationship to the environment can hardly be defined as being a perfect fit. Rather, it is characterized by processes of *creating* this world and *creating ourselves* within it. In order to comprehend these processes, in turn, we need to create "anti-environments,"[27] that is, we need to create rational distance between ourselves and our own situation. By way of example, McLuhan refers to the sailor in Edgar Allen Poe's story 'A Descent into the Maelstrom.' The latter, though spinning around in the maelstrom himself, "staved off disaster by understanding the action of the whirlpool" and thus, in effect, dragged himself out of danger by his own neck.[28] It is only from such a position that humans can yank themselves out of the water, so to speak, and master the laws of new media worlds like surfers on the ocean's crests (by this he meant media surfaces) who are ever in pursuit of the perfect (electromagnetic) wave.

In what follows, I will examine some additional aspects of the relationship among fish, people, and water – a reference system that has less to do with epistemologically valuable analogies than it does with media-technological analyses of an object of knowledge that is exemplary in both its visual and conceptual fuzziness. My point of departure is the following question: What happens when media theory no longer operates on the surface but rather jumps off its surfboard into the cold water and encounters swarming

25 McLuhan and Fiore, *War and Peace in the Global Village*, 175.

26 Ibid.

27 Ibid., 177.

28 McLuhan and Fiore, *The Medium is the Massage*, 150.

schools of fish under the sea? What happens when the surfing *overview* of its analytical anti-environment can thereby become effective *within* this very environment? Such questions are meant to spur an investigation of the specific medial relations that arise, *come in between*, and become effective when biological research attempts to understand the dynamic flurries of fish schools as objects of knowledge. How does an understanding of these dynamics constitute itself in a novel and different way in relation to certain media-technological parameters? In what ways do the destructive, distortive, unruly, and opaque factors of the 'immediate' water environment of fish (as McLuhan called it) limit the applicability of media-technological systems of observation? And thus how, beyond the limiting surface of the sea, do new surfaces of reflection or impermeability appear that themselves enable the appearance of mediality within this presumably immediate environment?

While it is true that nautical terms already provide us with some understanding of the scope of the area of knowledge in question, oceans nevertheless evoke a degree of horror about the unfathomable. They introduce a new level of 'blurriness' in its most literal sense. As Bernhard Siegert has observed, this horror is exemplified in Plato's remarks about the "hostility of the sea." Siegert, who appraises in his work not only the water of the Mississippi but also the expanse of the sea and especially the ships that course through it, emphasizes Plato's deep distrust of water and the sea in general. Such distrust, he thinks, is associated in an essential way with Plato's suspicious attitude toward writing, which is expressed in the *Phaedrus* dialogue. About this foundational text of media-critical thinking, Siegert remarks: "Water is the element of a fatherless writing that has fallen away from the logos. According to Socrates's words, we are to sow our knowledge of what is just, beautiful, and good 'in water with pen and ink' [...]. To write is to venture where there is no nomos."[29] By means of various media technologies designed to 'write in water,' biological fish-school research aims to discover regularities within the demarcational tendencies of its 'non-objects' and to fix such regularities later in relation to their environment.

Through the non-object of schooling fish, two interesting aspects are redoubled and brought into connection with a media history of underwater swarm research. From an epistemological perspective, this would be a doubling of disruptive phenomena: to the visual and acoustic distortion of the environmental medium of water are now added the kinetic dynamics of whirring and flurrying schools. As an object of knowledge, the fish school therefore becomes blurry by a twofold factor of intransparency. This,

29 Siegert, 'Der Nomos des Meeres,' 45. Siegert is here referring to *Phaedrus* 276b–276c.

moreover, is joined by a conceptual doubling: for in the case of schools of fish, which are multiplicities between collectivity and individuality that form and transform themselves in three spatial dimensions and in time, a second structured space is added to the space and topology of the sea, which has often been described as 'nomadic' by philosophers and media theorists. The spaces of fish schools are demarked by their constant flux, by the relentless tension that they exert on the geometrizing order of Euclidean grids. They are examples of momentary and spontaneous adaptions to internal and external factors. Schools organize themselves, so to speak, as *opportunistic* spatial structures. On the basis of local interactions, they create a specific space – a second environment within their environmental medium, the sea. And thus they become a second smooth space within the smooth space of the sea, which Gilles Deleuze and Félix Guattari famously defined as follows:

> In striated space, lines or trajectories tend to be subordinated to points: one goes from one point to another. In the smooth, it is the opposite: the points are subordinated to the trajectory. [...] In smooth space, the line is therefore a vector, a dimension and not a direction or metric determination. [...] [S]mooth space is directional rather than dimensional or metric. Smooth space is filled by events or haecceities, far more than by formed or perceived things.[30]

Yet, as smooth space *par excellence*, the sea also became, in Deleuze and Guattari's words, "the archetype of all striations of smooth space" and thus a reminder that "the smooth itself can be drawn and occupied by diabolical powers *of organization*" and that "there exist two nonsymmetrical movements, one of which striates the smooth, and one of which reimparts smooth space on the basis of the striated."[31] A media history of fish-school research is immediately related to this reciprocal relation between smooth and striated space, to the oscillation between Euclidean grids and Riemann's non-Euclidean, anexact, and morphological geometries.[32] This relationship is based on the topological structure of smooth space, within which every point can generate additional points in its immediate vicinity. Each of these vicinities thus forms a sort of local Euclidean space, yet the connections of

30 Deleuze and Guattari, *A Thousand Plateaus*, 478–79.

31 Ibid., 480 (emphasis original).

32 See Riemann, *On the Hypotheses Which Lie at the Bases of Geometry*; and Husserl, 'The Origins of Geometry,' 155–80.

the vicinities to one another are not predetermined and can come about in a variety of ways. As the mathematician Albert Lautman once noted, "Riemann space at its most general thus presents itself as an amorphous collection of pieces that are juxtaposed but not attached to each other."[33] This juxtaposition is reflected in the complex relationship between the local micro-dynamics within a fish school, which are based on simple parameters and which produce, as nonlinear interconnective processes, global dynamics that cannot be reduced to local properties. In order to describe the smooth space of the school, attempts have been made to striate the smooth space of the sea – to prescribe its water, so to speak. Attempts have been made to envelop them metrically or, put another way, to equip them with boundaries. Deleuze and Guattari elaborate:

> Heterogeneous, in continuous variation, it is a smooth space, in so far as smooth space is amorphous and not homogeneous. We can thus define two positive characteristics of smooth space in general: when there are determinations that are part of one another [i.e., Riemannian space and metrical space] and pertain to enveloped distances or ordered differences, independent of magnitude; when, independent of metrics, determinations arise that cannot be part of one another but are connected by processes of frequency or accumulation. These are the two aspects of the *nomos* of smooth space.[34]

This conceptual delineation of smooth space will serve as a point of departure for my effort to approach fish schools as *epistemic aggregations* – as the results, in Deleuze and Guattari's formulation, of processes of frequency or accumulation. Yet to observe the trajectories of schooling fish as a topological structure organized into vicinities within the likewise topological space of the open sea is often to fish in troubled waters. Observations of open water can fail on account of the very smoothness of the oceanic space, which in many respects can prove to be a locus of non-knowledge rather than knowledge – a space that can sabotage the fruitful production of knowledge by its sheer vastness, by its inhospitable nature, by the poor lighting conditions beneath the water, or by the whitecaps that form atop waves when the wind reaches level four on the Beaufort scale. The biologist Julia K. Parish and her colleagues have summarized the problem neatly:

33 Lautman, *Les schemas de structure*, 34–35. Quoted from Deleuze and Guattari, *A Thousand Plateaus*, 485.
34 Ibid.

> Following individual animals (or units of anything within a moving aggregation) in space and time turns out to be very difficult. Tracking requires a known frame of reference within which the object moves. If an object moves very fast, the rate at which its position is sampled must also be fast to accurately record changes in speed and direction. For confined objects, such as fish in a tank, this is relatively easy. However, tracking a fish in the ocean is more difficult, as it is likely to swim away.[35]

Not only is it difficult to define a frame of reference in the open water – to striate the space – but even when that is done, the 'objects' of investigation can simply swim away from it. Before the availability of technologies that could tolerate, at least to some extent, the hostility of fish schools' natural environment, the study of schools largely took place on land: in the tanks and aquaria housed by institutes of marine biology. This section will thus focus at first on these terrestrial observations and experiments before moving on to examine the research conducted in open water. The glass enclosures used for laboratory experiments happened to yield remarkable results: it was not until researchers had confined the smooth space of the sea within research aquaria and measured the activity of schooling fish in time and space that they were able to recognize that the sea is a smooth space to begin with. Such media-based samples and measurements were the first to reveal the eventfulness and intensities of this non-object and to create a vectorization of its movements *by means of* gridding itself. On account of these metrics, which were 'inspired' again and again by the smooth space of the school, accumulations and frequencies became legible for the first time. Out of striated space, smooth space emerged: trajectories that could be tracked.

Writing in Water

The exploration of the smooth space of swarms proved to be a rather frustrating occupation, though. Here is a telling quotation from a study directed by Julia Parrish:

> Assuming structure is advantageous, how is it maintained? Laboratory and field attempts to address this question in fish schools have been limited, in part because obtaining three-dimensional trajectories on specific individuals for a relevant period of time is difficult. Data that do exist are typically from highly artificial conditions (e.g., relatively small schools in

35 Parrish et al., 'Introduction – From Individuals to Aggregations,' 7.

> highly lit still-water tanks). Three-dimensional tracking techniques have not advanced to the stage where it is feasible to observe large schools (i.e., over 10), in three dimensions, over long times (i.e., for more than seconds).[36]

That such remarks can be found in a biological study from around the year 2000 is indicative of the persistent problems that fish schools pose for media-technological sampling. In this regard, everything had begun so hopefully after the end of the Second World War, when technical media of (slightly delayed) observation first came to be used in biological fish-school research and progress came to be made in minimizing the blurriness of observations and experimental structures for the sake of taking accurate measurements. The mechanization and automation of local processes in fish schools coincided with the mechanization of recording techniques in which, according to Étienne-Jules Marey, phenomena could be expressed with images formulated in their own language: as "images of objectivity" in Lorraine Daston and Peter Galison's sense of the term.[37]

Soviet researchers, for instance, thus began to use a new system for sampling and measuring schools of fish. It involved installing a camera orthogonally above the observation tank so that its visual area offered a two-dimensional, bird's-eye view. The field of vision itself was divided into quadratic cells of equal size by making use of the floor of the aquarium, which had been furnished with large, ten-by-ten centimeter tiles.[38] A net of cellular measuring fields thus extended across the aquarium floor – a system of coordinates that made it possible to capture the dynamics of schooling fish more effectively. The camera, which now took the place of human observers, was discretized in two different ways. On the one hand, the grid inscribed beneath the water allowed for individual schooling fish to be oriented more accurately in space; on the other, the film recordings of the movements of the school could be broken down into a precisely timed sequence of individual images. What such sequences of images brought to light was their inherent clock – a chronometer that lent a high degree of accuracy to the researchers' findings. Michel de Certeau has pointed out that, since the time of James Cook, the chronometer has been the authoritative device – "autonomous," "impervious to all alteration, inviolable" as it is – for enabling navigators to orient themselves on the open sea. It reconciles

36 Parrish et al., 'Self-Organized Fish Schools,' 297.

37 Marey, *La méthode graphique dan les sciences expérimentales*, iii–vi. Cited in Daston and Galison, 'The Image of Objectivity,' 81.

38 See Radakov, *Schooling in the Ecology of Fish*, 96.

the circle with the straight line and thus one could say that it likewise reconciles smooth and striated space. An example of such spatial orientation is provided by Phileas Fogg: "[T]his gentleman [...] did not travel, he described a circumference," according to Jules Verne's *Around the World in Eighty Days*. Fogg himself functions as a chronometer that relates any change of course to a referential time of departure and can thus rectify matters.[39] Thus the main interest of fish-school researchers in these recordings lay neither in the illusion of motion that such film strips evoked nor in the fascinating and hypnotic beauty of observing fish schools in motion but rather in analyzing and breaking down this movement into differential individual frames. It came down to establishing the relation of these individual images to one another. Of interest were the changing conditions of the school that could be captured by these images and measured according to the grid.

At the moment when images and image carriers became measuring instruments, thus bringing into closer association the blurs of the image and the inaccuracies of a measurement, a fundamental epistemic problem recurred concerning the analysis of movements, a problem that has been virulent since Marey's invention of and experiments with the chronophotographic method. From the beginning, Marey denied that the 'motion pictures' of film had any scientific relevance: "In the final analysis they show what the eye sees directly; they add nothing to the power of our sight, remove none of its illusions. But the true character of a scientific method is to supplement the weakness of our senses or to correct their errors."[40] Marey was probably not very interested in how moving bodies *appear*. Far more decisive for him was the *accurate analysis* of the motion of bodies, which could only be accomplished by breaking such motion down to its smallest possible elements. According to Joel Snyder, the use of technical media of observation in place of human observers did not merely induce a competition between the two. Rather, as Marey stressed, the dissection of motion sequences into precisely specified individual temporal units opened up an entirely novel level of analysis – a whole new realm of reality:

> Not only are these instruments sometimes designed to replace the observer, and in such circumstances to carry out their role with an incontestable superiority, but they also have their own domain where nothing can replace them. When the eye ceases to see, the ear to hear,

39 See Certeau, *Heterologies: Discourses of the Other*, 147–148 (in a chapter entitled 'Writing the Sea'). Certeau is here referring to Chapters 2 and 11 of Verne's novel.

40 Marey, 'Preface.' Quoted from Gunning, 'Never Seen this Picture Before,' 249.

> the touch to feel, or indeed when our senses give deceptive appearances, these instruments are like new senses of astonishing precision.[41]

They produced data, as Snyder further remarks, that did not exist apart from the graphical procedures to which they owed their existence. First and foremost, graphical methods and chronophotography *generate* the data pertaining to phenomena of motion: the latter presents motion as a relation between distances and the time required to cover them at any given moment in an image that, once placed within the context of its 'neighboring' images, becomes a sort of survey photograph. These images are not representations of motion that could have been perceived before, because they exceed the capacities of human vision by several factors: with their shutter speeds of 1/1000 of a second, which had been achieved by the end of the nineteenth century, they far surpassed the sensitivity of any form of biological sight and were able to measure "infinitely small lapses of time."[42] They do not, however, render visible the phenomenon itself – the *motion* of a horse, runner, or bird.[43] Rather, the *méthode graphique* makes use of appropriate devices to translate the complex motions of living beings into curves: as a connection of numerous individual points on an *x,y*-coordinate system in which the *y*-axis serves as a measure for spatial motion and the *x*-axis represents lapsing time. The graphical lines of the curves visualize and quantify the complex motion of bodies in time and make them accessible for mathematical analysis.[44]

If the place of motion pictures in fish-school research is to be evaluated, it is essentially a matter of subtracting motion out of the images, excluding obscurities by means of the high number of frames per second, and breaking down the motions of individuals into discrete spatio-temporal units. Motion pictures only gain their epistemic surplus at the moment in which they are frozen. It is only in such a way that they differentiate themselves from the sensory perception of human beings and thereby acquire scientific value – in contrast to a running film, which merely serves this perception and imitates its conventions.

The measurements of moving bodies, however, do differ essentially from the image-based measurements of schooling 'bodies.' The object of such analytic efforts is not, as in Marey's work, established sequences of motion with causally connected progressions that can be analyzed according to

41 Marey, *La méthode graphique*, 108. Quoted from Snyder, 'Visualization and Visibility,' 380.
42 Marey, La méthode graphique, iii.
43 See Vagt, 'Zeitkritische Bilder.'
44 See Brain, *The Graphic Method*.

the *span* of time that they have traversed. The motion of an individual bird's beating wing, for example, will ultimately provide the analytical timeframe whereby it will repeat itself after a certain time. Moreover, the elements of chronophotography's objects and bodies that interact in dynamic processes have relatively fixed relations to one another in that, for instance, they are connected with one another at fixed points of articulation. They thus generate regular curves in the recording system. The couplings of interactive 'elements' in a schooling body, on the contrary, are free and constantly changing, and the paths of schooling individuals are better described as trajectories resembling Brownian motion than as repetitive processes. In addition to the regular and inter-individual behavior in a school of fish, there are also, on account of rapid and constant situational changes and the multiplicity of interactive elements, constant deviations from typical behavior. This is something that a collection of individual paths and an analysis of image sequences will fail to detect in an effort to explain global sequences of motion, especially if the resolution amounts to no more than twenty-four frames per second. The novelty here, as far as Marey's ideas are concerned, is that the object of media-technologically supported investigations is not an 'animal locomotion' that is *divided* into stages of motion but rather a *coupled* 'school motion.' This is a different, nonlinear type of motion in which the form of the object to be observed changes as a whole. What comes into play is a particular sort of cinematics that Marey failed to take into consideration. Moreover, this cinematics of merely loose and repeatedly dissolving and merging couplings takes place in *three-dimensional* space. The two-dimensional media of chronophotography or film could not capture the direction of such motion, a fact that alone produced distortions in the analysis of the images.

The case of fish schools, however, requires another step to be taken after the detailed breakdown of blurry motion, and this is to reassemble the individual frames into new and sharpened sequences of images that visualize or (perhaps more fittingly) *animate*, in a new form, the previously unclear or unrepresented material in the filmed motion picture, such as the accurate trajectories of schooling individuals. These attempts at visualization will be discussed in greater detail later on; first, however, it is necessary to return to the difficulties involved with observing of schools of fish.

In the section devoted to laboratory experiments in his book *Schooling in the Ecology of Fish*, Dmitri Radakov, who made use of gridded aquaria for measuring schools in the 1950s and 60s, pointed out that, even under artificial laboratory conditions, it is essential to observe a large number of schooling individuals in order to make the transition from pure quantity to a new form of quality – in order, that is, to evaluate the individuals of a school as "a certain biological category,"[45] as his fellow countryman I. I. Mesyatsev once wrote:

> [A]n investigation of schooling behavior demands that we study not only the interrelations of two or a few specimens, but the general regularities inherent in a fairly large school as a unit, in which quantity goes over to quality. [...] [W]e generally experimented with schools composed of several scores or hundreds of specimens, even though this made it very difficult to set up the experiments and to evaluate their results.[46]

Radakov studied schools of small species such as minnows. In doing so, he hoped to circumvent the basic problem that he did not have access to an aquarium large enough for schools of mature herring or mackerel. He also stressed the importance of creating the most natural possible conditions, though this often led to technical problems.[47] One consequence of this was habituation effects, which could distort his research findings. Schools that had been in the aquarium for a long time, for instance, would tend to slow down their reaction speeds, to the extent that it would be necessary to replace them with fresh specimens. With small fish, at least, there was enough room in the tank to observe the execution of the collective maneuvers of a school composed of around a hundred individuals. As mentioned above, the quantitative evaluation of the experiments was conducted with the help of "motion picture photography":

> Filming makes it possible to see repeatedly on the screen fish performing actions which we wish to study, and this is often very important in view of the multitude of fish in a school and the speed of their movement. Moreover, this technique allows us to show the film in 'slow motion,' as is often done in the case of sports events. In addition, knowing the rate at which the film is channeled in the camera during shooting and the scale of the image, we can determine the speed and path of movement of fish and other objects whose image appears on the film.[48]

Here, swarms became media events whose eventfulness, like that of a sporting event, emerged only after the fact in slow motion. They became

45 Mesyatsev, 'Structure of Shoals of Schooling Fish.'
46 Radakov, *Schooling in the Ecology of Fish*, 54.
47 Ibid.
48 Ibid., 56–57.

events that are unthinkable apart from the media that created them.[49] The film exposure of twenty-four frames per second became the clock and, simultaneously, the grid of measurement in time and space (it did not, of course, achieve anything near to the detailed precision of Marey's chronophotography). Developed and projected one frame at a time, the movements of the schooling individuals were calculated by measuring, with changes in scale in mind, the changes of their positions from one image to the next (or, in more crude calculations, from one image to the *n*th) either on the projection screen or by arranging photographic prints.

With this combination of real-time recording, which could track the fish as they moved from one zone of the aquarium to the next, Radakov was able to create a map of a school's basic patterns of motion in two dimensions plus time. In his experiments, he was not only able to observe certain typical forms of schooling maneuvers; he was also able to record them graphically. This, moreover, he was able to do not only on a film-by-film basis but also cumulatively to create motion vectors within his tile grid. The measured movements of schooling individuals within a given zone of the grid could be aggregated and conveyed statistically to chart something like the intensity of motion within the quadrant at hand:

> A school of 'verkhovka' was put in an aquarium (1 × 2 × 0.4 m). A small cigar-shaped object (rough model of a predator) was trailed along on a string beneath the water surface. Upon this development, the school split into two parts which merged again after the object had passed. We filmed this process [...]. The method we proposed for a quantitative evaluation of the school's movement is as follows. [...] Any elementary part of a space to be investigated is characterized by a vector, the modulus of which is equal to the density of the school (that is, the number of fish in a given square), while the direction is determined by the predominant (average) direction of movement of the first fish in that square. A circle was arbitrarily divided into 12 sectors, so that the orientation of the fish was estimated to within 30°.[50]

Every film image thus yielded a corresponding vector field of the filmed school. These vector fields could then be aggregated to form a time-dependent "directional diagram of action." From this information, highly particular intensities could be established for each zone; from the striated space of the

49 On such phenomena, see Engell, 'Das Amedium – Grundbegriffe des Fernsehens in Auflösung.'

50 Radakov, *Schooling in the Ecology of Fish*, 96. *Verkhovka* are small schooling fish for which the English translator could not find an adequate equivalent.

tiled floor, in other words, emerged the dynamics of a smooth space. This brought to light its directionality, its events in time, its constant *becoming* – characteristics that, described by Deleuze and Guattari as antithetical, lead away from a topographical order toward a topological structure. In this process, directionality arises from zoning, from a determined localization, while time courses within and dependent upon this space, and in discrete steps. It is arranged according to the frequency of the film's images, to which the vector images are also related: "Each of the squares in which vectors are present may give rise to a function which is dependent on time and which offers an independent value for analysis."[51]

And yet a categorical transformation had taken place: the uniform and absolute space of the grid – a Euclidean and Newtonian stage – gave rise to a dynamic and temporalized quantity of tiles – a swarm of tiles characterized by their trajectories, lines, and positional relations. Measurable on the grid and yet freed from it, together they defined a new topological space: a space that could only arise from the respective positional and motion relations between the elements of the whole. The local informational value inscribed in each field made the relations of each individual zone amenable to mathematical analysis and allowed what was observed on film to be reduced to abstract and complexity-reducing models. "[W]e hope," Radakov remarked, "that our attempt to find such a method will be of some use in further work with mathematical models of schooling maneuvers, a method which offers obvious advantages."[52] The principle of *intensifying* local parameters by taking into account neighboring positional relations – that is, the creation of complex geometries on the basis of simple structures and rules – would again become a matter of interest in the context of cellular automata, which I will discuss in a later chapter.

For now, Radakov's gridded aquaria and his translation of film stills into vector images made it possible to record and operationalize two-dimensional images plus time. A few years later in the 1960s, the biologists J. Michael Cullen and Evelyn Shaw would devise an experimental system that implemented the temporal dimension in an entirely different way and could accommodate an additional spatial dimension.

The Linearity of the Doughnut: Swimming with the Current

Aquarium walls are not only boundary surfaces for measurements and visual surfaces for observation; they are likewise surfaces of distortion.

51 Ibid.

52 Ibid., 99.

Both Albert Parr and Dmitri Radakov mentioned the problem that, in laboratory experiments on schools in an aquarium, the schooling structure would often be disturbed. This would typically happen when the school under investigation was forced to change its direction abruptly as it came into contact with one of the walls of the aquarium. As Parr's experiments showed, this could lead to entirely different structures, such as the 'mill,' or, in case of certain species, it could lead to a total collapse of the school structure, especially when a school approached one of the corners of a traditional aquarium and suddenly had to deal with two walls. Anticipating this problem, Evelyn Shaw used a doughnut-shaped tank with a screen around it in order to minimize external influences.[53] Observations were made either from above or from the side through a one-way mirror. This form of aquarium enabled the schooling individuals to polarize and swim in a continuous forward motion without ever having to turn around. In a sense, the design of the aquarium is a materialization of Certeau's idea, mentioned above, about reconciling the circle with the straight line. By considering the factors of attraction and repulsion in schools, disturbances and distortions could be reduced in the experimental system and the undeceivable dimension of time could also be reflected architectonically in the schools' self-organization. Explicitly underscored in all of this was the internal processes that lead to the formation and cohesion of schools. External influences, in contrast, were minimized so that the results of the observations were not distorted.

In 1965, furthermore, Shaw and his colleagues J. Michael Cullen and Howard Baldwin described two systems that could allow for the three-dimensional observation of fish schools.[54] Both methods involved taking photographs from above an aquarium, each (as with Radakov's experiments) against a gridded background. In this case, the grid consisted of a white sheet of styrene plastic with black lines forming ten-by-ten-centimeter squares. One of the systems made use of a stereophotographic method in which the relative position of the two calibrated cameras enabled a spatial view that, at a known distance and angle, could make an accurate photogrammetric measurement of the distances and relative positions of the individual schooling fish to one another.

The second approach was the so-called 'shadow method,' which would be applied more widely in the 1970s. It required only a single camera, but the latter was flanked by a spotlight pointed in a particular angle. Each of

53 Shaw, 'The Schooling of Fishes,' 130.

54 Cullen et al., 'Methods for Measuring the Three-Dimensional Structure of Fish Schools.'

the observed fish thus cast a clear shadow on the floor of the aquarium. Given the size of the shadow in relation to the size of the corresponding fish – and given the angle of incident light and the depth of the water – one could likewise convey the three-dimensional position of individual fish. The disadvantage of this method was that, if the schools being examined were too large, the shadows of the fish would overlap with one another and thus be difficult to distinguish.

The circular form of the aquarium and the continuous swimming of the school also allowed the latter to be followed easily by the camera, such that the positions of the individual fish did not change in relation to the camera's perspective. The development was described by Brian L. Partridge and his collaborators: in order to achieve a detailed analysis of a school's structure and dynamics, a camera was attached to a gantry that rotated above the aquarium. Its speed was controlled from one observation station, while a second observer was situated on the gantry itself and "gave a continuous 'racetrack' commentary of the positions of each fish with respect to the rest. This was recorded on the videotapes, and at a later stage individual fish were identified in each film sequence."[55] Thus it was possible to record a fish school of twenty to thirty individuals over longer periods of time, during which the frame of reference moved along with the fish. The gridded space no longer enclosed the motion of the school but rather, in a sense, moved with time. Thus an additional spatial dimension was added to Radakov's observational system, and schools could be studied in light of the constitutive factors of *attraction* and *repulsion*.

In their analysis of the recordings, Partridge and his colleagues accomplished something rather amazing. Up until the turn of the century, their four-dimensional measurements, which were conducted in the middle of 1970s, remained by far the most thorough data set on the subject and even provided, according to Julia Parrish, a metric against which simulation outputs could be tested.[56] For, until the 1980s (that is, before the possibility of automated data processing), "researchers interested in collecting four-dimensional data sets had to repeatedly digitize hundreds, if not thousands, of points. Method sections from several fish schooling papers [...] are full of agonizing descriptions of the number of frames analyzed (e.g. Partridge et al. hand digitized over 1.2 million points). The endless hours of data collection were enough to turn anyone away."[57] Anyone, apparently, except Partridge

55 Partridge et al., 'The Three-Dimensional Structure of Fish Schools,' 278–279.
56 Parrish and Viscido, 'Traffic Rules of Fish Schools,' 67.
57 Parrish et al., 'Introduction – From Individuals to Aggregations,' 10.

and a few of his colleagues. At first, video sequences were chosen that were copied onto 35-millimeter film with a frequency of thirty frames per second:

> In all, nearly 12,000 frames of film were made for the experiments in 1975 and 18,000 were made for those in 1976. This corresponds to 184 and 214 separate film sequences, respectively. Once the films were made, the position of each fish's snout and its shadow in each frame of each sequence [...] was determined using an inexpensive online interactive coordinate plotter developed for the purpose.[58]

The plotting program was designed in such a way as to correct for optical errors, blurriness, lens aberrations, the "nonlinearity of the video system," and so on: "Final coordinates were accurate to +/- 0.25 cm."[59] The endless swimming in a circle inside the special tank led to input, to the digital storage of massive amounts of data points, and to output consisting of mountains of pen-writer trajectories. Socrates's 'writing in water' – the sowing of ink through the pen – was partially automated in this case: as knowledge of the just, beautiful, and good, writing was delegated to machines – computers, potentiometers, AD-converters – and to programming languages. In its four-dimensional analysis, the smooth space of the school that had emerged from the grid was fragmented into a bundle of trajectories – of partially parallel and partially intersecting lines – and manifested itself on the plots of the paper machine, once again, as a dimensional reduction in two dimensions plus time. Where the zones had previously been charged with intensities, they now gushed forth directly in rays of time and motion, each corresponding to an individual fish in the school. In the doughnut tank, circularity gave way to a renewed fragmentation of intensities into individual linearities as they simultaneously extended along the temporal axis. The accumulations of these linearities, the reconstructed course of time (immobilized in paths), and the conglomerating, intertwining, and relatively self-orienting lines of motion were thus the epistemic aggregations from which the basic functional parameters of the schooling individuals' interaction could be ascertained and quantified. On such grounds, the formation of certain adaptive patterns of motion – or the synchronization of inter-individual distances or speeds – could presumably be comprehended, and a degree of order was meant to be brought to drifts of data.

58 Partridge et al., 'The Three-Dimensional Structure of Fish Schools,' 279.
59 Ibid.

Hand Digitizing: Data Tablets

This development could only be achieved by compiling the fragmented positional data in a suitable way. The “low-cost interactive plotter” that Partridge used to this end possessed the charm of a hodgepodge device contrived in someone’s Silicon Valley garage, even though it happened to be built in an office at Oxford University: “Provided a user has access to a computer running ALGOL or FORTRAN (such as PDP-8, Digital Equipment Corporation) which is equipped with analog-digital (A-D) conversion, he can build the entire system for under $50, which is about 1/200 the cost of most commercially available plotters.”[60] The plotter consisted of a horizontal table supporting a plexiglass screen, beneath which a mirror was mounted that allowed film to be back-projected onto the screen itself. On both sides of the screen, a potentiometer was installed at a defined distance from one another. Each of the potentiometers was attached to a string, both of which were in turn mounted to a cursor. The latter was meant to determine the respective *x,y*-coordinates of a point. The strings were kept taut against the potentiometers by a set of weights attached to their ends, so that a triangle would be generated for each point to be plotted. Thus, from that triangle, the value of each *x* and *y* could be calculated geometrically.[61]

Because the position of the potentiometers was fixed and the voltages across them were proportional to the lengths of the strings creating the triangle, two potentiometer signals could be fed into one multiplexer, which would generate a signal that would in turn be directed to a computer through an analog-digital converter. By means of a control panel – in this case a rack with ten push buttons, at least five of which governing an independent procedure – various input and output modes could be calibrated. The five basic procedures of the program were the following:

> (1) Scales: calibrates the voltages across the potentiometers to distance moved by the cursor, scales the coordinates to correct for the size of the projected image, and initiates a dialogue to define various parameters. (2) Point: determines the current cursor position and subtracts the coordinates from a reference point in the picture, since no projector positions each frame in exactly the same place. (3) Error: deletes the previous point plotted by decrementing the counters. (4) Missing: outputs characteristic

60 Partridge and Cullen, ‘Computer Technology: A Low-Cost Interactive Coordinate Plotter,’ 473.

61 Ibid., 474.

> *x,y*-coordinates for missing data [...] so that the coordinates are left out of further analysis. (5) Frame: outputs the current frame of data to magnetic disk or paper tape, writes the frame number and current time (calibrated to original film speed), advances the projector, and increments the frame counter.[62]

In Partridge and Cullen's design, the computer and the interface formed the very same 'man-computer symbiosis' that J. C. R. Licklider had conceptualized as early as 1960.[63] In the initial dialogue with the program ('scales'), for instance, the operator could set a series of parameters concerning the expected number of coordinates, the number of frames to be plotted in the film sequence, or the maximum distance that a point might move from one frame to the next. Whenever the criteria were exceeded, the program would indicate that an error has occurred and allow the deviant point to be replotted. In this regard, the program's accuracy in detecting errors caused a bit of unease among the researchers: "Maddeningly, the computer is usually right when it suggests, on the basis of these criteria, that points have been plotted out of order."[64] To stick to the truth, moreover, the man-machine symbiosis would in fact become a *woman*-machine symbiosis: ENIAC girls and Ivan Sutherland's imagined secretaries beckon and pose the question about who had really been responsible for the agonizing task of data entry.[65]

In an article from 1974 (and thus a few years before Partridge designed his contraption), Sutherland had described his own advanced system for digitalizing three dimensional coordinates with up to seven 'pens.' This involved defining the relations between *x,y*- coordinates from a top view, defining a third, *z* coordinate from a side view, and then linking the three coordinates directly with one another. In Partridge and Cullen's system, two inputting steps were necessary to achieve this – fixing the *x,y*-coordinates and then measuring the position of these points in relation to the shadows created by the individual fish. To illustrate the capabilities of his tablet, Sutherland's examples included sketches of ships and castles, which, as true-to-scale *bodies with surfaces*, are of course far easier to convert into computer graphics than the coordinates of schooling fish, not to mention

62 Ibid., 475.

63 See Licklider, 'Man-Computer Symbiosis.'

64 Partridge and Cullen, 'Computer Technology,' 476.

65 About the ability of computer-illiterate people to use interactive graphical user interfaces, Sutherland reported that even a secretary could work at a computer without having to understand how it functioned. See Pias, *Computer Game Worlds*, 95. Pias is here referring to Sutherland, *Sketchpad: A Man-Machine Graphical Communication System*, 33.

that their frames of reference are much easier to reproduce. Nevertheless, Sutherland also underscored the utility of his system for comparing 'perspective views' and photographs, especially for making automatic corrections when the axes of the two views are not perfectly aligned.[66]

In an article from 1981, two Chicago-based researchers introduced a "computerized film analyzer" called GALATEA, which had first been designed in 1973 but was now specified to assist fish-school research. This system supplemented Partridge's data tables in that it used a "digitizing area" as a graphical display upon which film images of the process under investigation could be projected. With a lightpen, users could mark points on this graphics board, and these would be processed into so-called "kinegrams" by a PDP-11/40 mini-computer. By means of an additional kinescope, the kinegrams could in turn be superimposed onto the film images and synchronized to them with a clock. Thus, it was possible to mark objects on the film images in real time, and these digitalized markings could lead to an ongoing analysis of the animation (kinegrams) being produced – an 'analysis by synthesis':[67]

> The resulting set of projected points, overlaid directly on the source image, provides all the advantages of tracing paper in verifying entries and avoiding omission and duplication of points. [...] With such a system, it is possible to record accurately the *x, y* positions of hundreds of points in an hour [...] and the various moving points can be connected with lines to create animated stick figure representations of the objects under study.[68]

GALATEA was described by its developers as a system that could be applied in biological and biophysical laboratories like "dynamic tracing paper for film," whereby the user – like a piece of pattern-recognition technology – contributes to the image analysis through an interface: "[T]he user directly transcribes the features he discerns using his own sophisticated interpretive pattern recognition capabilities."[69] Moreover, as soon as a grid of at least six coplanar reference points could be constructed, the system would also make it possible to reconstruct a three-dimensional image out of two stereo-photogrammetrically measured two-dimensional images. The reference grid would even make it possible to film a given situation with moving

66 Sutherland, 'Three-Dimensional Data Input by Tablet,' 453–454.

67 Potel and Sayre, 'Interacting with the GALATEA Film Analysis System,' 52.

68 Potel and Wassersug, 'Computer Tools for the Analysis of Schooling,' 16–17.

69 Potel and Sayre, 'Interacting with GALATEA,' 52.

cameras, so long as all the reference points in the sequence were visible (the position of the objects in relation to the camera could be computed later on for each individual image): "Perhaps the most important feature of the system is that it does not require persistent fixed reference surfaces, as in the shadow method or methods involving mirrors. The system is unique in providing an arbitrary perspective of the final data on the video screen."[70] This would be especially interesting, or so the authors thought, for analyzing studies conducted in the open sea, where the controlled conditions of research aquaria are of course not a possibility. Because of a general lack of comprehensive fish-school research in the following years, however, more than a decade would pass before the system was used to that end. In the meantime, lured by more powerful computer programs and a general interest in self-organization processes revealed by chaos research and complexity studies, biologists versed in computer science turned their attention instead to programming various simulation processes (to be discussed in Chapter V).

That said, all of the information that was transformed into computer graphics by means of such tablet-digitalization and image-analysis methods, which filled up Partridge and Cullen's plots and magnetic storage belts with three-dimensional positional data, heighten the following suspicion: When the process of determining relative positions with the help of computers – always over just short periods of time and with just a few schooling individuals – is confronted with so much data and changes in data that it becomes extremely time-consuming to make meaningful calculations with them, then this seems to indicate quite strongly that the schooling individuals themselves were hardly capable of dealing with such time-intensive and calculation-intensive processes (as regards the act of processing so many sensory perceptions). Thus it is probably the case that schooling fish can possess only a limited amount of knowledge about the school as a global structure. Rather, the early side effects of the first computer-based epistemic approaches to fish schools suggest that their global structure depends on the dynamic interaction of relatively few individuals in local vicinities. In the conversion of the biological system of the fish school into computer-supported recording systems, the media-technological functionality of the digitalization process sheds light on the possible limitations of the system being converted. If both are regarded as information-processing systems, the technical arrangement can – depending on the computing capacity at hand – provide certain insight into the decentralized organization of fish schools as a biological system.

70 Potel and Wassersug, 'Computer Tools for the Analysis of Schooling,' 17.

The use of data tablets and other computer-supported analytical tools at the time in biological fish-school research is also indicative of the fact that the relevance of visual technologies persisted without interruption even during the era of electronic data processing and the production of mathematical models of schools. In fact, it was only by way of such methods that the first step could be taken toward a comprehensive visual analysis of the movements of schools in four dimensions. This step, moreover, involved making the first direct link between the study of schooling and the computer-based methods of a conceptually novel form of graphic design – a link that would later lead to the development of unique and effective models in biological fish-school research.

3. Fishmen[71]

> The most important limiting factor, for any in-site study on marine population, is the fact that water is an environment that can be considered as opaque.[72]

In the case of laboratory aquaria and their artificial conditions, that which is advantageous (namely for analyzing the internal processes of fish-school organization) can also be seen as a limitation. Under such conditions, it is questionable to what extent observations of this sort are applicable to the behavior of very large schools in their natural environment. Moreover, it might also be the case that the fish in question can only develop certain abilities to adapt when interacting with the many external influences encountered in this environment – the conditions of undersea currents, the availability of food, the threat of predators, and so on. Like the ornithologists discussed in my second chapter, scientists studying schools of fish likewise entered the habitat of their 'object' of research, not least because the knowledge they might attain there could be of commercial interest to the fishing industry.

The *in-situ* observation of fish schools in the open sea, which can be thought of as a complementary strategy to aquarium-based research, posed its own diverse set of problems. If schools are approached from a bird's-eye perspective, for instance, they are only visible as vague, amoeba-like *surfaces*. Yet from the late 1940s to today, aerial photographs have been

71 Such is the title of the opening chapter in Jacques-Yves Cousteau's *The Silent World*.

72 Gerlotto, 'Gregariousness and School Behaviour of Pelagic Fish,' 239.

used to follow fish schools along their migratory routes or to determine their distribution within a given area. These efforts, too, probably derive first of all from the desire to generate results of commercial interest. After all, such photographs and recordings have hardly provided any information at all about the internal structure or three-dimensional form of schools and the way these structures and forms change in response to external or internal factors. All that can be seen (or not seen) in such images is the top fifteen to twenty centimeters of the school in question. What is more, the degree of visibility depends on the opacity of the water and the serenity of its surface. Even when there is little wind, the ocean's waves create a carpet of distortion that precedes even the intransparency of the school. On top of this, it is necessary to have a sufficiently strong source of light and there is always the risk that the 'object' under investigation will react to the noise of the airplane. From the air, too, it is only possible to make approximate measurements of a school's size and speed.[73] Unlike the observational systems in the laboratory aquaria at land-based marine research institutes, an external view of the sea permits no more than a limited view of the activity taking place below its surface. Only on rare occasions does the smooth space of the sea present a smooth and highly transparent surface.

For a long time, something similar was also true of underwater observations in the open sea, for instance by divers, automatic cameras, or from submarines. In this case, the only schooling individuals that could be photographed or filmed were those nearest to the recording media. Only in exceptional circumstances was it possible to determine the size, form, and behavior of large schools, because the visual range amounted to no more than fifty meters and it was difficult to capture a clear image of anything more than twenty meters away. Moreover, the technological equipment was highly expensive, especially given the fact that it enabled such limited access to the 'object' being studied.[74] Also, the coloration of most fish schools ensures that they stand out as little as possible against a dark background. Finally, there was one more complicating issue that the (aptly named) marine biologist Wolfgang Fischer raised in an article from 1973:

> From my own experiences, it seems especially important to register the gained data as quickly as possible. When working underwater, the amount of information lost through *forgetting* is particularly great. The ideal solution to this problem is for the diver to have a wireless radio

73 See Radakov, *Schooling in the Ecology of Fish*, 46.

74 Gerlotto, 'Gregariousness and School Behaviour of Pelagic Fish,' 239.

> connection to the surface and for him to use a TV recorder. In this way, the acoustic information from the diver and the corresponding optic information can be recorded synchronously.[75]

Just as in the case of Partridge's studies with the toroidal aquarium, running commentaries were also used under water to contextualize and enrich the visual data with an additional level of information. And Dimitri Radakov added yet another point that might have been important for prototypical scientists to keep in mind: "For short-sighted persons it is very important that the mask be made with depressions in the viewing part (made of plastic) to correspond in curvature to the lenses of the glasses worn (account being taken of the refraction of rays with the transition from one medium to the other)."[76] Despite such difficulties, open-water observations played a significant role within the field of biological fish-school research. At the very least, they made it possible to circumvent the abiotic factors that influenced laboratory studies and experiments conducted with aquaria. Because the presence of divers – so long as they kept a sufficient distance from the schooling individuals – did not affect the latter's behavior, the "light diving method" (as Fischer called it) made it possible to observe schools in their natural environment in a sort of participatory manner.[77] In doing so, it was recommended that divers should use the lightest possible equipment, which would guarantee their freedom of movement and allow them, according to Hans Hass, "to live like a fish in the realm of fish and, through such close contact, to study far better than ever before the manifold life in the sea."[78] Here the main authority is not the surfer, who according to Marshall McLuhan governed the laws of the counter-environment while elegantly operating on the surface of sea.[79] It is rather the diver who, submerged beneath the water's surface to unexplored depths, augments the media history of fish-school research by monitoring his object of knowledge with new media techniques.

As was the case during the nascent years of land-based behavioral research, during the early stages of producing images of underwater life, university-based zoologists were not on the forefront in the development

75 Fischer, 'Methodik und Ergebnisse der Erforschung des Schwarmverhaltens von Fischen,' 393 (my emphasis).

76 Radakov, *Schooling in the Ecology of Fish*, 49.

77 Fischer, 'Methodik und Ergebnisse der Erforschung des Schwarmverhaltens von Fischen,' 392.

78 Hass, *Fotojagd am Meeresgrund*, 11.

79 See McLuhan and Fiore, *War and Peace in the Global Village*, 175.

of new media-technological processes. Here, too, it was amateurs with various backgrounds and intentions who attempted, by means of technical apparatuses and novel methods of observation, to cast a new perspective on the behavior of marine organisms – techniques of observation that they both tested and optimized. Life beneath the sea was no longer fascinating simply as a classic subject of science-fiction literature or the early cinema (such as Georges Méliès's silent film *Under the Seas* from 1907). In the reflections of the water's surface, the films of late 1920s became above all self-reflexive: the relations among science, fiction, and film were renegotiated when the underwater film pioneers Jean Painlevé, Hans Hass, and Jacques-Yves Cousteau began to dive into aquatic worlds with the aim of making recordings that were as true to reality as possible. Their bold efforts to film previously unseen forms of life meant they had to develop the necessary technologies themselves in order to be able to move around in the water along with their recording equipment. These three underwater filmmakers not only developed revolutionary camera techniques; they were also protagonists in the revolution of diving techniques. Whereas Painlevé also worked in the aquaria housed at his "Institute in the Cellar" (as Léo Sauvage called it) and dreamed of having a complete underwater studio, the self-proclaimed adventurers Hass and Cousteau took to the open sea. With his friends and later with his wife Lotte, Hass explored sharks and coral reefs, while Cousteau – the "subaquatic astronaut"[80] – began, after the war, to use futuristic technology on his ever more spectacular expeditions.

All three directors can tentatively be classified as makers of 'science films' in which biological life and cinematographic apparatuses are aesthetically combined. As André Bazin remarked:

> When Muybridge and Marey made the first scientific research films, they not only invented the technology of cinema but also created its purest aesthetic. For this is the miracle of the science film, its inexhaustible paradox. At the far extreme of inquisitive, utilitarian research, in the most absolute proscription of aesthetic intentions, cinematic beauty develops as an additional, supernatural gift. [...] The camera alone possesses the secret key to this universe where supreme beauty is identified at once with nature and chance [...].[81]

80 McDougall, 'Introduction: Hybrid Roots,' xvii.

81 Bazin, 'Science Film: Accidental Beauty,' 146. Originally published as 'Le film scientifique: Beauté du hasard,' *L'Écran français* 121 (1947), 10.

In this light, there are at least three important aspects to the relationship among film, technology, and aquatic life between the 1930s and 1950s. *First*, the status of the science film as a legitimately scientific way of accessing biological phenomena and observing 'life' was a matter of debate. In particular, underwater films were suspected of using tricks and deception and were thus considered little more than 'entertainment for the uneducated.' In 1928, for instance, members of the French Academy of Sciences reacted to a screening of Painlevé's *The Stickleback's Egg* with skepticism and even indignation.[82] Hans Hass always placed great value in demonstrating the authenticity of his underwater recordings 'in the image.' According to Bazin, the epistemic potential of science films was permanently in danger of being blurred by the illusion-making potential of the 'cinema of imagination.' *Second*, these science films reflected their own mediality. In their attempts to create the illusion of having immediate access to objects living and moving under water, the filmmakers had to be aware of the mediality of their equipment and the features of the environmental medium in which they were recording. The task of filming underwater was and is one of eliminating or at least reducing distortions. In a vivid way, science films introduced a theory of media that understood transmission in terms of negating the parasitic noise that would always be in the transmission's path ahead of time. *Third*, in science films, cinema gained access to an 'apparatus-based aesthetics' in the interplay between anthropomorphism and the uncanny metamorphoses of life under water. In the unseen and unforeseeable biological life *processes* made visible by these films, their technical aesthetics emerged in relation to coincidental and potential events: in this medial arrangement, life attested to its vitality *by means of* its movement. The counterpart to this cinematic way of thinking about movement was stasis, which would enable of clear view of things. When, in the final image of Painlevé's 1954 film *Sea Urchins*, the urchins in question rearrange themselves to form the static letters *FIN*, it becomes clear that, in the words of one critic, "[w]hen movement ceases, [...] the show is over."[83]

Already at this point, it was innovations from the entertainment sector (in the broad sense) that demonstrated and prefigured the applicability of technical media in biological contexts as well – a transference of technical knowledge that, at the end of the 1980s, would recur in a similar manner in the context of computer graphic imagery in multi-agent systems (this process will be discussed in the next chapter). These innovations also provided an important impulse for producing images of fish schools under the water.

82 See Berg, 'Contradictory Forces: Jean Painlevé 1902–1989,' 17.
83 Rugoff, 'Fluid Mechanics,' 56.

From the 'Institute in the Cellar' to the Open Sea

The French surrealist and Bugatti driver Jean Painlevé, who collaborated, for instance, as the "chief ant handler" on the ant sequences in Luis Buñuel and Salvador Dali's short film *Un Chien Andalou* (1929),[84] was perhaps the first amateur who attempted to use photographic and film techniques – which he developed himself – for the scientific analysis of subaquatic life. In 1923, having abandoned his coursework in mathematics and medicine, Painlevé began to study zoology at the Sorbonne, where he enrolled in courses at the affiliated Roscoff Marine Biological Station on the Breton coast. There he not only became acquainted with his later life companion and scientific collaborator Geneviève Hanon; he was also introduced to certain experimental-biological approaches to marine life.[85] Beginning in the late 1920s, he worked alongside Hanon and the cinematographer André Raymond at the so-called Institute of Scientific Cinema, which was housed in a Parisian basement, to produce realistic film recordings of underwater organisms kept in aquaria. In 1935, the journalist Léo Sauvage visited Painlevé's private institute and was amazed by how sophisticated it was compared to traditional research institutions:

> The filming room offers a spectacle as colorful as it is diverse. There is something bohemian about Jean Painlevé's Institute, something fresh, youthful, spirited, bustling and unconventional that challenges the mummified sciences of the Academy in the most insolent way. The walls are white, covered with buttons, switches, levers, meters. How do they know what's what? And of these countless, inextricable wires that go in every direction, come back, entangle and separate, which goes to a projector, which to a camera, which to a socket? [...] I am stopped in my tracks, stunned, before a new apparatus for filming in slow motion [...]. Painlevé explains how everything is made out of old things, refurbished and transformed. Thus one of the elements in the camera is a mechanism from a clock, bought somewhere at a discount. But it has been modified, a system of spare cogs adapted to it, allowing the recording speed to be changed at will. The camera is completely automatic.[86]

That said, each of these high-tech apparatuses had its own set of problems. It was not only that the glass walls of Painlevé's saltwater aquaria would

84 Winter, 'Science is Fiction: The Films of Jean Painlevé.'
85 See Berg, 'Contradictory Forces,' 9.
86 Sauvage, 'The Institute in the Cellar,' 126–127.

occasionally crack on account of the heat produced by his six to seven studio lights. The filmmakers also had to deal with resistant objects of observation: "These animals," as Painlevé reported, "are mobile, capricious, and completely unconcerned with the way you wish to film them. So you must simply yield to them, bow to their whims, and then, be patient."[87] In comparison with fish schools, the creatures that he was filming at the time – sea urchins, crabs, octopuses, and sea horses, for instance – were admittedly far less dynamic organisms (and, above all, they could be observed individually). Still, it was not exactly easy for him to get the right shot. So as not to miss the birth of a sea horse, for instance, Painlevé installed a device on the brim of his hat that would emit an electrical shock if he happened to doze off and rest his head against the camera.[88] As a technical solution to the conditions of observation, this act of integrating the researcher into the media-technological arrangement in fact ensured that he did not sleep through the decisive moment and that the camera recorded an event that had never before been filmed.

At the beginning of the 1930s, Painlevé made his first attempts to film *in situ*. To do so, he used a Debrie Sept camera with a handmade watertight housing, the front of which was made of glass. With room for no more than seven meters of 35-millimeter film, however, the camera could only record for a few seconds, and thus Painlevé routinely had to resurface to change the film.[89] Beyond that, he also had to struggle with the deficiencies of the diving equipment, which had a mechanical pump to supply him with air from the surface: "The goggles were pressing against my eyes, which, at a given depth, triggers an acceleration of the heart by oculo-cardiac reflex. But what bothered me most was that at one point I was no longer getting any air. I rose hurriedly up to the surface only to find the two seamen quarreling over the pace at which the wheel should be turned."[90] In 1933, the marine captain Yves Le Prieur introduced a diving suit that did not need to be connected to the surface with any hoses – "a self-contained underwater breathing apparatus that combined a high-pressure air tank with a specially designed demand valve."[91] The air tanks were made by the French tire company Michelin. Around the same time, flippers were developed and marketed as "swimming propellers" by Louis de Corlieu. Enthused by Le

87 Ibid., 128.

88 See Berg, 'Contradictory Forces,' 25.

89 See ibid., 23 and 25.

90 Ibid., 25 (quoted from one of Painlevé's unpublished documents).

91 Ibid., 27.

Prieur's invention, Painlevé cofounded with him a diving club called the Club des sous-l'eau, which would go on to organize diving exercises in the Mediterranean for its nearly fifty members and host elaborate parties at its swimming pool in Paris.[92] For Painlevé, this opened up a whole new world: "Indeed, he dreamed of one day creating a studio – complete with film equipment, scientific apparatus, and technicians – entirely underwater."[93] Yet in the shadow of the coming war, underwater scientific explorations were far from most people's minds, and Le Prieur's diving apparatus would soon be used primarily by the French navy instead of the amateur pioneers of freediving.[94]

In a text from 1935 titled "Feet in the Water," Painlevé indicated how seriously he pursued scientific interests with his media-technological observational ensemble, how important it was for him to use cutting-edge technology, and how difficult it was to set up and manage this ensemble: "In choosing the aquatic world as a field of investigation, we have encountered two problems, nonexistent elsewhere: 1. Establishing the basis for the study of aquatic animals which, unlike that of land and air animals, has so far been conducted in a summary and backward fashion. 2. Obtaining photographs that are as clear and illustrative as possible under the most realistic conditions."[95] According to Painlevé, this meant that the observational ensemble had to be modified for each new organism that he wanted to film; in a sense, he had to adapt to the conditions of the object of knowledge. Here, in the intermediary space between epistemic and technical things, the researcher had to make adjustments until he found an arrangement that would allow new and never-before-seen events to be seen and, above all, to be recorded:

> Whether shooting in freshwater or saltwater, light poses a delicate problem. As in all studios, various light sources – ambient and spot – are necessary to illuminate the specific area. After compensating for the reflections and refractions through the water of the aquarium's glass, the correct amount of light must be determined: there must be enough light to be visible on film without, however, bathing the animal in so much light as to affect its behavior. [...] When the lighting is changed – increased or decreased – some animals will switch directions, for example, descend

92 See ibid.
93 Ibid., 29.
94 See ibid.
95 Painlevé, 'Feet in the Water,' 131.

> when they had been climbing. Or a shrimp might vomit in front of the lens just when one expected the most ethereal ballet from it. [...] Or an octopus who constantly lifts everything that is around it, clouding the water with its groping tentacles, might, when one's back is turned, escape from the tank, flatten itself out, slip under the studio door and tumble out the window onto the embankment below to the surprise of bathers.[96]

In these behavioral studies, it was not only the exclusion of abiotic factors that was problematic but also the very behavior of many of the organisms themselves. They either moved hardly at all or so abruptly that it proved difficult to focus on them. Moreover, filming with a high-speed camera – to slow down the bodily movements in the film, for instance – could only really function in the case of fixed life forms. It was hardly realistic for the camera to follow the unpredictable directions of swimming fish.

In the case of *in situ* recordings, as mentioned above, the technical devices were largely inadequate for working under water, and this was true of both the filming and the diving techniques. Working on the coast made something else clear as well. Conversations with fishermen there often revealed their exaggerated and mythologizing view of animal behavior, which was grounded in the traditional anthropomorphizing perspective that the field of ethology hoped to overcome. "There are," as Painlevé wrote, "so many myths to shatter! The most preposterous anthropomorphism reigns in this field: everything has been made for Man and in the image of Man and can only explained in the terms of Man, otherwise 'What's the use?' This leads to observations that are inaccurate."[97] With their careful and bio-logically conceived technical arrangements, however, his films were meant to provide a more accurate view of things. The latter were no longer based on an anthropomorphic perspective but rather profited from (and was limited by) the technical components of his observational ensemble. Painlevé's goal was to achieve a symbiosis between his biological object of research and his technical apparatuses – or perhaps rather to instigate a sort of co-evolution in which the observational setting adhered to the objects of knowledge and thus made processes and types of behavior visible in such a way that his audience would perceive these objects of knowledge – these marine organisms – in a new and different way. This co-evolution functioned in the interplay of disruption, anticipation, and the suppression of interference between the object of knowledge and the

96 Ibid., 131–32.
97 Ibid., 136.

technical system or, in Painlevé's optimistic words: "So, in sum, just when you think you have finally perfected a technique, you are forced to change it. We now use color in some of our documentaries, just as cartoons do. And we now bring spotlights into the water with us. Through it all, however, we have kept the pioneers of film in mind: they exemplify the desire to press on, regardless."[98]

Just as novel methods and their technical specifics alter the objects of knowledge under examination, the objects of knowledge also transform the technical specifications being used. Biological and technical knowledge – or, more pointedly, the behavior of organisms and the behavior of media-technological apparatuses – blended together in the advanced methods used to produce moving images of aquatic life in motion. The noise produced by the objects of observation made it necessary to engage with technical apparatuses and their constant modification. This represented the condition of possibility for joining the perspective of the researcher to the dynamic processes of animal behavior beneath the sea. Media techniques are not simply extensions of the human senses; in that they work along with them to exclude the disruptions produced by the objects being observed, they are rather indicative of the conditions and limitations of our perceptive faculties. The researcher ultimately proves to be a bricoleur who tinkers and experiments between biological objects, technical media, and his or her human perception.

"Half Tarzan, Half Grzimek"[99]

Beginning in the late 1930s, another pioneer of underwater filming and innovative developer of undersea cameras and breathing apparatuses was the Austrian Hans Hass, who in his 1947 film *Menschen unter Haien* (People among Sharks) described the significance of having lightweight equipment as follows: "We have devoted ourselves to this mysterious environment with body and soul, and in doing so we ourselves have almost become fish-like beings. We truly feel at home beneath the waves. Every now and then will one of our heads appear silently above the water's surface for a breath of air, only to re-submerge into the unfathomable depths to face ever new miracles and adventures."[100] A few scenes later, the ease with which he was able to move through the water was contrasted with the sluggishness necessitated

98 Ibid., 139.
99 Ralf Bülow, 'Halb Tarzan, halb Grzimek.'
100 *Menschen unter Haien*, directed by Hans Hass.

by the typical diving equipment at the time, which made it nearly impossible to come into close contact with any marine life. Hass's initial motivation for improving diving equipment came from a meeting with the American author Guy Gilpatric in the French Riviera, where Hass was vacationing after finishing secondary school. Gilpatric first infected the eighteen-year-old with his enthusiasm for spear fishing, and soon thereafter their expeditions became events with cameras and film recordings. For such fishing trips, Gilpatric had already designed watertight glasses (out of modified pilot goggles) that allowed him to see clearly under the water. Moreover, he had written a sort of guidebook for divers that was published in 1938 as *The Compleat Goggler*.[101] In this book, Gilpatric exuberantly described his first experience with his goggles as a sort of diving *flight*:

> I was unprepared for the breathtaking sensation of free flight which swimming goggles gave me. [...] The bottom was fifteen feet below me, now, but every pebble and blade of grass was distinct as though there were only air between. The light was a soft bluish-green – even, restful, and somehow wholly appropriate to the aching silence which lay upon those gently waving meadows.[102]

Because Vienna, of course, is not exactly close to the sea, Hass tested his diving apparatus – before an audience of bathing beauties – in a public swimming pool in the city's nineteenth district before taking it out on an expedition (this scene is staged in Hass's first film, *Pirsch unter Wasser* [Underwater Stalking], which was released in 1942). And over the course of the next few years he developed and tested various apparatuses that would enable him to dive as a 'frogman' – for instance by redesigning the rescue devices meant for submarine passengers. On the grounds of these types of experiences, Hass was assigned to the Wehrmacht's diving unit from 1943 to 1945.

Because Hass could not find any adequate equipment for filming and photographing under water, he built his own watertight housings for cameras. In this arena, too, he alluded to his pioneering achievements in a subtle way:

> From catalogues I learned that there were supposedly certain cameras in America for use under water, but they did not seem to be especially

101 On this book, see Hannah and Mustard, *Tauchen Ultimativ*, 25.

102 Gilpatric, *The Compleat Goggler*, 3.

> reliable, or else I would have encountered good and authentic underwater photos in one publication or another. This was not the case, however, and here I want to make a point to stress this. So far, what has been presented to the public with the label 'underwater photo' or 'underwater film' were, in the majority of cases, taken in aquaria or swimming pools or, in the best case, from diving bells in clear lakes. Often, too, they are just trick shots taken in a studio.[103]

His 1942 book *Fotojagd am Meeresgrund* (Photo Shoot on the Bottom of the Sea) contains 'photographic evidence' of such forgeries. Hass himself took great care that his pictures included certain elements to ensure that they would undoubtedly be regarded as 'authentic' and as having been taken in the open sea. He aimed to discredit any accusations of trickery in advance, and he described the construction of his technical apparatuses in detail. In order to make cameras suitable for underwater use, he at first made their housings out of brass, and later out of plexiglas, "that elastic and unbreakable glass from which the transparent cabins of our bombers were manufactured"[104] – military know-how thus also played an important role the improvement of underwater media techniques. Hass initially used an automatic Robot II camera to take pictures – equipped with a frame viewfinder with cross hairs to take accurate shots – and he used a 16-millimeter Movikon K 16 to make films. He conducted his own series of tests with different exposures and focuses in order to get a feel for the proper settings and to optimize the results of his work. As early as 1949, his experiments with photographic and technical equipment resulted in the so-called 'Rolleimarin' underwater camera housing, which was commercially produced by the company Franke & Heidecke and which, outfitted with the 'Rolleiflex' camera made by the same producer, served for years as the standard equipment for underwater photography. The housing, which weighed 5.3 kilograms, enabled pictures to be taken at a depth of up to one hundred meters.[105]

In his publications, Hass repeatedly discussed his experiences with underwater photography and filming. He described, for instance, the optical effects when using various focal lengths and flashes, and he catalogued the types of distortion produced by the environmental medium of water.

103 Hass, *Fotojagd am Meeresgrund*, 70.

104 Ibid., 74.

105 See the company's brochure titled '90 Jahre Rollei – 90 Jahre Fotogeschichte' (p. 3), which can be viewed online at http://www.90jahre-rollei.com/de/presse/90_Jahre_Rollei.pdf (accessed 19 March 2018).

Such insights were summarized in *Der Hans Hass Tauchführer* (The Hans Hass Guide to Diving) – as the title makes clear, Hass had long been a brand name by the time the book appeared in the 1970s. He also made sure to underscore the aspects of underwater recording that were of scientific interest: "And not least, the underwater photographer can make valuable contributions to science. [...] Such images, however, are only truly valuable when the photographer takes note of all the circumstances surrounding the observation. These include – to name a few – the duration of the process, the time of day, and the exact location and depth."[106] In order to make precise measurements of fish schools – of the distances between the individual fish, for instance – it would be necessary, just as it was in research aquaria, to install a frame of reference as a matrix of orientation. In the case of open-water observations, however, the installation of such a frame is complicated by the mobility of the camera and the changing fields of measurement. It would be extremely rare for a school of fish in the open sea to swim past some sort of grid-shaped background structure. Later studies would therefore employ a mobile control frame in which cameras were oriented relative to one another and in an absolute relation to certain features of the environment (rock formations, for instance): "The control frame photography is used to calculate the interior, relative, and one component (scale) of the absolute orientation of the two cameras."[107] For every setting, the control frame had to be reestablished; if the schooling individuals under observation happened to swim out of the object space, all of these recalibrations would have been in vain.

In the area of film, too, Hass made fundamental contributions, and in the 1940s he began to introduce Europeans to never-before-seen moving images of the world beneath the sea. Beginning in the 1950s, much of the action in his films, which combined features from animal documentaries with more campy elements, was driven by the somewhat chaotic appearances of his wife and assistant Lotte. This led one television critic from the *Spiegel* magazine to coin the clever rhyme "Keine Grotte ohne Lotte" ('No grotto without Lotte'), and the *Hessische Nachrichten* newspaper referred to one of the couple's research trips to the Red Sea as a "pin-up expedition."[108] In these entertainment-oriented films, the arduous task of relaying information about underwater experiences, which in previous movies had typically involved dull voice-overs, was replaced by lively conversations on the floor of the sea.

106 Hass and Katzmann, *Der Hans Hass Tauchführer*, 39.
107 Osborn, 'Analytical and Digital Photogrammetry,' 54–55.
108 See Bülow, 'Halb Tarzan, halb Grzimek,' n.p.

The mouthpieces used by Hass functioned not only as breathing devices but also as microphones. In these scenes, facts about science and science fiction appear side by side and have – at least from today's perspective – a rather comic effect. In such paradisiacal communication conditions, the filmmaker was hardly concerned, as the scientist Wolfgang Fischer had been, with forgetting or failing to record important information.

Yet Hass did not shy away from discussing the typical problems involved with underwater filming. These included, for instance, the difficulty of knowing the proper distance to shoot from and the challenge of tracking moving fish without jerking the camera around. Over time, he developed a hands-free swimming technique that allowed him to film smooth sequences under water, where he was now able to take "weightless" tracking shots that "not even the best Hollywood studios could offer."[109] As Wolfgang Fischer would point out, however, it is necessary for the camera to be in a fixed position in order to produce accurate recordings (or at least non-misleading recordings) of the behavior of schooling fish. Though Hass was far more interested in introducing the broad public, in the most entertaining way possible, to the hitherto unfamiliar flora and fauna of the oceans than he was in conducting any specific sort of fish-school research, he nevertheless used a camera stand for certain shots, which he planted into the seabed, or he would film from a fixed standing or sitting position. While it is true that Hass earned a doctoral degree in zoology from the Humboldt University in Berlin, which was awarded in 1944, and that he often worked closely with scientists on his many expeditions (one noteworthy collaborator was the behavioral researcher Irenäus Eibl-Ebesfeld), the goal of his research trips was not to analyze the behavior of fish schools, despite the fact that he holds the honor of having taken some of the first photographs of schools in the open sea.

The Subaquatic Astronaut

Of course, as far as studying and visualizing marine life under water is concerned, the most famous pioneer is Jacques-Yves Cousteau, the French diver, oceanic explorer, and 'adventurer' (as he was fond of calling himself). In more than a hundred films and fifty publications, he was largely responsible for producing images of (and stoking the world's fascination for) the 'silent world' beneath the surface of lakes and oceans. And like Painlevé and Hass, Cousteau was also a busy developer of media-technological apparatuses.

109 Hass and Katzmann, *Der Hans Hass Tauchführer*, 39.

As a young naval gunner and extreme swimmer in the mid-1930s, Cousteau was likewise stationed on the Mediterranean coast. In order to optimize his crawl style of swimming and protect his eyes from saltwater, he acquired a modern set of Fernez goggles.[110] Since the 1920s, Maurice Fernez's goggles had been the eyewear of choice among sponge divers in the Mediterranean, and in Cousteau's case they would irrevocably change the entire course of his life. In his book *The Silent World*, which was published in 1953, he described his first diving experience as follows:

> One Sunday morning in 1936 [...] I waded into the Mediterranean and looked into it through Fernez goggles. [...] I was astounded by what I saw in the shallow shingle at Le Mourillon, rocks covered with green, brown and silver forests of algae and fishes unknown to me, swimming in crystalline water. Standing up to breathe I saw a trolley car, people, electric-light poles. I put my eyes under again and civilization vanished with one last bow. I was in a jungle never seen by those who floated on the opaque roof.[111]

During the next two years, Cousteau snorkeled off the coast and studied the physiology of diving, especially the problems involved with maintaining body temperature. After a few failed experiments with coating his skin in grease, he tailored a vulcanized rubber suit that could be inflated with a layer of air for additional insulation. This, too, proved to be problematic. Cousteau did not only have to cope with the extra buoyancy created by the suit; "[a]nother weakness of this dress," he noted, "was that the air would rush to the feet, leaving me in a stationary, head-down position."[112] These issues would not be resolved until 1946. Before that, in the early 1940s, Cousteau began to work with the engineer Émile Gagnan to improve Le Prieur's compressed-air diving apparatus. In an effort to regulate the use of cooking gas in automobile engines, Gagnan had developed a 'demand valve' that could also be used for automatic air intake when diving.[113]

Like Hass, Cousteau's work was initially greeted with skepticism by the public, and it was this that motivated him to produce technical images of his underwater experiences: "Our friends ashore listened to our undersea reports with maddening boredom. We were driven to making photographs

110 See Cousteau, *The Silent World*, 8.

111 Ibid., 9.

112 Ibid., 13–14.

113 Ibid., 19.

to reveal what we had seen. Since we were always on the move downstairs, we began with motion pictures."[114] At first, Cousteau and his team used an "obsolete" Kinamo I camera, for which a machinist named Léon Veche constructed a watertight case. Because no 35-millimeter film was available during the war, they resorted to buying fifty-foot rolls of Leica film, the negatives of which had to be spliced together in a darkroom.[115] Later he would film with a 'bathygraph,' the focus of which was easy to adjust on its pistol-shaped handle. Over the years, Cousteau's expeditions under water became more and more spectacular, and the technologies that he used themselves became a common element in his films. Underwater vehicles such as the diving scooter (which had been developed for the navy), small submarines with striking designs such as the SP-350 Denise 'diving saucer' from 1959, and automated diving robots would come to feature in his ever more sensational images of the world beneath the surfaces of the sea. The life under water being observed with technical media was accompanied by human submarine life supported by technical apparatuses, be it in the form of a 'subaquatic astronaut' or in a diving saucer. Only those who can move "like a fish among fish" (as Hass put it) – regardless of whether this movement is assisted by simple rubber flippers and aqualungs or by advanced mini-submarines with integrated cameras and spotlights – are able to adapt sufficiently to the realities of the opaque subaquatic world and be granted access to observe the behavior of the unfamiliar organisms living there. Part adventurers and part researchers, only subaquatic astronauts are fit to discover the 'outer space' of the seven seas.

Techniques for observing underwater life have continued to be developed to the present day and have resulted in fascinating film recordings of fish schools, as is well known from recent productions. Consider, for instance, the almost unbelievable sequences in the BBC's films *The Blue Planet* (2001) and *Deep Blue* (2003) or in the French movie *Oceans* (2010). The two-part television series *Swarms: Nature's Incredible Invasions* (2009), which was likewise produced by the BBC, even presents images from the middle of a flying swarm of locusts, in which case the boundaries blur between the images caught on camera and computer animation. There is one common feature, however, in the tradition of nature documentaries from Hass or Cousteau until now: although they all provide audiences with a vague *idea* of how individual swarming organisms behave in relation to each other and how entire swarms behave in response to predators (for instance), a valid

114 Ibid., 22.
115 Ibid.

and precise evaluation of the *exact* behavior of swarming individuals can hardly be made on the basis of such footage. That said, films of this sort, with images that have only become more impressive over the years, do renew and strengthen our fascination with the intransparent organization of swarm collectives. The intransparency and dynamics of schooling fish are presented through observational media that allow the opacity of the environmental medium of water to become transparent.

Regarding underwater biological 'fieldwork' and the study of fish schools, however, there are additional factors to consider, namely those that make it possible to quantify and compare the results of one's observations. It will therefore be necessary to take a closer look at the methods of free diving and their contributions to science.

Swarm Research in the Open Water

Whereas the initial impulse for using diving apparatuses and developing technical observation and recording devices for underwater conditions may not have come from the scientific side of things, the use of such media techniques for scientific ends was nevertheless discussed throughout the 1960s. In a survey article, for instance, the marine biologist Ruper Riedl outlined the potential applications of diving in response to its growing popularity among scientists since the 1950s. According to Riedl, the aim of this submarine fieldwork was to gain a clearer view of complex communication, which was especially relevant to certain subdisciplines of behavioral research and ecology. For here, before any causal analyses could be conducted, it was first necessary to ascertain the "correlation between complex phenomena."[116] In this context, 'gaining a clearer view' meant the ability to save data – usually on film – concerning dynamic behavioral processes in their natural habitat, which would later be available for analysis. It goes without saying, then, that (media-) technical apparatuses lay at the heart of such endeavors: "Beyond the narrow confines in which the diver is placed in time and space," wrote Riedl, "the area of perception can be expanded at will by means of apparatuses." Before it was possible to examine things in an experimental setting, it was first necessary, he thought, to investigate the phenomena of interest in the "location of their activity."[117]

Riedl was of the opinion that underwater research, which was still in its early stages at the time, had to be scientifically bolstered by a combination

116 Riedl, 'Die Tauchmethode, ihre Aufgaben und Leistungen bei der Erforschung des Litorals.'
117 Ibid., 337.

of "experimentation and causal analysis." In his article published in 1973, Wolfgang Fischer discussed the potential and problems of diving methods for fish-school research in particular, and he addressed Riedl's concerns. Fischer stressed that divers, when 'tagging along' with a school, should attempt to move on a level parallel to it in order to ensure that the any changes of position made by the individual fish could be reliably tracked on the recorded images. Of course, as far as making somewhat precise measurements is concerned, it was likewise necessary when taking photographs in a laboratory setting to install a stable frame of reference in order to place the movements of schooling individuals in relation to a geometric grid.[118] Fischer noted that, for the sake of data analysis, he himself had used a thousand meters of film and seven hundred photographs. With a special projection device, the films were processed as a series individual images, each of which was traced onto a transparency: "By overlapping the traced individual images, it was possible to follow with accuracy the course of reactions that took place within a school of fish and to measure them in milliseconds, given that the precise temporal distance between the images was known. The use of high recording speeds made it possible to ascertain and make precise measurements of the types of reactions occurring within a social grouping of fish."[119] In the case of the photographs, his analysis was partially based on the negatives, which were examined through a projector or in an enlarged format of 18 × 24 centimeters. As with his analysis of films, he concentrated above all on the following factors: "the distances between individuals, the fish at the head of the school and the length of time spent in that position, the form of the school, its dimensions, the number of individuals, the beginning of a reaction, the manner in which the so-called 'reaction wave' spread, the combination of species within the school, their affiliation to certain forms of groupings, and reinforcing effects."[120]

As in the studies conducted by Partridge and his team, the marine biologists in Kiel had to analyze individual technical images by hand. The respective positions of schooling fish were marked on the projected image and compared to the positions in the neighboring frames. This visualized data set also made it possible to analyze a school's reaction waves and evasive maneuvers in response to environmental factors from a new perspective and independent of disruptive abiotic stimuli. Here the reaction speed of a fish school was

118 See Fischer, 'Methodik und Ergebnisse der Erforschung des Schwarmverhaltens von Fischen,' 393.
119 Ibid.
120 Ibid.

estimated to be between 600 and 1,500 milliseconds. In other experiments, the researchers also tracked the tendency of different schools to combine and the behavior of individual fish in relation to congregations. It was revealed that the fish under examination typically joined individuals of the same size and that larger schools 'attract' smaller schools of the same species, which tended to join larger structures. In all of these studies, nets placed in the water prevented the schools from swimming out of the area of observation. In the end, the main results of Fischer's experiments were the following:

1. Marine schooling fish form a genuine school only when in the pelagic zone. At night and while eating, the school disbands.
2. The reactions within a school spread according to certain rules and can be measured in time.
3. A reaction within a school can be instigated from various places.
4. There are no leading fish within a school.
5. The reactions of freshwater schools correspond to those in salt water, though freshwater schools exhibit certain different social forms.
6. The freshwater species *Alburnus alburnus* proves to be a model fish for studying the reactions of schools to fishing gear and for the further investigation of the rules governing schooling.[121]

In addition to providing further evidence that there are no leading individuals within schools – an issue discussed again and again throughout the development of biological fish-school research – Fischer's research yielded another interesting finding: he demonstrated that the behavior of a freshwater fish could serve as a model for further studies of the principles of schooling and for different species of fish. Because of the difficulties involved with keeping and studying saltwater fish, this realization promised to facilitate further research, which would be applicable to oceanic fishing as well. For, as mentioned above, Fischer's main concern was commercial fishing, and he was hopeful that it might soon be possible to influence the behavior of entire schools in order to make it easier to catch them:

> Of particular value to the fishing industry seems to be the reinforcing effect that operates within a school [...]. [...It] would be entirely possible to steer positively conditioned fish in such a way that they could initiate a school-wide reaction in the direction of the fishing net. In order to improve fishing tactics and techniques, the reaction times of the schools

121 Ibid., 400.

> that were measured in these investigations would have to be considered in light of the swimming speeds of the fish in question, which are in part already known. When trawling, fishermen would also have to keep in mind the fact that reactions within a school can be initiated from any given part of it.[122]

This is indicative of a way of thinking that has already been influenced by systemic or systematic approaches. Just as in other areas of biological research, efforts to simplify the experimental setting led scientists to look for a model organism that would allow them to draw general conclusions about the relational system of fish schools.[123] A few decades later, moreover, the use of 'conditioned' individuals to influence entire schools under different technical conditions would yet again become a popular approach. In the meantime, however, attempts to measure and quantify the structures of fish schools in their natural habitat remained problematic and susceptible to technical errors.

In a study published in 1977, the American biologist John Graves attempted to use photographs to measure the spacing and density of fish schools swimming in the open sea, a method that was still underdeveloped at the time (in contrast, as we have seen, to experiments of this sort conducted with aquaria).[124] To do so, a camera system was lowered slowly from the surface into the middle of a school, where it took automatic photographs at set intervals. The system consisted of a compact camera in a watertight case made of plexiglass and aluminum. The housing was equipped with a depth release, floats, and a signal flag. As soon as the system was submerged, the camera righted itself and initiated an electric timer to activate the camera shutter and a strobe light. The camera took fourteen photographs per drop at intervals of twenty-four or forty-eight seconds while it sank at a rate of ten meters per second. According to Graves, the fish were not much disturbed by the flashing light, though they kept a distance of about two meters from the system.

With the help of enlarged photographs and an *x,y*-coordinate reader, which probably resembled the data tablets used by Partridge and his colleagues,

122 Ibid., 399.

123 What comes to mind here is the fruit fly *Drosophila*. Beginning in 1909, it was held up by Thomas Hunt Morgan as a model organism during his rediscovery of Mendel's laws of inheritance. By the 1930s, however, it was all but forgotten by biologists. Yet during the rise of genetics during the 1970s, it was ultimately rediscovered as a model organism. See Jacob, *Of Flies, Mice, and Men*.

124 Graves, 'Photographic Method for Measuring Spacing and Density Within Pelagic Fish Schools at Sea.'

Graves measured the lengths of the fish caught on film. In order to reduce photographic distortion and computer-processing time, he measured only those fish that were within a diameter of six to ten centimeters from the center of the images. When measuring the distance between the fish, he also simplified the procedure by assuming that every fish was the same size and that they were all oriented perpendicularly to the camera. In his analysis, differences in the size of fish were dependent only on their distance from the camera lens. The distance between the individual fish and the camera was determined by calculating the ratio of the standard fish size to the size of the negative image and by substituting this value into the camera's underwater calibration equation.[125] This would only function, however, up to a certain distance from the camera, at which point it became difficult to see images of 'smaller' fish on account of overlap, the murkiness of the water, and a loss of penetrating light. Finally, once Graves established the minimum fish size to be included in his program, a three-dimensional model of the photograph was built by calculating a third coordinate z, which was based on the fish-image size and certain adjustments to the x and y coordinates (to account for distance from the camera), and with this model it was possible to estimate the density of the school (the number of individual fish per cubic meter).

Besides, Graves provided some information about his sample size: "Anchovy schools appeared on 16 of the 230 photographs taken. For the 10 photographs in which the fish seemed perpendicular to the camera, the mean density of the school was 114.8 fish/m^3 [...] and the mean distance was 1.2 body lengths."[126] Thus, as Jules S. Jaffe later noted in a survey article on the methods used for recording three-dimensional congregations, the empirical basis of Grave's experiments was relatively flimsy, and their results were skewed by the assumption that all the fish were the same size. As far as studying fish schools is concerned, the main disadvantage of optical methods is the simple fact that such techniques cannot see through the bodies of the fish, which of course overlap.[127] This was a problem, moreover, that could not be overcome stereographically but only by means of acoustic methods. That said, optical data remained indispensable because, among other things, it was needed to provide points of reference for tuning acoustic systems. In an article titled 'Analytical and Digital Photogrammetry,' Jon Osborn has described, with reference to a system used by New Zealand's

125 Ibid., 231–232.
126 Ibid., 232.
127 Jaffe, 'Methods for Three-Dimensional Sensing of Animals,' 26.

Ministry of Agriculture and Fisheries, how intricate it can be to calibrate a stereo camera system that can reliably determine *z*-coordinates. The system in question used two Lobsiger DS 3000 cameras with 28-millimeter Nikor underwater lenses that were mounted behind a flat acrylic port, each with a reseau grid in its focal plane. The distance between the cameras was variable, and a control frame was photographed before and after each outing and was brought on the ship in case any of the cameras were disturbed at sea and needed to be recalibrated. The frame itself contained approximately sixty control points that had been coordinated to submillimeter accuracy with traditional surveying techniques. To calibrate the camera, the image coordinates of the control points and the corners of the reseau grids were measured with a stereocomparator.[128] At the same time, Osborn was quick to stress how difficult it is to track a large number of simultaneous movements and to convert these into an accurate three-dimensional model:

> Much of the current research in object recognition and matching relies on defining targets in terms of generalized geometric objects, such as lines, planes, and cylinders. While this is appropriate for industrial applications such as quality assurance for manufactured parts, these methods are less well suited to biological tasks where animals move and their shape changes as a function of animal distance and orientation with respect to the cameras.[129]

In the 1990s, after all, automated digitalization processes were still in their infancy:

> Accurate and reliable fully automated target recognition, correlation and tracking is still in the developmental state. [...] Most commercial automatic recognition systems rely on feature recognition for binary images (e.g. a high-contrast edge recognized as black and white). Therefore, image recognition of aggregating ants filmed against a high-contrast background is relatively simple. But automatic identification of every fish in a school, let alone recognition of the head and tail of every individual, becomes a problem of difficulty.[130]

Until the development of more advanced systems toward the end of the 1990s, the focus of such types of observation was not directed so much

128 See Osborn, 'Analytical and Digital Photogrammetry,' 54–55.
129 Ibid., 50.
130 Ibid., 51–52.

toward inter-individual connections and communication but rather toward such things as ecological parameters – toward reactions, for instance, to food sources, predators, and other external stimuli. It remained somewhat problematic to apply and compare the results of laboratory experiments to those of fieldwork conducted in the open water. As the French biologist François Gerlotto has stressed, however, it has been possible since the 1990s to observe and analyze fish schools in their natural habitat and to draw sound conclusions about their structure and behavior.[131] That said, Gerlotto and his contemporaries relied on a media-based method that made use of an entirely different type of technology to overcome the dual intransparency beneath the sea. With the help of digital technologies, this method – which, synesthetically, implemented *sound* in order to *see* things – was able to make fish schools visible in a new and specific way.

4. Acoustic Visualization[132]

> Despite their several shortcomings, these devices are justifiably regarded as 'eyes' by those who are engaged in exploratory fishing and who are studying the distribution and behavior of fish (particularly schooling forms) in natural conditions, especially in the sea. [...] In accordance with their usefulness in these investigations, hydrostatic devices are rightly compared to a microscope, used to study the most important biological questions.[133]

> The impression given by most schools observed by the author is more that of a huge amoeba moving along, including the outthrusting of heavy and bluntish pseudopodia.[134]

When the RMS *Titanic*, which was then the largest passenger ship of its time, sank after colliding with an iceberg on April 14, 1912, this not only came as a great shock to the progress-oriented societies on both side of the Atlantic. The catastrophe also spurred the implementation of wide-ranging measures to improve safety technologies at sea. Thus, inspired by this event, the

131 Gerlotto, 'Gregariousness and School Behaviour,' 234.

132 On this term, see Greene and Wiebe, 'Acoustic Visualization of Three-Dimensional Animal Aggregations in the Ocean,' 62.

133 Radakov, *Schooling in the Ecology of Fish*, 12.

134 Breder, 'Studies on the Structure of the Fish School,' 22.

Canadian radio pioneer and inventor Reginald A. Fessenden demonstrated, in March of 1914, the ability of his so-called Fessenden Oscillator to function as an acoustic echo ranging device. Aboard a ship operated by the U.S. Coast Guard, he tested his oscillator off the coast of Newfoundland, where it successfully detected the presence of an iceberg that was 3,200 meters away and where it provided accurate measurements of the depth of the sea. A few months earlier, as an employee of the Submarine Signal Company in Boston, he had already shown that this electromagnetic sonar device – "a 540-Hz air-backed electrodynamically driven clamped-edge circular plate" – could be used as an underwater communication tool on submarines (a Morse code carrier was used to modulate the oscillator).[135]

Fessenden's sonar method introduced an entirely different way to monitor fish schools, a method that began to gain popularity in the 1950s – though it was not unknown in the 1930s[136] – and had little in common with the optical approach to observing, quantifying, and tracking schools. The history of studying the transmission of sound under water, however, can be traced back to Leonardo da Vinci, who, in a notebook from the year 1490, mentioned the possibility of using an immersed tube to hear the activity of distant ships: "If you cause your ship to stop, and place the head of a long tube in the water and place the outer extremity to your ear, you will hear ships at a great distance from you."[137] This history also includes the experiments conducted by Daniel Colladon and Charles Sturm, who in 1826 attempted to measure the speed of sound as it travelled through Lake Geneva, and Lord Rayleigh's work *Theory of Sound*, which was published in 1877. And it would also have to include the various electro-technical innovations and their underlying theories of electromagnetism, which made it possible to construct the Fessenden Oscillator in the first place and which, a few years later, would also serve as the basis for Paul Langevin's ultrasonic submarine detector and Alexander Behm's ocean echosounder.[138] Yet however enlightening it might be to delve into this understudied area of media history, this is not the aim of the present section. Though I would like to do justice to the technological history of sonar, which has of course played a prominent role in military history since the First World War, my focus here will be

135 See Riley, 'Reginald Aubrey Fessenden (1866–1932).' Just five days after the sinking of the Titanic, the British engineer L. F. Richard patented a similar system, but he never brought it to market. See Burdic, *Underwater Acoustics Systems Analysis*, 3 (footnote 1).

136 See, for instance, Sund, 'Echo Sounding in Fishery Research.'

137 MacCurdy, ed., *The Notebooks of Leonardo da Vinci*, 268. Quoted from Burdic, *Underwater Acoustics Systems Analysis*, 1.

138 See, for instance, Simmonds and MacLennan, *Fisheries Acoustics*, 2.

on its use in overcoming the specific technical challenges of echolocating and monitoring schools of fish. These acoustic methods were and remain closely connected to visualization techniques – with acoustic images that have to be deciphered and interpreted in particular ways.

Because sound waves travel through water much better than light does, namely at a speed of around 1,450 ms^{-1}, echosounders (which transmit vertical pulses) and, as of 1929, multibeam sonar (which transmits pulses in multiple directions) were developed to generate images of underwater fish schools out of acoustic reflections. Such acoustic methods for locating and visualizing things belong to the general category of echolocation. Unlike the optical methods discussed above, these acoustic methods were used specifically to address the question of the organization and governing factors of fish schools as dynamic structures. The idea was that it should be possible to deduce the nature of local structures from a scan of a school's global structure (partially with the help of visual methods; see the section 'Oriented Particles' below). In that schools could now be surveyed in their natural habitat, the research perspective was able to expand from isolating internal factors to examining external factors as well when analyzing the structure and morphology of a given school. Well into the 1980s, attempts to track and analyze fish schools with sonar technologies were motivated above all by the paltry set of empirical data that existed concerning their behavior in open water. As with optical experiments conducted at sea, a further motivation was the question of the validity of any fish-school behavior that had been observed in aquaria. One aim, that is, was to exclude any abiotic factors that might influence the behavior and structure of schools. In the case of acoustic visualization, the dynamic and adaptive structural changes and modifications of fish schools were brought to light with novel methods of visualization.

Instead of the laboratory approaches of using running commentaries, fish shadows, stereoscopy, or data-tablets to identify the individuals of a school and their communicative function within the school as a whole, so-called 'acoustic eyes' were now implemented to examine schools from the outside. The use of hydroacoustic methods can thus be characterized as another effort to adapt media-based analytic strategies to overcome the limitations posed by the observational environment of open water and the inadequacies of optical methods in such conditions. Yet these acoustic technologies also resulted in new implications regarding the conceptual convergence of unity and multiplicity in swarms. These arose from the genuinely technical realities of the systems in question as well as from the specific status of the images generated by the analog and (later) digital imaging technologies that were used in conjunction with the acoustic devices.

Both in laboratory aquaria and in optical approximations of fish schools *in situ*, their organizational principles always evade any view of the system as a whole. For a long time, their dynamic order engendered a sort of *visual* intransparency that could only be resolved with a laborious amount of effort, and this, in turn, thwarted any efforts to analyze their *control-technical* intransparency. Acoustic methods, however, follow a different logic of resolution and thus created a new perspective on schools of fish. Instead of conceptualizing schools as collective structures that arise from the sum of relations between individual schooling fish, hydroacoustic methods were (and are) mainly concerned with identifying the specific features of an entire school in relation to its natural environment. In this case, the total structure of a school was scanned, and the individual fish as such were typically not recognized but rather postulated *a posteriori* by means of statistical methods. A fish-school morphology based on sonar technology could therefore never investigate the psychomechanics of interindividual communication; rather, it conceptualized its own form of margin or periphery that addressed the nexus of relations between the particular and the global in a new and different way. François Gerlotto and his colleagues, for instance, have defined fish schools as follows: "[T]he definition of a school relies more on the geometric properties of the fish aggregation, the existence, for example, of clear edges around a group of thousands or up to millions of fish concentrated in a reduced volume. In the same sense, an acoustic school is defined as 'a multiple aggregation of acoustically unresolved fish'."[139]

The acoustic tracking of fish schools also foregrounded a different set of experimental variables. Rather than addressing the details concerning the internal self-organization and behavior of individual schooling fish, these methods were used to approximate the broader changes that took place in the density and structure of a school. It was possible to study its dynamics of contraction and diffusion in relation to its natural environment. Its external morphology could be scanned along with its crude internal structure (and density). This dual view therefore conceptualized schools as *interfaces* between the schooling individuals and their environment by concentrating on their formations and maneuvers in response to external influences – on their capacity to adapt: "Fish will adopt the most favorable shape of aggregation depending on the local environment."[140] By forming

139 Gerlotto et al., 'Waves of Agitation Inside Anchovy Schools Observed with Multibeam Sonar,' 1405. Here the authors refer to Kieser et al., 'Bias Correction of Rockfish School Cross-Section Widths from Digitized Echosounder Data.'

140 Paramo et al., 'A Three-Dimensional Approach to School Typology,' 171–172.

a self-organizing 'manifold function-nexus' (in Sanford Kwinter's terms), fish schools thus evinced a relationship between the environment and schooling 'bodies' that could henceforth only be conceived in relation to and with itself and whose various forms of disruptive potential interfered with one another.

Noisy Targets: Copulating Shrimp and Flatulent Herring

At a sonar-technology conference held in the mid-1980s, the British physicist and writer Mark Denny met a French scientist who was interested in using sonar to track fish. Denny was somewhat surprised by this ambition. Even by that time, the use of sonar for biological research (instead of for military purposes) was still relatively rare, and thus the French researcher must have felt, in Denny's words, like "a fish out of water." The *target* that interested this scientist – that is, the detection of fish schools – was one of the very things that the military hoped to avoid. To military scientists, fish schools were no more than acoustic *interference* that hindered the detection of their own targets, which were typically submarines.[141] For them, schools of fish belonged – along with other types of interference such as the copulating shrimp that Denny could not help but mention in his book – to the 'external noise' of the ocean.[142] In the context of attempts to observe and locate other things, schools of fish represented a specific form of interference, and at first the idea must have seemed counterintuitive to those engaged in military research to regard them as a potential *target* of sonar technologies. In hydroacoustic systems, sound waves are typically transmitted in discrete pulses at a frequency between 12 and 500 kHz. By way of a vibrating membrane, the electric oscillations of the pulse generator are mechanically transformed into pressure oscillations that then spread through the water. Every object within the range of these acoustic signals reflects a part of this sound back to the pulse generator, which in turn converts the incoming sound waves into electric signals. Present objects will thus cause higher voltage in the receiver and thereby distinguish themselves quite clearly from the background noise in the system (the 'self-noise'). Accordingly, larger objects increase the amounts of electric voltage all the more. These

141 See Denny, *Blip, Ping & Buzz*, 164.

142 Denny claims that, during certain seasons, large amounts of copulating shrimp represent a genuine problem for submarine operators and that certain suspicious noises, which the Swedish navy once interpreted as signs of a nearby Soviet submarine, had in fact been produced by immense schools of farting herring (see ibid., 6 and 57).

latter instances are then amplified and represented, in a form of their own and to various degrees of detail, on optic displays. From the strength of the reflected noise and the temporal interval between the transmission and reception of the sound signals, it is thus possible to determine the distance to the reflecting object and its approximate density.[143]

When the very entities that cause interference, which fish schools represent *in themselves*, become the target of sonar-technical and acoustic-optical resolutions, then attention must be paid both to the 'self-noise' of the system as well to the 'external noise' (the focus of military applications). This is because the basic equation (and thus the correct determination of the distances, speeds, and movements of subaquatic objects) for calculating the speed of sound under water ($c = \sqrt{B/p}$) is not constant but rather always dependent on inconstant variables, namely elasticity (B) and density (p). The speed of sound depends on the temperature, depth, and salinity of the water in question, and the temperature of the water is itself dependent on depth and location, as well as on the weather conditions and the time of day. Even the structure of the water's surface can vary drastically and reflect sound in highly divergent and haphazard ways. Moreover, the floor of the sea – with all of its irregularities and unevenness – can also influence the transmission of sound.[144] A typical estimation of the speed of sound under water combines theoretical considerations with empirical results and thus looks something like the following equation:

$$c = 1449 + 4.6T - 0.055T^2 + 0.0003T^3 + (1.39 - 0.012T)(S - 35) + 0.017z,$$

where c = the speed of sound in m/s, T = temperature in °C, S = salinity in ppt, and z = depth in meters. Even in a sea of noise, the methods of underwater echo ranging can identify certain patterns. That said, these methods generate a sort of knowledge that is highly dependent on "experience and theoretical considerations" about the interfering factors involved, as William S. Burdic has observed:

> In spite of this rich variety of detailed characteristics, it is possible to recognize predictable patterns related to environmental conditions and geographic locations. Thus typical sound speed profiles are often available

143 For a convenient overview of the scientific and mathematical principles behind such techniques, see Bazigos, *A Manual on Acoustic Surveys*. Unless specified otherwise, I have here used the term *sonar* to denote both echograms as well as multibeam technologies.
144 See Burdic, *Underwater Acoustics Systems Analysis*, 11.

> for a given geographical location and season. Acoustic loss data for the boundaries, derived from a combination of experience and theoretical considerations, cover the expected ranges of wind speeds, grazing angles, bottom characteristics, and frequencies. Using this type of information, propagation prediction routines based on ray acoustics or more rigorous techniques are available to produce the average, or expected, transmission characteristics for a given situation. [...] However, if accurate detailed results are required, it is necessary to measure the sound speed profile carefully and establish the surface and bottom conditions at the actual location and time of the acoustic test.[145]

Perhaps to avoid such difficulties when tuning their sonar system to detect external noise – after all, dealing with the noise of fish schools posed enough challenges on its own – the first scientists to use an acoustic location system in this area of research decided to test it in an artificial pond. As in the case of simulation systems, which were developed later on and will be discussed below, it was the Japanese fishing industry that made the first strides in the development of sound localization systems. The first successful experiment was presented by K. Kimura in 1929.[146] In one corner of a fish-cultivation pond, Kimura installed a sound transmitter, and in the other corner he installed a separate receiver. The transmitter sent a ray at a twenty-degree beam with a frequency of 200 kHz and with a modulated and audible amplitude of 1 kHz, and this beam was reflected by the receiver on the opposite side of the pond. When a school of twenty-five *Pagrosomus major*, which are forty to fifty centimeters long, swam through the sound beam, the amplitude of the received signal would noticeably fluctuate. Kimura recorded this interference by photographing the wave forms as they were displayed on an oscilloscope.[147]

A good five years later, A. B. Wood and his colleagues would make further advancements with their so-called echosounder, which produced echograms on paper.[148] A few fishermen, such as the Englishman Ronnie Balls and the Norwegian Reinert Bokn, conducted experiments with these devices early on. Whereas Balls would not report about his experiments with the echosounder until after the Second World War (his aim was to discover schools of herring in the North Sea), Bokn is credited with publishing the first example of a fish

145 Ibid.
146 Kimura, 'On the Detection of Fish-Groups by an Acoustic Method.'
147 See Simmonds and MacLennan, *Fisheries Acoustics*, 3.
148 Wood et al., 'A Magnetostriction Echo Depth-Recorder.'

echogram.[149] In 1935, a Norwegian named Oscar Sund published a letter to the editor in the journal *Nature* that bore the title 'Echo Sounding in Fishery Research.' Here he provided an overview of his own findings by presenting photographic reproductions of four paper echograms. Produced by a 16 kHz echosounder, these images demonstrated the unexpected fact that the observed schools of cod were no more than ten to twelve meters wide and that they maintained a constant depth throughout the period of observation. In subsequent studies, Sund was also able to map the distribution of cod in the sea.[150] These studies also revealed another source of interference: "The bottom right-hand record," Sund noted, "is somewhat disfigured by the oscillations set up by excessive shaking of the ship's motor" – an observation indicating that schools of fish were not the only source of interference that had to be operationalized for acoustic scanning. Rather, additional technical sources of interference had to be taken into account beyond the self-noise of the sonar and the external noise of the ocean. These included – to name just a few – the sounds and turbulence produced by a ship's motors and propellers.

The rapid military advancements in echosounder technology made during the Second World War for the sake of detecting submarines – advancements spurred above all by the intensive ASDIC programs in Great Britain and the United States – led to systems that, by the beginning of the 1950s, were also adapted to be used by the fishing industry and fish-school researchers.[151] Although these expensive and technically refined systems were now available to scientists studying fish, it was still impossible for the latter to quantify the sound reflections at their disposal. That is to say, it was still impossible to calculate the total number of fish being tracked. Calculations of this sort would not be made until 1957, when L. Middtun and G. Sætersdal managed to estimate, from echogram images on paper, the individual reflections of individual fish that they had scanned.[152] Two years later, L. F. Richardson and his colleagues improved this method somewhat by (manually) counting an echosounder's signals from a calibrated cathode ray tube display. In the 1960s, experiments were finally conducted with automatic counting systems, in which so-called echo-counting devices – one a "pulse counter," the other a "cycle counter" – were connected to a transducer.[153] With this system, however,

149 See Simmonds and MacLennan, *Fisheries Acoustics*, 3.

150 See Runnstrøm, 'A Review of the Norwegian Herring Investigations in Recent Years.'

151 On the ASDIC programs – the acronym stands for "anti-submarine detection investigation committee" – see Hackmann, 'ASDICs at War.'

152 Middtun and Sætersdal, 'On the Use of Echo-Sounder Observations for Estimating Fish Abundance.'

153 See Mitson and Wood, 'An Automatic Method of Counting Fish Echoes.'

it was only possible to estimate the signals of individual fish when they were swimming at a sufficient distance from one another. Further progress was made in 1965 with the invention of the so-called echo integrator, which measured the relation between a fish school's density and the voltage of its echo squared – a method that would be used widely throughout the 1980s. Today, multibeam and split-beam technologies have made it possible to determine the precise position of individual fish within a sound beam and to track the movements of entire fish schools with three-dimensional models.[154]

Pings

> Acoustic surveys of aquatic organisms are notorious for large data sets.[155]

Unfortunately, it would be beyond the scope of this book to provide a detailed media-historical discussion and analysis of hydroacoustic methods; a comprehensive media history of sonar remains to be written.[156] Here I will therefore concentrate only on those hydroacoustic media-technological systems that have been used in fish-school research. At the heart of my discussion, moreover, will be the implications and complications that this shift in perspective brought about, especially in comparison with the optical systems surveyed above. It will be shown how scientists working with these acoustic systems managed to deduce, from the global structures and morphologies of fish schools, a new 'essence' of schooling that likewise differed from the various 'essences' that had been brought to light by previous analytical systems.

In fish, the swim bladder is responsible for backscattering 90–95 percent of sound. Species that lack this organ, such as mackerel, thus reflect only around one tenth of an acoustic signal. On account of their biological blueprint, it is thus possible to ascertain specific signatures for certain species of fish by measuring their respective 'target strength' – that is, the reflective potential of the fish of a given species – and relating it to reflections produced by a school of the same species. Or the reflections can be correlated to a sort of standard, which is usually a material mass with a defined density. Both methods make it possible to determine the approximate density of a

154 See Levenez et al., 'Acoustics in Halieutic Research.'

155 Towler et al., 'Visualizing Fish Movement, Behavior, and Acoustic Backscatter,' 277.

156 For thorough technical introductions to sonar, see Simmonds and MacLennan, *Fisheries Acoustics*; and Kalikhman and Yudanov, *Acoustic Fish Reconnaissance*. From the perspectives of media studies and the history of science, see also Oreskes, *Science and Technology in the Global Cold War*, 141–187; Shiga, 'Ping and the Material Meanings of Ocean Sound'; idem, 'Sonar and the Channelization of the Ocean'; and idem, 'Sonar: Empire, Media, and the Politics of Underwater Sound.'

school and the number of fish within it.[157] The original application of this method was, again, motivated by economic interests. With the help of sonar devices, fishing boats were able to detect schools, scan their form, calculate their approximate density, and study their reactions as global structures in order to maximize their catches and optimize future fishing methods. Sonar technologies thus created a genuinely new view of fish schools that was based on the perspective of fishermen:

> As congregation is, in many aspects, a most attractive and familiar animal behaviour, descriptions and definitions have been given by many authors since the pioneering works of Radakov. In much of the literature, schools are seen as groups of fish that are characterized by polarized, equally spaced individuals swimming synchronously, where the inter-individual distance is usually less than one body length. Partridge defines a school as a group of three or more fish in which each member constantly adjusts its speed and direction according to the actions of the other school members. In contrast, we may note that fishers have a different notion of school, which is for them a large aggregation of fish [...] in a small area, allowing their capture with appropriate gear [...]. As we can see, there is a difference in definitions between what a fisher considers a school to be, and that of the ethologist.[158]

Traditional single-beam echosounders, which were originally used to measure water depths, operate with pulses of sound called 'pings.' The latter are emitted almost perpendicularly from the bottom of a ship, and their relatively slow intervals mean that they can generate continuous two-dimensional acoustic images of objects vertically, but only discontinuous images horizontally (the process is similar to stacking cross sections of an object on top of each other). In short, fish are here represented as 'chips.' The signal is transmitted at an angle between five and ten degrees, so that, even at a short distance, the area being 'pinged' is quite large and the resolution is accordingly low. Radakov remarked as follows about the problems involved with visualizing two-dimensional sonar data, which during the 1960s often offered no advantage over the techniques of optical observation:

> [H]ydroacoustic devices [...] give imprecise images of [...] fish schools, merely symbols which have to be deciphered (and this can be sufficiently

157 See Ole Arve Misund, 'Underwater Acoustics in Marine Fisheries and Fisheries Research,' *Reviews in Fish Biology and Fisheries* 7 (1997), 1–34, esp. 8–12.
158 Paramo et al., 'A Three-Dimensional Approach to School Typology,' 178.

> done only up to a certain point); these instruments are incapable of characterizing the relative positions of the fish in space or their movements in the school at the time of performing a particular maneuver [...], and thus the most important regularities of schooling behaviour have still not been rendered accessible for study.[159]

It is difficult to compare acoustically generated images with corresponding optical observations, and they provide only imperfect information about events taking place within the global structure of schools. This is because acoustic signals are shaped like a cone. Acoustic images are thus distorted by the fact that the reflective surfaces become larger as the signal transducer moves farther away and because the emitted and reflected sound becomes inversely weaker as this distance grows as well.[160] At first, these effects had to be subtracted in order to give a smooth and undistorted shape to the collected data. Before multibeam sonar, which has been used for military purposes since the early 1970s, came to be implemented in fish-school research at the end of that decade, single-beam echosounders had gradually been refined "from primitive instruments that could barely discern a faint echo from the seafloor to sophisticated systems with complex signal-to-noise ratios and target resolution."[161] Higher resolutions, however, always entailed a more limited scanning area. Not until the development of multibeam sonar, which combines several beams with narrow emission angles, was it possible to scan large areas at a high resolution. Thus the ability to smooth out images was and remains dependent on the available capacities for processing incoming raw data. Accordingly, the rise in data density that accompanied the transition from single- to multibeam sonar technology necessitated an increase in data-processing capacity simply to extrapolate relevant information from the data at hand. In the field of fish-school research, advanced software and hardware were thus the enabling conditions for the refinement of acoustic analytical methods. At the same time, they also created space for new methods of visualization to assist in the analysis and interpretation of data.

With their presentation of the three-dimensional structures of fish schools, the visualizations produced by today's multibeam sonars are of course far more detailed than the early echograms. In a study examining the advantages of three-dimensional scanning technologies over

159 Radakov, *Schooling in the Ecology of Fish*, 41.
160 See Gerlotto, 'Gregariousness and School Behaviour,' 242.
161 Mayer et al., '3D Visualization for Pelagic Fisheries Research and Assessment,' 217.

two-dimensional methods, François Gerlotto and his colleagues have identified the following relevant factors for generating a realistic three-dimensional scan: "a large set of very narrow beams (less than 2°), covering ideally 180° and at least 90°, a very short pulse length, and a high ping rate."[162] Even this approach, however, has certain negative side effects: the use of frequencies higher than 200 kHz, for instance, limits the maximum range of the scan to just a few hundred meters. With a pulse length of 0.06 milliseconds, the sixty beams of their system were simultaneously updated seven times per second. Their sonar was equipped with two outlets – one analog, from which reconstructed images are recorded on video tapes, and the other digital, which produces a graphical matrix for each ping.

The post-processing of the videos was either done by eye or with the help of image-analysis software, namely the so-called MOVIES-B program developed by the Institut Français de Recherche pour l'Exploitation de la Mer (IFREMER). To use the latter, however, it was first necessary to convert the acoustic visualizations into matrices of 240 × 320 pixels in order to extract the most relevant geometric data. Regarding the digital output, the matrix that it generated for every ping consisted of 60 columns and 2,040 lines, thereby promising a far better resolution than that of the video images. At the time of the study, which was published in 1999, it was still an enormous challenge to collect and process such large amounts of data (roughly 50 MB per minute).[163] The quality of visualized acoustic images has thus been directly correlated to the storage and processing capabilities of the available computer equipment: As hardware has become more powerful, the approximations of the observed structures of schools have become more and more precise; as optical resolutions have improved, the display of the digital images has become closer to real-time processing; finally, the conversion of raw data into three-dimensional digital images has also made it possible to draw more accurate conclusions about the 'object' of study. Thus, according to Larry Mayer and his colleagues, "the combination of digital data, large areal coverage, and high resolution will allow us to develop tracking algorithms that can quantitatively monitor changing behavior."[164] Without visualization tools, the sheer mass of data seems unmanageable.

As Jorge Paramo and his coauthors have demonstrated, the quantitative and qualitative differences between two-dimensional and three-dimensional sonar are striking. Their comparison of two-dimensional

162 Gerlotto et al., 'From Two Dimensions to Three,' 7.

163 Ibid.

164 Mayer et al., '3D Visualization for Pelagic Fisheries Research,' 223–224.

and three-dimensional acoustic observations not only revealed that the two-dimensional perspective had distinguished more schools than were actually there. The observed dimensions of the two-dimensional and three-dimensional objects were also completely different, with the mean length of one school measuring almost twice as long in three dimensions than in two.[165] Moreover, whereas two-dimensional graphics still have to be processed in one's mind into 'life-like' images, three-dimensional visualizations can be dealt with somewhat naturally and intuitively. The latter's graphical processing allows access to the behavior of fish, the dynamics of schools, and to possible sources of interference. Finally, they also make it possible to adjust experimental arrangements and compare various models of behavior: "With such an approach, researchers would be able to quickly determine if their experimental strategy is appropriate for a given set of circumstances."[166]

The aforementioned need to smooth out the data material, from which these amorphous two- and three-dimensional digital images are formed, necessarily and always draws attention to that which has to be smoothed – to the multiple types of interference that fish schools cause for hydroacoustic methods as well. On the one hand, acoustic signals often reflect off of so many individual schooling fish that their delayed echoes can make it seem as though the school in question is far larger than it actually is. Visually, this multiple scattering of signals comes to be presented as a "downward tail."[167] On the other hand, especially dense areas of a school can conceal parts of the environment and parts of the school itself. Such so-called "acoustic shadowing" can cause the size of a given school to be underestimated by up to 50 percent.[168]

What is more, the observational systems themselves can cause interference, for instance by being attached to the bottom of a ship. Noise is thus coming in from two fronts: On one side of the system, the movements of the fish schools being scanned represent a constant source of interference. On the other side of the system, the presence and motion of the ship can influence the behavior and structure of nearby schools.[169] And not least, the acoustic signal itself can incite an evasive reaction on the part of the schooling fish. As soon as it hits a schooling structure, the individuals will flee away from the

165 See Paramo et al., 'A Three-Dimensional Approach to School Typology,' 172–173.
166 Mayer et al., '3D Visualization for Pelagic Fisheries Research,' 220.
167 See Weill et al., 'MOVIES-B: An Acoustic Detection Description Software,' 262.
168 See Misund, 'Underwater Acoustics,' 13.
169 See Gerlotto et al., 'From Two Dimensions to Three,' 6–7.

cone-shaped sonar signal. In doing so, the school expands to its maximum volume, as the individual fish begin to swim apart from one another at the maximum 'nearest neighbor distance.' It can even happen that the school will splinter into separate parts. In other words, the acoustic signal, which is intended to scan the original form of a school, can bring about significant mutations in the school itself. As the self-noise of the system as whole begins to magnify, the observation no longer simply constructs an object out of all the interference but also simultaneously causes the object to become more diffuse.[170] Thus, in a paradoxical development, the very scanning technology to which schools of flatulent herring once represented a disruptive source of acoustic clutter has become a source of interference for the clutter itself.

All of these parameters have to be taken into account when transforming acoustic signals into optical displays. At the intersection between acoustic and optical media, it is revealed how the fuzziness of digital technologies and their filtering and compression algorithms have made it possible to view the objects under observation at a far higher resolution – how they are able to construct a sharply delineated object where analog methods have fallen short. This has typically involved using image-filtering algorithms in conjunction with manual methods of image segmentation (thresholding) in order best to process the backscatter of acoustic signals: "Following thresholding, a single morphological filter pass was used to eliminate all small objects and smooth the outline of larger ones. Previous experience has shown that this combination best preserved school morphology and biomass. The remaining detected objects are identified as schools."[171]

Blobs

Over the course of these acoustic and computer-graphic processes, however, fish schools, which are prototypical non-objects without surfaces, are generated into bodies *with* surfaces. Here the vagueness, opacity, and whirring quality of the nevertheless stable edge of a school's surface are visualized only approximately, while the sound signals reflecting off the school's *interior*, which is always in a state of erratic motion, continues to be a source of interference. The edge – the simultaneously permeable, dynamic, and yet stable 'center' of fish schools – appears here as a sort of acoustic packaging, as a reflective approximated surface that simultaneously

170 Soria et al., 'Effects of External Factors (Environment and Survey Vessel) on Fish School Characteristics Observed by Echosounder and Multibeam Sonar in the Mediterranean Sea,' 154.
171 Beare et al., 'Spatio-Temporal Patterns in Herring (*Clupea harengus* L.),' 470.

encases the congregation. In the scientific processes for visualizing acoustic data, it is these approximated surfaces that make the forms of schools visible; they create a boundary surface that separates the school from the noise of its environment and smooths over the noise of the school itself. As a consequence, however, it has had to be accepted that the processes of smoothing distort the object of investigation. On the one hand, such processes entail that information is lost about individual schooling fish; on the other hand, the development of appropriate algorithms is hardly uncomplicated and requires that simplified assumptions be made about the data, which can in turn lead to the introduction of artifacts and to the concealment of relevant discrepancies and data-points.[172]

Yet this artificially generated surface is more than just a casing. Through a process called 'echo-integration,' by which the density of a given school can be calculated quite precisely on the basis of the received signal,[173] the acoustically scanned surface also extends into the deep. This is accomplished by means of a statistical method for estimating the average number of individual fish per unit of volume. Images from multibeam sonars do not differentiate the 'outer shell' of a fish school from its 'content.' Rather, the 'noise surface' of the sonar results in a sort of *all-surface* in which the particularity of the school cannot be represented.

The methods of acoustic focusing, the reductionism inherent to the visualization tools of multibeam sonars, and the computer-graphical representation of schools as density distributions all make it difficult to describe fish schools in geometric terms: "In general, the schools had an amoeboid shape which are not well described with standard geometric volumetrics like spheres or ellipsoids."[174] Schools scanned with hydroacoustic methods thus come to resemble that shapeless creature of B-movie fame: the Blob. The idea in the movies is simple: a gelatinous extraterrestrial mass reaches earth on a meteor and develops an active life of its own, over the course of which it ends the lives of human beings and other absorbable beings by incorporating them through its surface. What is interesting is how the notion of the blob was received by so-called postmodern thinkers. In the middle of the 1990s, for instance, Greg Lynn adopted this figure as a metaphor in order to formulate a new theory of architecture based on non-Euclidean

172 See Li, *Oriented Particles for Scientific Visualization*, 2, 38. In his thesis, Li develops an oriented particle system for fish schools that avoids making approximate calculations of artificial surfaces in acoustic visualizations.

173 See Gerlotto, 'Gregariousness and School Behaviour,' 239.

174 Paramo et al., 'A Three-Dimensional Approach to School Typology,' 175.

geometry. Here he made sure to underscore the very oscillation between unity and multiplicity that makes the status of the swarm so precarious: "The term blob connotes a thing which is neither singular nor multiple but an intelligence that behaves as if it were singular and networked but in its form can become virtually infinitely multiplied and distributed."[175]

Blobs, too, are conceptualized in terms their edge or surface. In an off-hand way, they really are what they eat. According to Lynn, blobs are organisms that are topologically inverted; they do not swallow things into an internal cavity but rather, like amoebas, they cling to things and gradually incorporate them through their surface.[176] Blobs have no need for a mouth because, in practical terms, their digestive system is simply turned inside out:

> The blob is all surface, not pictorial or flat, but sticky, thick and mutable. In virtually every instance, a B-film blob is a gelatinous surface with no depth per se; its interior and exterior are continuous. [...] These blobs are neither singular nor multiple since they have no discrete envelope. Essentially, a blob is a surface so massive that it becomes a proto-object. [...] Blobs possess the ability to move through space as if space were aqueous. Blob form is determined not only by the environment, but also by movement. [...] [They] are defined not as static but as trajectories.[177]

The gelatinous blob mass is always mutating in relation to its dynamic environment; a blob does not possess an ideal static form that can be detached from the specific conditions in which it finds itself. Yet its form is influenced not only by the environment but also by its own movements and trajectories. In at least two respects, then, blobs share a connection with hydroacoustic visualizations of fish schools.

For one, it is impossible to distinguish between the interior and the exterior of their respective surfaces. Fish schools, too, reflect sound as a 'mass' of sound surfaces back to the receiver, and these are processed and visualized as three-dimensional proto-objects. This process may obscure the dimension of the particular, but, according to Lynn, blob-like surfaces point to new and flexible relations between homogeneity and heterogeneity:

> [B]lobs intervene on the level of form, but they promise to seep into those gaps in representation where the particular and the general have been

175 Lynn, 'Why Tectonics is Square and Topology is Groovy,' 172.
176 Ibid., 170.
177 Ibid., 171.

> forced to reconcile – not to suture those gaps with their sticky surfaces, but to call attention to the necessary strategies of structural organization and construction that provide intricate and complex new ways of relating the homogeneous or general to the heterogeneous and particular.[178]

In fact, the data gathered by means of acoustic methods have led to the conclusion that fish schools in their natural habitat exhibit far more flexible structures than they do in research aquaria. The postulate that schooling fish tend rigidly to maintain an 'ideal' distance from their nearest neighbors had to be rejected. Echo-integration images of density distributions within fish schools have revealed, among other things, more and more large open spaces within their structures. Neither rigid psychomechanics nor inflexible models have been able to provide an explanation for these so-called vacuoles.[179] It seems somewhat ironic, then, that it was Charles Breder, of all people, who made the following comment in his article 'Fish Schools as Operational Structures': "Theoretically at least, fish schools could take any shape. Considered as three dimensional 'blobs', they have been described and photographed in a wide variety of shapes."[180]

For visualizations generated by means of echo-integration, the digital data of every ping are recorded and, after a school has been completely scanned, used to produce a three-dimensional representation of its structure and to calculate its dimensions. The software used by Paramo and his research group extracts all of the volumetric pixels ('voxels,' for short) inside a school and presents them as a four-column data set, which contains three spatial coordinates and the observed density for each voxel. This makes it possible to calculate the two- or three-dimensional features within the school in question. If a given voxel's density is less than the threshold set, then it is registered and represented as a 'hole' by the software. So-called vacuoles – the large empty spaces within the global structure mentioned above – thus arise when many neighboring voxels happen to be 'empty.' Essentially, densities are calculated as the squared voltage in each voxel along a regular scale of 256 grades, from 0 (for a density lower than the threshold value) to 255 (the maximum density that can be measured).[181]

Blobs and swarms defy comparison with any model that idealizes an eidetic relationship between bodies and the symmetries or proportions of

178 Ibid., 169.

179 See Gerlotto, 'Gregariousness and School Behavior,' 234–236.

180 Breder, 'Fish Schools as Operational Structures,' 483.

181 See Paramo et al., 'A Three-Dimensional Approach to School Typology,' 173.

their parts – an idealized relationship that Vitruvius described long ago in his *Ten Books on Architecture*:

> The design of a temple depends on symmetry, the principles of which must be most carefully observed by the architect. They are due to proportion [...]. Proportion is a correspondence among the measures of the members of an entire work, and of the whole to a certain part selected as standard. From this result the principles of symmetry. Without symmetry or proportion there can be no principles in the design of any temple; that is, if there is no precise relation between its members, as in the case of a well-shaped man.[182]

At first glance, Roman temples may seem to have nothing to do with fish schools, but the way that scientists would later use symmetric models to evaluate a school's proportions – the relation between a school's parts and the school as a whole – will in fact need to be discussed at length in my next section. That said, a 'morphology of the amorphous' (in Benoît Mandelbrot's terms) yields a flexible perspective that is the exact opposite of these classical concepts: blobs and fish schools do not simply represent a higher grade of complexity in comparison with Euclidean 'standard geometry.' They also describe an entirely different level of forms that Euclid, Vitruvius, and Da Vinci had excluded for being utterly formless.[183] As Greg Lynn has remarked:

> Because of its desire for a holistic model of the body – one that is essentially static – only bodies that can be ideally reduced through a process of division to whole numbers are acknowledged in architecture. The proportional correspondence between a temple and a *well-shaped man* are based first on a single organization regulating all parts to the whole and second on the presence of a common module. This formulation of the body as a closed system in which all parts are regulated by the whole is organized from the top down. Proportional orders impose the global order of the whole on the particular parts. This whole architectural concept ignores the intricate local behaviors of matter and their contribution to the composition of bodies.[184]

182 Vitruvius, *The Ten Books on Architecture*, 72. See also Wittkower, *Architectural Principles in the Age of Humanism*.
183 See Mandelbrot, *The Fractal Geometry of Nature*, 1.
184 Lynn, 'Body Matters,' 143.

With the figure of the blob, it is possible to reconceive this dictum about holistic models in terms of productive and *integrative* differentiations – differentiations that allow a local level to emerge from bodies from the bottom up.

Blobs and hydroacoustically scanned fish schools, however, share something else in common as well. Both are connected to their environment by the blurry dividing line that separates them from the environment itself. This means that the boundaries between the inside and the outside of a 'body' always has to be redefined, given that internal areas are constantly being influenced by external forces out of their control while the areas in question are pushing outwardly to expand and reconfigure themselves. Thus deterritorialization, in Deleuze and Guattari's sense of the term, not only forces interiors outward but also – and at the same time – unifies a constantly changing interior by internalizing external forces. In short, it is a process that combines involution with evolution.[185]

These movements are always subject to the provision of viewing bodies not as closed and stable entities but rather as multiplicities. Such 'bodies' develop new qualities; blobs are not simply the whole of their parts. Because they remain open to external influences, they are always something less than a whole. Moreover, as a topological 'both-and,' they oppose Euclidean geometry and a dualistic, Cartesian view of the world. It must also be noted, however, that they are a multiplicity without any visible particularity. They are a dynamic jelly, an acoustic jelly that can be conceptually extended to the 'blob-becoming' of fish schools that, by means of acoustic visualization processes, take on the form of a continuous surface where their interior and exterior fold into one.

The operationalization of the 'fisherman's perspective' by means of statistical methods and reductive filtering processes points to a certain constitutive scientific stratum, to a *dispositif* of the science of data production that is even more fundamental than laboratory experiments: this statistical 'smoothing' and analysis of data evokes the trend of bureaucratization that has increasingly defined scientific activity since the eighteenth century and has striven to contain that which can no longer be determined positivistically and to manage blurry or vague phenomena by means of statistical averages. The data-processing of acoustic signals is an example of "epistemological accounting."[186] In this manner, fish schools may indeed still be defined spatially, but they are no longer defined territorially. The numerical analysis

185 Ibid., 135.

186 Schäffner, 'Nicht-Wissen um 1800: Buchführung und Statistik,' 124.

of their morphology and structure does not describe a topographical form but rather a quantity of data that is infused with a dynamic temporal function and thus generates a "neutral space of events."[187] In their orientation toward dynamic time – as Stefan Rieger would put it – the futures of fish schools as an object of knowledge are cybernetically operationalized. The future as an interplay among probabilities, sequences of events, and environments applies both to the structure of perceptible objects as well as to acts of technical and human perception. Of central importance, then, are the conditions and modalities of this data-processing. For, regardless of whether the problem at hand concerns the determination, storage, or transmission of information, it is only when data have been 'designed' that they allow us to have any mathematical and (at least since Claude Shannon's time) information-theoretical knowledge about their data events:

> It is only this conjectural knowledge about the design of data that allows what in technical contexts is called data smoothing and what underlies perhaps its most consequential application: data compression. As of this point, at the latest, information theory, probability theory, and data calculus lost their specialist status and began to address, as a phenomenon, a current lifeworld in its totality, that is, the foundations of its data-based constitution.[188]

Acoustic analysis and the processing of such data modulate fish schools as amoeba-like all-surfaces. Here again, their data drifts are treated quite differently from the way they were treated in laboratory experiments, namely as epistemic aggregations that, with their detailed density distributions, are optimally prepared for the nets of the fishing industry. The tightly knit nets of fishermen are in hot pursuit of the ever mutating 'sieve' of the sonar, whose "mesh will transmute from point to point" (to borrow a line from Deleuze).[189] Here, technologies of 'state' control are implemented for the sake of studying the non-object of fish schools, and they rely on the same methods that had allowed states themselves to be transformed from territories into volumes of data.

Oriented Particles

As I hope to have shown, the visualization of acoustically collected data about fish schools has been of crucial significance for their operationality.

187 Ibid., 128.
188 Rieger, Kybernetische Anthropologie, 42–43.
189 Deleuze, 'Postscript on the Societies of Control,' 4.

Moreover, the media-historical origins of this technique in the act of tracking objects with sonar represent a starting point in a technical-historical line of development that leads to computer-graphical imaging processes.[190] For the latter, it was convenient to understand schools as three-dimensional objects with a surface in order to make their morphology evident. Yet this method was somewhat superficial itself because, as mentioned above, the process of smoothing data can obscure relevant information and cause information about individual fish to be lost altogether:

> Fish schools could be displayed as a connected closed surface so that its shape could be perceived. However, this display method has two obvious disadvantages. First, the information about individual fish is lost. Secondly the algorithms to generate a closed surface can be complicated. Such surface approximation algorithms inevitably involve making simplifying assumptions about the data. These algorithms also tend to obscure artifacts resulting from the surveying method.[191]

By the middle of the 1990s, when the image-processing and storage capacities of certain computer hardware were sufficient to handle large volumes of data produced by multibeam sonars and to visualize these data in real time (before then, data were collected in the open water and then processed on land), the question was raised about how it might be possible, with minimal effort, to produce more realistic visualizations of the scanned non-object. A project undertaken at the University of New Brunswick, for instance, aimed to develop a visualization tool that could interpret the data from multibeam sonars and simultaneously integrate navigation information in order to identify the geographical coordinates of particular soundings. This information would then be displayed in an animated 3D representation.[192] Instead of generating the relatively static, potato-shaped blob of traditional three-dimensional graphics, a dynamic quantity of particles could accurately *present* the scanned school of fish. This approach had other the advantages as well. On the one hand, it made it possible to delineate the form of an entire school from a global perspective; on the other hand, it also made it possible to zoom in on the positions of individual schooling fish.

190 Friedrich Kittler, for instance, traced the media-historical genealogy of computer graphics back to the military's use of radar to track enemy airplanes – a line of development that can also be tied to the sonar scans of fish schools and their representation in computer visualizations. See Kittler, 'Computer Graphics: A Semi-Technical Introduction.'

191 Li, *Oriented Particles for Scientific Visualization*, 38.

192 Ibid.

In designing his system, the mathematician and computer scientist Yanchao Li relied on concepts from the field of digital graphic design, which, since the early 1980s, had been developing so-called 'particle systems' or 'oriented particle systems' for visualizing 'bodies without surfaces' such as noise, fire, and clouds of dust.[193] The latter systems create dynamic objects by means of the defined behavior of many individual, virtual particles. By using a so-called distributive behavioral model, which is an explicit reference to the findings of biological swarm research,[194] they have produced convincing simulation models of natural swarming behavior that are in turn cited in almost every recent article devoted to fish-school research (this chiasmus will be discussed at greater length in the next chapter).

Li used squares to represent individual fish. It should be stressed that, on account of disruptive influences, these squares only rarely visualize the precise positions of the scanned fish. Rather, they represent a graphically generated resolution – in individual particles – of a school's density within a particular volume of water. It is no longer the acoustically scanned edge or surface of a school that extends into the depths; rather, a school's density parameters are resolved into (virtual) particles on the basis of its stochastically determined and algorithmically smoothed form. The resulting form is thus conceptualized in terms of its edge. In the particle system, every particle receives the same orientation parameters regardless of whether it is on the edge or in the center:

> We investigated two methods for specifying the particle orientation. The first is borrowed directly from Newton's law of gravitational attraction. We assume that fish are attracted to each other as if there are forces between them. We use the combined force vector generated with respect to all the other fish in the school to determine the particle orientation for a particular fish. [...] The second method we evaluated involves using a force function such that the attraction between fish was proportional to the inverse of distance.[195]

Perhaps even more interesting is what Li mentions a few lines later, especially because it closely resembles a remark made Charles Breder about his own mathematical model of swarms from the 1950s: "Note this is not intended as a model of fish behavior. It is simply a visualization method"[196] – a com-

193 See Reeves, 'Particle Systems: A Technique for Modeling a Class of Fuzzy Objects.'
194 See Reynolds, 'Flocks, Herds, and Schools.'
195 Li, Oriented Particles for Scientific Visualization, 39.
196 Ibid., 39.

ment that hints at the interconnection of 'fish and chips' to be discussed in my final chapter. This interconnection is reflected both in the 'realistic' animation used for special effects in films as well as in biological studies that rely on such visualizations in order to infer potential biological structures on the basis of underlying program structures. Both sides draw upon a set of simple rules from which complex structures can arise – one side to minimize computing time and programming effort, the other because it is dealing with relatively simply arranged schooling individuals that can be animated in large numbers.

With the implementation of oriented particle systems for scientific visualizations, representations of biological fish schools were made more 'realistic.' Stochastically smoothed within a mass of data, the edge was thus once again visualized as an edge of particles composed of stochastically determined schooling individuals. In short, the 'realistic' representation of scanned data functions only by means of a simulation, and the biological foundations of the non-object that it serves to represent are inherent to the simulation's model of visualization. The combination of advanced acoustic methods of fish-school analysis in the open sea and object-oriented digital modeling has since been taken even further. The motivation for this was not only to make the abundance of collected data easier to interpret by means of three-dimensional visualizations. Three-dimensional models have also been used to improve the tuning of acoustic devices. In a 2003 study conducted by researchers at the University of Washington, for instance, the authors' stated goal was as follows: "Through the visualization of empirical and simulated data, our goal is to understand how fish anatomy and behavior influence acoustic backscatter and to incorporate this information in acoustic data analyses."[197] Thus the visual processing of sonar data has not only helped scientists to acquire a clearer understanding of the dynamic behavior of fish schools; conversely, it has also been used to produce simulation models in order to determine how their behavior and their movements are reflected in changes to individual target strengths and sound echoes. Equipped with such information, researchers are then able to modify their filtering and analytical techniques. On this matter, the authors of the study refer to Edward R. Tufte's foundational work on designing quantitative data:

> One approach used to examine how biological factors influence echo amplitudes integrates modeling of organism behavior with acoustic measurements of fish distributions in computer visualizations. These

197 Towler et al., 'Visualizing Fish Movement,' 277.

> visualizations should present large data sets in a coherent and comprehensive manner; reveal several levels of detail within data sets; avoid distortion of measurements; prompt the viewer to think about mechanisms that cause observed patterns; and encourage comparisons among data sets.[198]

To this end, they integrated three components into an object-oriented programming environment and assigned each to its own object class: "Our goal is to integrate echosounder properties with fish anatomy, backscatter model predictions, and fish trajectories to visualize factors that influence patterns in backscatter data."[199] The 'echosounder class' includes all the properties of the echosounder and the transducer – its position, direction, range, frequency, and so on. The data processed here are meant to determine the target strength and the orientation of the observed fish. The 'echo fish class' includes behavioral and anatomical data about the fish of a given species, such as its swimming velocity. This information is used to produce a three-dimensional model of individual schooling fish that vividly indicates how, for instance, a change in its position in relation to the transducer will alter the backscatter of its acoustic signal. Finally, the 'trajectory class' simulates the behavioral repertoire of the fish by means of an agent-based schooling model. With this model, scenarios can be created that visualize the dynamic changes in the backscatter produced by the individual schooling fish according to their movements relative to the sonar system. This makes it possible, for instance, to track the fluctuating relation between the 'acoustic size' and the actual length of a fish over longer periods of time.

In this case, too, imaging techniques merge with a horizon of possibility – concerning their technical optimization – that is provided by the images themselves. It is only by being combined with the operative images of agent-based computer simulations that fish schools become describable both on the global and the individual level. Therefore, in addition to gathering data about the global behavior and the movements of entire schools in the sea, acoustic studies are also valuable for those interested in combining this broader perspective with the level of individual behavior. At the same time, moreover, another operative factor is inherent to the visualizations produced by computer simulations: by sharpening the instrumental focus of the sonar's acoustic 'eyes,' they function like feedback mechanisms for generating more accurate empirical data. The output being generated (data images) is itself fed back as input into the process of generating data images.

198 Ibid., 277–278.
199 Ibid., 278.

As I hope to have shown, the acoustic methods used to visualize fish schools are also subject to various sorts of noise and interference. Yet despite the self-noise of the sonar system, the external noise of the sea, and the noise produced by the schooling itself (what military operators of sonar referred to as 'clutter'), adequate methods were eventually developed for processing the essential data. For without appropriate filtering and smoothing processes, it was impossible to extract any relevant information about the behavior of fish schools from the volumes of complex acoustic data scanned by sonar systems. Fish schools have thus become the object of statistical operations; they have become quantities of data in which the edges of the aggregation and the individual fish themselves exist only as approximations or averages – as stochastically determined areas, in any case, of that which is visually undeterminable. They have been made compatible with a bureaucratic epistemology that can make calculations on the basis of non-knowledge. Rather than being hindered by the intransparency of schooling organizations, this approach operationalizes this opacity by means of approximations and thereby sacrifices detailed knowledge about the inter-individual connections among the fish. That said, acoustic methods are indeed able to track fish schools over longer periods of time, and they also make it possible, on the basis of advanced visualization processes, to study the typical global actions and reactions taking place within large schools. Beyond simply producing an ontological, representative congruency between that which a school might be and that which is visualized in the digital image, acoustic methods operationalize the visualizations themselves. Approximations and formal congruency, that is, are sufficient for the operational needs of fishermen.

As I have already suggested, however, this new operative epistemic strategy only could have been developed in conjunction with digital imaging processes. The latter converted the imprecise images of older sonar systems into a world of three-dimensional and time-sensitive survey images in which the complex acoustic data could be analyzed in a systematic way. Ultimately, this process brings to light an interconnection that has been a theme throughout this book, namely the use of biologically inspired processes of computer simulation and their visualization tools in biological research that has itself been inspired by computer technology. Here, tools developed with information from fish schools are recursively used to study fish schools themselves. Under these conditions, schools are transformed from mysterious non-things into data-based objects of knowledge – into *epistemic aggregations* of a technically informed media history.

IV. Formulas

Abstract
Under the term *formulas*, this chapter investigates complementary strategies in order to describe the dynamics and functions of biological collectives. It examines how, on the basis of patchy empirical data, attempts were made to construct mathematical models concerned with the geometric form of fish schools or with the algorithms of the local behavior of swarm individuals. It thereby follows traces which link biological swarm research to cybernetic ideas of 'communication' or 'information transmission.' Equipped with a new technical vocabulary, researchers began to describe swarms as 'systems' and were able to conceive of them in new ways. Nevertheless, the first approaches to simulating swarm dynamics in the 1970s received little attention, a fact that was likely due to the inability at the time to display dynamic processes visually.

Keywords: cybernetics, fish school, mathematical model, geometry, sensory systems, models as mediators

1. Models as Media

As demonstrated in the previous chapters, Charles Breder was right when, in an early article, he referred to fish schools as 'notoriously difficult laboratory materials' – a characterization that applies just as well, if not more so, to studying them in the open sea.[1] Alongside the efforts being made to gain empirical data about the control logic and organization of schools by means of various observational and analytic systems, a second (and complementary) strategy explored models and mathematical *formulas* in order to describe the dynamics and functions of biological collectives. Models, after all, provide the opportunity for drawing connections between different scales of observation and between different influential variables. They offer access to

1 Breder, 'Studies on the Structure of the Fish School,' 7.

Vehlken, S., *Zootechnologies. A Media History of Swarm Research.*
Taylor & Francis Group, 2019
DOI 10.5117/9789462986206_CH04

levels that the technical methods of visual or acoustic analysis cannot access, and they combine theoretical considerations and empirical data. However, two further sets of problems have to be considered in this context which exceed the difficulties that I mentioned in the previous chapter – namely the question of viewpoint, the feasibility of convincing experiments in research aquariums, and the revealing of visual 'invisibilities' by acoustic methods. One issue concerns the ways in which the relationality of schools could be converted into models and how to measure the epistemic value of such models in conjunction with other forms of knowledge production in the field of biology. How, that is, did the specific mediality of the models take shape? How did researchers proceed in their attempts to model fish schools? What were the central functions that the models were intended to depict? And in addition, I will eventually turn to the question of why the first simulation models, which were developed in Japan starting in the mid-1970s, were largely inconsequential to later biological research.

Given that schools exist neither on the level of the individual nor on that of the collective but rather on a third level where multiplicity and relationality intersect, the question of scaling has been of great interest in the field of fish-school research.[2] The overall behavior of a school is not simply the sum of the behavior of its parts; rather, some of its patterns of motion and other properties can only be observed on the level of the total system. Efforts to perceive and measure these patterns and structures are influenced by the perspectives that result from choosing a particular scale of resolution. In making this choice, one accepts that every minute detail on a microscopic level will not be crucial to processes taking place on a less detailed level. The question that always has to be asked is *which* factors will continue to play a role when the perspective shifts to another level of resolution:

> The problem of how information is transferred across scales cannot be addressed without modeling. In relating behaviors on one scale to those on others, one is often dealing with processes operating on radically different time scales, in which much of the detail on faster and finer scales must be irrelevant to those on slower or broader scales. Because decisions about what one can ignore require a quantitative evaluation of the manifestations of processes across scales, a quantitative approach is both unavoidable and powerful.[3]

2 See Thacker, 'Networks, Swarms, Multitudes,' n.p.

3 Levin, 'Conceptual and Methodological Issues in the Modeling of Biological Aggregations,' 245.

As we have seen, it is extremely problematic to produce a detailed optical-acoustic analysis of fish schools when dealing with a large number of individual fish; with such methods, a quantitative approach is only possible in limited circumstances. Here, models offer the possibility of supplementing such approaches with otherwise inaccessible levels of investigation.

According to Margaret Morrison and Mary S. Morgan, models can be regarded as autonomous scientific instruments that can generate both theoretical as well as empirical knowledge. If models are viewed as 'mediating instruments' between these two domains, then they are liberated from the auxiliary role that they have long played in the formation of theory and they gain, from this new position of autonomy, a specific operational role of their own as research instruments.[4] What is the basis of this autonomy? Morrison and Morgan argue that models are not mere derivatives from either theory or data material; rather, they combine elements from both sides – "elements of theories and empirical evidence, as well as stories and objects which could form the basis for modeling decisions."[5] These elements are integrated into a common formal (mathematical) system that can express the fundamental relationships between the variables under consideration. This interaction does not necessarily take place in such a way that empirical material is mathematically represented *by means of theory*; rather, models make representations conceivable that would not be conceivable within existing theories. Models, then, provide a new theoretical understanding of the phenomena under investigation.[6] Accordingly, theory is not an algorithm for the construction of models, because the latter always contain simplifications and approximations that are determined according to the data at hand. The use of analogies, moreover, represents another creative element in the construction of models. This not only involves the introduction of neutral features into them but can also involve the introduction of negative features, which can in turn lead to new theoretical results and unexpected empirical explanations. Models should not, therefore, be located somewhere in the hierarchical middle ground between theory and the empirical world but rather *along the way* (so to speak) toward the epistemic strategy of computer simulation, in which the theoretical and empirical spheres come to overlap at various points throughout the construction of the model in question.

If such partial autonomy can be accepted, then this raises the question of the independent existence of models as research instruments in the

4 See Morrison and Morgan, 'Models as Mediating Instruments.'

5 Ibid., 13.

6 Ibid.

production of knowledge. What is their effective contribution to such activity? It seems obvious to regard them as aids in the formulation of theory if certain theories fail to offer valid explanations for the phenomena that they are attempting to describe. Often, models serve to test existing theories or to experiment with them within a particular theoretical framework. They are used to study the implications of theories in concrete situations. And what is especially interesting is their use for testing theories that cannot be applied in any other way. That said, models can also create a direct link between the instrument and the object of an experiment, given that they can be manipulated to equate to the physical properties (for instance) that they are meant to represent. For example, if the physical laws of gravity are applied as a model for the formation of fish schools, their parameters can be transferred to quantify the schooling individuals' psychomechanical impulse to be attracted to one another. These quantifications thus allow the model to measure something that would be impossible to measure in the object of investigation itself.[7]

Yet it is important to stress that models produce representations, and that these representations are of a particular sort. In this case, a representation should not be understood as an exact mirror of a phenomenon but rather, according to Morrison and Morgan, "as a kind of rendering – a partial representation that either abstracts from, or translates into another form, the real nature of the system or theory, or one that is capable of embodying only a portion of a system."[8] Despite this limitation, however, it is possible to model each of the partial aspects of a given system or to combine different models for the sake of representing a single system. In the construction, modification, and use of models – regardless of whether they are mathematical equations, diagrams, computer programs, or something else – knowledge can be generated about a theoretical framework, about existing empirical data, and finally about the model itself.

In what follows, I will examine various approaches to modeling the organization of fish schools. Each of my case studies drew upon the suspicion – already formulated in laboratory and aquarium studies – that schools operate according to a simple set of rules, and each of them also introduced mathematical approaches that would pave the way for later simulation processes. Although they were intermediary technologies between visual analytic processes and time-based simulation scenarios, they were nevertheless used to develop explanatory models for schooling collectives that went

7 See ibid., 19–23.

8 Ibid., 27.

beyond what could be observed and beyond purely biological concepts. Chief among these approaches are those that aimed – quite early on – to furnish observational and experimental data with a mathematical underpinning and sought to supplant such data with basic formal parameters. Since the late 1970s, it has become increasingly difficult to draw a sharp line between models, simulations, and 'computer experiments.' Many (mathematical) models have been temporalized with the help of the computer and tested in simulations, so that the boundaries between these epistemic strategies have become more and more fluid. Accordingly, this section will end with two early computer simulation models that – as will be shown – were simply developed *too early* and thus hardly received any attention at the time. They were unable to benefit from the momentum behind agent-based computer simulation, which was only able to be maintained in conjunction with advanced graphical visualizations (this momentum would not begin to gain steam until the end of the 1980s). Before then, the epistemic potential of computer simulations for fish-school and other types of swarm research was hardly explicit.

Both the models and computer simulations of fish schools shifted the focus of media-theoretical analysis away from the problems of tracking, gridding out, and analyzing their behavior toward the issues involved with their visual representation. The present section will thus concentrate on models of fish schools that made their underlying mathematical structures (geometrically) conceivable and that attempted, with these graphical representations, to make the dynamics of their respective schooling systems at least approximately accessible. In contrast to the aims of experimental and empirical research, however, the goal here was to describe the sort of geometry that could be derived from the order of the school itself. Whereas the earliest methods aimed to geometricize the non-object of the school and to situate and order it in time and space within a prescribed system of coordinates, subsequent models represented approaches that asked which geometries the schools themselves produced. Though predating the technical advances that led to performative and dynamic visualization methods of computer simulation, they still indicated a new way of visualizing and formalizing biological fish schools. They are part of a gradual development to eliminate *nature* from the school – to eliminate its observable (but ever elusive) 'naturalness' in favor of the formally describable calculability that would ultimately result in the operative use of swarm intelligence as a technical process.

Most of the diverse attempts to define fish schools with models focused on partial aspects of their formation and behavior so that schooling collectives could be conceptualized along the lines of such things as magnetized iron filings, geometric space lattices, selfish herds, or integrated sensor systems. For their part, the models entailed a departure from ontological definitions and they gave preference to similarity relations over optically or acoustically visible phenomena. In short, they defined and described formal relations.

2. Synchronization Projects

Elementary Operations

> Imagine a space alien looking at rush hour traffic on the L.A. freeway. It thinks the cars are organisms and wonders how they are moving in a polarized way without collisions. The reason is that there is a set of rules everyone knows. We are the space aliens looking at fish, and we don't have the driver's manual.[9]

By the 1970s at the latest, the insufficiencies of optical and acoustic methods for observing, analyzing, and measuring fish schools in laboratory settings and in open water began to incite scientists to try a new epistemic strategy by constructing models. By importing models from other scientific disciplines, fish-school researchers attempted to explain the parameters that governed such things as the specific 'nearest neighbor distance' between the fish, their polarized orientation, their clear edge, and their tendency to take certain forms and make coordinated movements at a global level. Models were thus implemented as media for generating knowledge; in a sense, they were linked together in a feedback loop with the insufficient empirical data that had already been collected about schools of fish. Most of the first attempts adopted the so-called 'individual-based modeling' approach,[10] and they were largely based on the early studies by Albert Parr and Charles Breder, who were the first to present a mathematically grounded description of the (geometric) structure of the schools that they had observed in the laboratory. Laboratory work and modeling have never been mutually exclusive; on the contrary, they need one another. Evelyn Fox Keller was therefore right to criticize a report issued by Princeton's Institute for Advanced Studies for containing the following remarks: "While the physical sciences have had this mathematical/theoretical tradition from their beginnings, biology has

9 Julia K. Parrish, as interviewed by Klarreich, 'The Mind of the Swarm.'
10 Gerlotto et al., 'Waves of Agitation,' 1405.

had a different history. [...] [It] has been more focused on laboratory work. However, several areas of biology have gradually developed an understanding of the important role that mathematical approaches can play."[11]

These comments say less about the apparently deficient use of mathematical approaches in biology than they do about their authors' outdated, Popperian notions concerning the strict separation of theory and experiment. In the field of fish-school research, at least, scientists began rather early on to construct mathematical and physical models, and this activity did nothing to preclude their serious engagement with experimental systems. This ongoing use of computer technology did more than enhance the possibilities for collecting, inputting, and storing data. User-friendly software and increasing computer literacy also enabled scientists to process these data and to develop their own models without any 'professional' help: the net effect is the beginning of a new culture in biology, at once theoretical and experimental, and growing directly out of the efforts of workers who Dearden and Akam describe as "a breed of biologist-mathematicians as familiar with handling differential equations as with the limitations of messy experimental data."[12] This generation of researchers practiced in the field of biology what Galison referred to (regarding the history of physics) as the third epistemological way, for they were no longer simply content to work with the partially redundant (and occasionally chaotic) data generated by experiments or fieldwork.

In their edited volume *Animal Groups in Three Dimensions*, Julia Parrish and William Hamner made a conscious effort to integrate new and interdisciplinary approaches into the study of fish schools. They argue that research of this sort requires more than just a biologically motivated approach; otherwise, scientists would never be able to free themselves from the 'technological morass' entailed by previous research methods employed in laboratories and in the open water. Their anthology of interdisciplinary approaches was meant to serve as a springboard for creative scientific work in the future:

> Furthermore, we are neophytes. Our measuring devices, our computers, our words, and our graphics may never let us adequately describe the aesthetic beauty of a turning flock of starlings or a school of anchovies exploding away from an oncoming tuna. What we see as apparent simplicity we now know is a complex layering of physiology and behavior, both

11 Quoted from Keller, *Making Sense of Life*, 254.

12 Ibid., 258. Here Keller refers to Dearden and Akam, 'Segmentation *in Silico*.'

> mechanistically and functionally. It is our sincere belief that an interactive, multidisciplinary approach will take us farther in understanding how and why animals aggregate than merely pursuing a strictly biological investigation. It is also more fun.[13]

Thus, it is also possible to regard fish-school researchers like Parr and Breder as early representatives of a branch of mathematics whose applications throughout the history of science have included molecular biology as well. With their idea of schooling machines, they conceptualized above all their intrinsic processes and were more concerned with the psycho-physical manner in which schools function than with their biological functionality. Furthermore, Breder endeavored to separate physical from biological parameters. His interest lay in mechanistic epiphenomena that were independent of biological, adaptive functions. He therefore viewed schools of fish as the algebraic sum of numerous behavioral units that interacted in a complex way and determined the activity of the school.[14]

Because it was (and remains) problematic to test mathematical models in the case of schools, and because computer-based simulations were still years ahead, Breder proposed drawing a connection to physical phenomena that exhibit similar structures and dynamics. Mathematical theories of physics were thus introduced to fish-school research as "models *for* the construction of objects" (in Keller's terms).[15] In his 1951 study on the structure of fish schools, Breder relied on the physical laws of gravity to describe the process of multiple schools coming together to form a large collective. Casual observations had revealed that individual schools of fish can exert influences on one another. This is reflected in their tendency to blend together, whereby smaller schools will increase their speed to join a larger one. Breder regretted that there were still no quantitative studies of this phenomenon, and because no such research was foreseeable, he suggested a theoretical approach that he figured would be fruitful if it could separate the purely biological aspects of the phenomenon from its purely physical aspects. His proposal was to

13 Parrish et al., 'Introduction – From Individuals to Aggregations,' 13.

14 Breder, 'Studies on the Structure of the Fish School,' 24.

15 See Keller, 'Models of and Models for: Theory and Practice in Contemporary Biology,' 72–82 (my emphasis). Here Keller distinguishes between the role of 'models *for*,' which, more than simply representative of something, are instruments for scientific activity such as the development of new concepts and theories, from 'models *of*' things that already exist. Models do not, therefore, simply stand for something else but themselves become active 'autonomous agents' in the processes of generating knowledge. See also Griesemer, 'Three-Dimensional Models in Philosophical Perspective,' 435.

calculate this especially pronounced aggregational behavior by means of the Newtonian laws of gravity: "The above remarks naturally suggest at once the possible applicability of the common gravitational formula, which in addition is equally useful in studies of magnetism and electrostatics as well as a variety of other less well-known fields."[16]

The main reason why Breder deemed it legitimate to discuss the attraction of fish to other fish in terms of gravitational formulas was the latter's pure applicability: because a uniform distribution of mutually attractive elements in space is gravitationally unstable, it should follow from this that a uniform distribution of fish within a given volume of water would result in 'clots' of fish in dense masses. Whereas Albert Parr had focused on the fish at the edge of a school to explain their collective structure, Breder postulated that the density, cohesiveness, and stability of a school should be proportional to the power of attraction operating between all the schooling individuals. In the case of fish, moreover, he believed that these characteristics were not automatically 'biological' features in themselves but were rather equivalent to any similar physical situation. It was only in the wake of such physical processes, he thought, that genuinely biological processes of selection would then come into play: "Selection would then operate on the elements (fishes in this case), modifying the basic homotrophic attitude in a manner concordant with the survival of the groups of elements. As can be observed and should be expected on such a basis, many 'answers' that are obviously adequate to long-continued survival have been made by differing social groups of fishes."[17] On the basis of this approach, it was even possible to model fish schools by comparing their activity to that of iron filings within a dynamic magnetic field; their twists, turns, and various orientations would then indicate the influence of external environmental factors. The algebraic sum of such a magnetic field could be resolved by vector analysis as soon as all the magnitudes and their nature were understood. At that point, however, such factors remained little more than indications of the non-uniform influences that shaped a given school.[18]

At the time, biologically and empirically oriented researchers sharply criticized the mechanistic and analogy-based approach of the 'Parr-Breder School' (pun intended?) for being less than substantial.[19] One reason for this was that Parr and Breder compared rotational phenomena produced by the

16 Breder, 'Studies on the Structure of the Fish School,' 22.

17 Ibid., 23.

18 Ibid., 22.

19 Radakov, *Schooling in the Ecology of Fish*, 18–19.

weather and inanimate things – hurricanes, tornados, whirlpools, locks of hair, for instance – with the 'milling' behavior of (living) fish without taking into account specific environmental or material factors.[20] That said, their approach nevertheless expanded the repertoire of biological fish-school research in that it did not restrict itself to specific biological and ecological realities. Rather, its aim was to discover basic principles of aggregation whose interdisciplinary application contained the *possibility* – no promises were given – of providing insight into fish schools as an object of knowledge. After all, they were fully aware that they were dealing with entirely different things when they ventured to compare the behavior of fish to that of inanimate objects: "That is, the notation describes the observed schooling without postulating the precise nature of the attractive forces."[21]

In the following years, Breder advanced his physical research, in which he continued to seek similarities in aggregational behavior, by incorporating more detailed mathematical models of fish schools. His article titled 'Equations Descriptive of Fish Schools and Other Animal Aggregations,' which appeared in 1954, presents a system of algebraic equations for determining the cohesion of schools and then compares these results to the observational data produced by empirical fish-school research.[22] Breder's system of equations – somewhat like Parr's before him – models a dynamic equilibrium based on the forces of attraction and repulsion in relation to the distance between the individual fish. The typical schooling structure for a given species of fish would form when a balance was achieved and the schooling fish established a standard distance between one another. At shorter distances, the variable of *repulsion* would increase exponentially to prevent bodily contact; at greater distances, the variable of *attraction* would be stronger in order to enable the school to form in the first place.

Because Breder did not have any adequate empirical data at his disposal to serve as the basis for his model, he modified the numerical values in his formula at random until it produced a family of curves that could – or so he postulated – be used to describe various forms of fish-school aggregations. Whether a diffuse structure or a dense school swimming in one direction. both forms could be generated by altering the two variables. The model became problematic, however, because not every individual fish is visible to all others within a school. Thus the elements of a school could not be treated as mass particles that each have an identical effect

20 Ibid., 27.

21 Breder, 'Equations Descriptive of Fish Schools and Other Animal Aggregations,' 362.

22 Ibid., 361.

on one another. It would be better, Breder thought, to think of them as having surface effects that could extent around six fish deep. Beyond this limit, according to his model, it would be irrelevant whether there were just a few more hidden rows of fish or a thousand, for the latter would be out of sight in any case.

Breder attempted to reduce the inter-individual relations within fish schools to a simple formula, and to do so he necessarily ignored such things as environmental influences or disruptions. What is interesting is that he went directly to the abstract level of mathematical operations and continued to work on that level until his graphs began to indicate similarities with the patchy empirical data that existed at the time – until the mathematical model began to resemble (insufficiently observed) reality. Before comparing his model to photographs of real fish schools, he expressed two reservations. The first concerned the problem – so common in open-water research – of establishing an accurate frame of reference for taking measurements. The distances between the individual fish could not be measured in absolute units but rather had to be expressed as a decimal part of the mean length of the fish pictured in the images. And second – "since we are here not dealing with mathematical models, but with actual physical and biological imperfection" – mean values and ranges had to be employed in each of his case studies.[23] (This way of operating would be prototypical for the epistemic strategies behind computer simulation, which will be discussed in my next chapter).

The goal of Breder's mathematically formalizable model was not only to express similarities between physical and biological collective phenomena on a phenomenological level but also to create a level of description by means of mathematical formalization that would, in principle, allow the functionality of self-organization to appear *avant la lettre*. Thus it could be said that the application of physical laws to fish-school research was based on the suspicion that it might be possible to make general statements about the behavior of aggregations without having to be concerned with the ontological status of the elements that constitute such collectives. In an entirely un-ecological manner, swarm behavior was formally conceptualized as a pure epiphenomenon of concentration.

Beyond all of this, it is interesting to note that Breder also proposed that information was transferred within fish schools by way of a 'shock wave' phenomenon, and that this suggestion opened up a new research horizon that led from the mechanical modeling of schools to modeling them on the

23 Ibid., 364.

basis of information theory. In schools, he thought, information was passed along so quickly by the behavior of the individual fish that its transmission had to be described in explosive terms. Unlike the case of 'explosive collectives' in mass psychology, however, this did not result in uncontrolled and uncontrollable behavior but rather to a sort of optimized reorganization. According to Breder, it is especially noteworthy that the individual schooling fish at no time have access to *complete* knowledge about the total behavior of their school, and that this limited knowledge is in fact a condition that enables such behavior to function in the first place:

> [R]emarks [...] on imitative behavior and information transmission are clearly related to these considerations on the size of a school necessary to prevent each individual from being completely informed about the behavior of every other individual. Relating such matters to the study of the nature of fish schools and other animal groupings connects this work clearly with modern theories of communication through both animate and inanimate systems. See Shannon and Weaver for a pertinent exposition of the trends of current communication theories.[24]

According to the quantitative understanding of information at the heart of Claude Shannon and Warren Weaver's model of communication, what is important is not the *content* of a transmitted message but rather the amount of data that is lost (ideally, as little as possible) in a given transmission. "Shannon's information theory," as Claus Pias has argued, "is thus a communication theory in the broadest sense; it treats words, writing, music, or painting on the same logical level as communication between multiple devices or communication between humans and devices."[25] Or, in Shannon's own words:

> In communication engineering we regard information perhaps a little differently than some of the rest of you do. In particular, we are not at all interested in semantics or the meaning implications of information. Information for the communication engineer is something he transmits from one point to another as it is given to him, and it may not have meaning at all. It might, for example, be a random sequence of digits, or it might be information for a guided missile or a television signal.[26]

24 Ibid., 368.
25 Pias, 'Zeit der Kybernetik,' 428.
26 Shannon, 'The Redundancy of English,' 248.

Technical information theory may indeed be "just like a very proper and discreet girl accepting your telegram," in Weaver's own words,[27] but perhaps it is for this very reason that it is so interesting in the context of fish schools and other swarms. Here, after all, the quantified transmission of information becomes *visible*. The convergence of motion and information transfers – and, to some extent, the channel and the message – is reflected in the global movements and structural changes of the school. From this point on, moreover, the ability of schools to transmit information efficiently and rapidly under the condition of limited knowledge became an interesting problem for the new field of mathematical information theory and an interesting problem in its own right. Biological fish-school research and the burgeoning science of cybernetics were suddenly united by the epistemic process of mathematical modeling.[28]

Synchronized Swimming

In March of 1951, things were well underway at the eighth conference sponsored by the Josiah Macy, Jr. Foundation in New York. One topic on that year's agenda was 'Communication Between Animals,' and the speaker invited to address this issue was the psychologist Herbert G. Birch. Birch surveyed the results from a number of experiments conducted with various animals. Not only did he describe the behavior of amoebas that feed on potassium permanganate and starfish that consume mussels: in even greater detail, he also engaged with Karl von Frisch's pioneering research on the ability of bees to process symbols,[29] and discussed Theodore Schneirla's experiments on the collective behavior of army ants.[30] Throughout his talk, he pondered the question of whether the behavior that he was describing qualified as *communication* or whether it might be more accurate to assign it to the concept of *interrelations*. Suddenly, however, the discussion turned to the topic of synchronization, and it was at this point that Charles Breder's research on fish schools came into play. At issue was the trigger

27 Weaver, 'Recent Contributions to the Mathematical Theory of Communication,' 27.

28 Here it should be kept in mind that Shannon and Weaver's focus was on linear communication between one transmitter and one receiver, both of which were thought to be relatively closed entities. Of course, the situation is different with swarms, in which many swarming individuals interact with one another simultaneously while also being subject to external environmental influences.

29 See Von Frisch, *Bees: Their Vision, Chemical Senses, and Language.*

30 See Schneirla, 'Social Organization in Insects'; and Maier and Schneirla, *Principles of Animal Psychology.*

mechanisms of collective behavior and the transmission of information through movement:

> [I]sn't it true that if you fired off a gun in this room, everybody would jump within a millisecond? – We wouldn't within a millisecond. I doubt if we should all jump. – There would be an appropriate electrical pickup on every person. – Perhaps within a millisecond of each other, not within a millisecond of the gunshot. – Yes, within a millisecond of each other. That is just what I doubt. – How long does it take a man to respond to a shot like that? How long from the gunshot to his response? – Two-fifths of a second. – Well, a couple of hundred milliseconds, so it implies a synchronization of a half per cent or so. – That is so, and that is an extremely abrupt and vigorous stimulus [...].[31]

At the heart of the ensuing discussion was the question of how the collective reactions of similar agents could be synchronized in time and space. And this raised the additional question about the extent to which the observed simultaneity could be ascribed to systemic processes or whether the impression of synchronous activity was simply evoked by the inadequate perception of the processes. Rather than being understood as the rhythmical alignment of the clock rates of different systems, synchronization was here regarded, first, as a process of connecting in space and time the specific event structures of an environmental system with those of a collective within it and, second, in terms of the synchronization phenomena that take place *within* this collective, which enable it to adapt to the external system.

In Birch's opinion, the term *communication* applied quite well to the swarming behavior of fish and birds, and this conclusion was in line with Norbert Wiener's understanding of cybernetics as the science of "communication and control in the animal and the machine."[32] This statement quickly caused the group to focus on the theoretical considerations concerning the sort of communicative feedback mechanisms that would have to be operational in fish schools in order to explain their collective maneuvers. The neurophysiologist Ralph W. Gerard, for instance, reported about observations that he had made off the pier at the Marine Biological Laboratory in Woods Hole. What had fascinated him the most were the highly-involved signals and the precise timing that enabled the fish schools swimming below him to change direction so quickly – and to do so without any leading individual

31 Birch, 'Communications Between Animals,' 468.

32 Ibid., 461.

or group hierarchy: "As far as I could tell from observing them visually, there was no wave of change from a leader or any other fish. They all moved simultaneously."[33] As Herbert Birch claimed, moreover, such phenomena occur in species whose individual behavior is highly determined.[34]

According to the cybernetic way of thinking, synchronization within a swarm collective was based on a 'process,' on 'communication,' and thus on an exchange of signals. For this reason it is difficult to distinguish between the concepts of synchronization and simultaneity as they were used by the participants at the Macy Conference. For, with his *procedural* understanding of simultaneity and his theory of relativity, Einstein had already dissolved the Newtonian conception of absolute time into many different times. It is only in a limited way that modern physics can comprehend simultaneity as a process of synchronizing different 'clocks' at different locations. Thus the hour of synchronization strikes when simultaneity is said to be defined as a process.[35] In short, synchronization is the creation of simultaneity in the process of making a relative and relational comparison of events at different locations.[36]

According to the state of research at the time, swarms evidenced something like a leap in scale from the level of individual capability to the level collective behavior, and this leap was based on a process of exchanging signals. Because a school of fish – "a spatial relation of a neat, orderly, a repetitive kind"[37] – arises from the individual and disorderly act of schooling, it was believed that there must be functional relations among the fish to synchronize the movements of the individuals in such a way as to produce the global movement of the school as a whole. Schools and other swarms could thus be regarded as open-ended *synchronization projects* in which a multitude of asynchronous individual movements become synchronous movements without the help of any master clock. These global movements, in turn, synchronize the internal processes of the school with events from its environment, thereby enabling an adaptive behavior that instantaneously becomes visible in the morphology of the school's spatial structure.

In this synchronization process, it could be argued, *first*, that 'swarm time' and 'swarm space' come together in an inseparable, reciprocal relationship. Internal and external stimuli, which influence a swarm as a whole, are synchronized through neighboring connections in a way that is

33 Ibid., 468.

34 Ibid., 461.

35 See Galison, *Einstein's Clocks, Poincaré's Maps: Empires of Time*, 13–41.

36 For a detailed discussion of the concept of simultaneity, see Jammer, *Concepts of Simultaneity: From Antiquity to Einstein and Beyond*.

37 Birch, 'Communication Between Animals,' 461.

then reflected in specific patterns of motion. Movement in swarms, as the participants in the 1951 Macy Conference already knew, always entails the transmission of 'information.' The elaborate signaling and timing observed by Gerard thus converged into a single dynamic process. *Second*, swarms discredit Norbert Wiener's rather dismissive assessment, in *The Human Use of Human Beings*, of collectives made up of 'simple' (instead of 'higher') living organisms.[38] Through their modes of synchronization, that is, swarms can be associated with that peculiar aspect of intelligence which Wiener himself would recognize when considering the alpha rhythm of the brain, namely the fact that the ability to think is itself based to some extent on synchronization phenomena.[39] Is it not possible to draw a line from here to the diffuse concept of swarm intelligence, which, since the end of the 1990s, has been implemented with computer technology to serve such a wide variety of disciplines? *Third* and finally, swarms can be conceptualized as systems that always create a robust and flexible global structure in response to ever-changing environmental influences. They exist in a process-based state of dynamic equilibrium. More recent swarm research has referred to this phenomenon with the label "sensory integration systems,"[40] thus calling to mind the concept of the homeostat – the "machina sopora," in Grey Walter's terms – that the British cybernetician W. Ross Ashby had developed in the 1950s.[41] To borrow a term from Andrew Pickering, both arrangements can be regarded as 'philosophical machines' that make difficult sets of problems visible and thus comprehensible through their processing.[42]

My main contention in this section is that the concept of synchronization connects, in various ways, a certain epistemic view of swarms and with its prehistory, which was strongly influenced by cybernetics. It was a particular set of cultural techniques that made it possible to analyze the synchronization phenomena in biological swarms. These analyses can in turn be associated with today's processes of swarm intelligence, in which the spatio-temporal operations of biological collectives are themselves transformed into medial processes in order to address particular problems of regulation and control. And this, in turn, raises the question of the extent to which the computer-scientific conceptualization of swarms, for instance, is in sync with its biological models.

38 Wiener, *The Human Use of Human Beings*, 48–73.

39 Wiener, *Cybernetics*, 181–204.

40 Schilt and Norris, 'Perspectives on Sensory Integration Systems,' 229.

41 Ashby, *Design for a Brain: The Origin of Adaptive Behaviour*.

42 Pickering, 'A Gallery of Monsters,' 238.

Alpha Rhythm

Travel reports dating back to the sixteenth century repeatedly offer accounts of wondrous and even unsettling collective phenomena. In his log of an expedition undertaken from 1577 to 1580, for instance, the British sea captain Francis Drake recorded the following observations about a giant swarm of "fiery worms":

> Our General [...] sayled to the certain little Island to the Southwards of Celebes [...] thoroughly growen with wood of a large and high growth [...]. Among these trees night by night, through the whole land, did shew themselves an infinite swarme of fiery worms flying in the ayre, whose bodies being no bigger than our common English flies, make such a shew of light, as if every twigge or tree had been a burning candle.[43]

Fascination with such uncanny phenomena tends to multiply even more when 'swarms' exhibit a sort of collective unity. While traveling down a river in Thailand, the German physician Engelbert Kaempfer noticed an impressive example of synchronicity – of collective, rhythmic illuminations – that reminded him of the contractions and relaxations of the human heart:

> The Glowworms [...] represent another shew, which settle on some Trees, like a fiery cloud, with this surprising circumstance, that a whole swarm of these insects, having taken possession of one Tree, and spread themselves over its branches, sometimes hide their Light all at once, and a moment after make it appear again with the utmost regularity and exactness, as if they were in perpetual Systole and Diastole.[44]

Infinite swarms and fiery clouds blinking in rhythm: as aggregations, fireflies exhibit manners of behavior that have led people to presume that they are governed by a mysterious organizational authority that somehow coordinates their activity. Between 1915 and 1935, there was a veritable boom in scientific publications on the synchronicity of fireflies, some relying on observations made around the turn of the century. Theories were proposed about a central leader that set the rhythm, or the synchronicity

43 Haklyut, A Selection of the Principal Voyages, Traffiques and Discoveries of the English Nation, 151. Quoted from Strogatz, Sync: How Order Emerges from Chaos in the Universe, Nature, and Daily Life, 11 (note).

44 Kaempfer, *The History of Japan*, vol. 1, 45. Quoted from Strogatz, *Sync*, 11 (note).

was dismissed outright as a mere observer effect. In the journal *Science*, for instance, the biologist Philip Laurent was confident enough to offer the following solution to the puzzle: "The apparent phenomenon was caused by the twitching or sudden lowering or raising of my eyelids. The insects had nothing whatsoever to do with it."[45] Needless to say, Laurent's opinion would remain an outlier. During the 1960s, however, and thus in a different, cybernetic and information-theoretical episteme, this synchronous and rhythmic illumination was regarded as a process of self-organization in which fireflies, like resettable oscillators, adjusted the blinking rhythm of their signals in response to the perceived signals of other fireflies.[46]

Norbert Wiener himself was of the opinion that synchronicity and processes of synchronization should be regarded as fundamental aspects of self-organization. In a chapter entitled 'Brain Waves and Self-Organizing Systems,' he cited the rhythmic oscillations of fireflies as an analogue to the organizational function of brain waves in the coordination of human neural activity. Ever since Alessandro Volta's and Luigi Galvani's experiments in the eighteenth century, physiologists have been aware of the phenomenon that a weak electrical connection forms between two electrodes fixed to different areas of the scalp, and that the strength of this voltage fluctuates over time. In the twentieth century, electroencephalographers were able to record these fluctuations on so-called strip charts, which yielded insight into human brain activity. A characteristic curve on such a chart is the so-called alpha rhythm, which is a pattern that oscillates by approximately 10 Hz. The alpha rhythm occurs in people who are in a relaxed and conscious state with their eyes closed.[47]

Wiener went about studying this alpha rhythm in far greater detail, for he suspected it functioned like a sort of pacemaker for the brain. He knew already that such a pacemaker did not follow the rhythm of a 'master clock,' for scientists were aware at the time that neurons are imprecise oscillators. Rather, he believed that a sophisticated synchronization process was responsible for this rhythm, a process that enabled a collective pace to arise from an immense number of imprecise neurons. Neurons with a faster rhythm would adapt to those that are slower, and vice versa, until they achieved an approximate rate of ten cycles per second. According to Wiener's hypothesis, this process of 'frequency pulling' would have to leave behind

45 Laurent, 'The Supposed Synchronal Flashing of Fireflies,' 44. Quoted from Strogatz, *Sync*, 11 (note).

46 See Buck and Buck, 'Mechanism of Rhythmic Synchronous Flashing of Fireflies.'

47 See Strogatz, *Sync*, 41–42.

a telltale signature in the alpha rhythm: whereas it might be expected that the measured frequencies of neurons should resemble a bell curve, with the majority of neurons near the middle (10 Hz) and the rest gradually tailing off to one side or the other depending on their speed, if the neurons were in fact able to influence each other's frequency, this would change the shape of the curve considerably, for a far greater number of neurons – almost all of them – would be adjusted toward the middle. That said, neurons with extremely divergent frequencies would probably remain outliers and not adapt to the rhythm of the majority. The result would a distribution curve with a tall spike in the middle and a double dip on either side.[48]

With the help of Walter Rosenblith and his method for making electronic recordings of brain waves, Wiener set out to test this hypothesis with the greatest possible accuracy. He announced the results of the experiment in his book *Nonlinear Problems in Random Theory*, though his overall presentation was rather vague.[49] Even later experiments on the alpha rhythm failed to verify his theory. Nevertheless, his approach was pioneering for the study of synchronization phenomena in large collectives. "Whereas earlier mathematicians," as Steven Strogatz has pointed out, "had been content to work on problems involving two coupled oscillators, Wiener tackled problems involving millions of them. Perhaps even more important, he was the first to point out the pervasiveness of synchronicity in the universe."[50] In the wake of Wiener's hypothesis, the mutual adaptation of rhythmic oscillations indeed came to be noticed in a variety of different areas, including biology and the case of fireflies, which Wiener himself had put on his research agenda:

> It has often been supposed that the fireflies in a tree flash in unison, and this apparent phenomenon has been put down to a human optical illusion. I have heard it stated that in the case of some fireflies of Southeastern Asia this phenomenon is so marked that it can scarcely be put down to illusion. Now the firefly has a double action. On the one hand it is an emitter of more or less periodical impulses, and on the other hand it possesses receptors for these impulses. Could not the same supposed phenomenon of the pulling together of frequencies take place? For this work, accurate records of the flashings are necessary which are good enough to subject to an accurate harmonic analysis. Moreover, the fireflies

48 Ibid., 42–44.
49 Wiener, *Nonlinear Problems in Random Theory.*
50 Strogatz, *Sync*, 41.

> should be subjected to periodic light, as for example from a flashing neon tube, and we should determine whether this has a tendency to pull them into frequency with itself. If this should be the case, we should try to obtain an accurate record of these spontaneous flashes to subject to an auto-correlation analysis similar to that which we have made in the case of the brain waves.[51]

Norbert Wiener placed synchronization phenomena at the heart of every process of self-organization. The latter do not require any external or centralized pacemaker. Instead, they are constituted by the constant adjustments made by their own elements, regardless of whether the 'nature' of these elements is technical or biological.[52] Wiener's idea that the self-organized synchronization of neurons within the brain enabled thinking to take place could thus be extended to other biological collectives in which self-organization plays a role – not least to swarms. A sort of temporal organization, for instance, is reflected in the visual arrangements of fireflies or in the spatial structures of fish schools and flocks of birds. Above all, however, his idea was applicable to technical implementations, in particular to a machine called the homeostat, which W. Ross Ashby had introduced in the late 1940s.[53] With this apparatus, Ashby intended not only to reproduce the self-organizational structure of brains (and complex organisms); he also hoped to equip it with a feature that happened to be especially significant to the spatio-temporal organization of swarms, namely the ability to adapt to external disruptions and environmental influences.

A Race for Relaxation

In his book *Design for a Brain*, Ashby approached the topic of adaptation from the perspective of neurophysiology, and he described his vision for the homeostat by referring to the results of some truly gruesome experiments:

> A remarkable property of the nervous system is its ability to adapt itself to surgical alterations. […] Over forty years ago, Marina severed the attachments of the internal and external recti muscles of a monkey's eyeball and re-attached them in crossed position so that a contraction of the external rectus would cause the eyeball to turn not outwards but inwards.

51 Wiener, *Cybernetics*, 200–201.

52 Ibid., 201–203.

53 Ashby, *Design for a Brain*, 100–121.

> When the wound had healed, he was surprised to discover that the two eyeballs still moved together, so that binocular vision was preserved. [...] The nervous system provides many illustrations of such a series of events: first the established reaction, then an alteration made in the environment by the experimenter, and finally a reorganization within the nervous system, compensating for the experimental alteration. The Homeostat can thus show, in elementary form, this power of self-reorganization.[54]

Ashby's homeostat was meant to serve as a mechanical model for this sort of independent adaptation to altered environmental conditions. It had a singular purpose: to maintain equilibrium while converting electrical inputs into outputs. The apparatus consisted of four identical units that were connected to one another with electrical wires. On the top of each unit was a rotating electromagnet with a needle indicating its position. Each unit emitted output to the other three, and thus each received input from the others (causing their respective needles to swing). Each of the electromagnets, in turn, was part of an additional internal circuit belonging to its unit, and this circuit was connected with a wire to a semicircular water trough, the contents of which were kept at a constant voltage by means of a battery. The varying position of the magnet's needle controlled the strength of the current that flowed through the wire. This individual current was the output of each of the homeostat's units, and thus this output current was proportional to the deflection of the magnet from the center: "The angular deviations of the four magnets form the central position provide the four main variables," as Ashby maintained.[55] Each unit thus had four synchronous inputs influencing the position of its needle: "As soon as the system is switched on, the magnets are moved by the currents from the other units, but these movements change the currents, which modify the movements, and so on."[56] This arrangement could adopt two basic modes depending on how the input parameters – the currents – were manipulated; it could either be stable or unstable:

> If the field is stable, the four magnets move to the central position, where they actively resist any attempt to displace them. If displaced, a *co-ordinated* activity brings them back to the centre. Other parameter settings may, however, give instability; in which case a 'runaway' occurs

54 Ibid., 104, 107.
55 Ibid., 100.
56 Ibid., 102.

> and the magnets diverge from the central positions with high increasing velocity – till they hit the ends of the troughs.[57]

In the case of instability, however, the homeostat did not at first behave adaptively; rather, it firmly remained in the position that its needle – now pushed to the far end of the scale – indicated as unstable. Ashby therefore implemented an additional feedback mechanism. If the output current from one of the units exceeded a certain limit, a relay would then trigger a process that would alter the unit's internal parameters (that is, the magnitude or sign of the currents being fed back to it) in discrete steps. These changes took place at random; they were, in Ashby's words, "deliberately randomised by taking the actual numerical values from Fisher and Yates' Table of Random Numbers."[58] Through this process of random resettings, the machine could thus reconfigure itself until it achieved stability once again and its needle settled down in the middle of its range. The homeostat, according to Andrew Pickering, "was a machine for staying the same: whatever you did to the magnets, or however you tinkered with its internal connections, it would reconfigure itself to achieve stability."[59] It was for this reason that William Grey Walter famously called it a *machina sopora* – a 'drowsy machine' whose sole purpose was to rest and relax.

In an article on the materiality of cybernetics, Pickering has discussed the essential implications of this machine, and several of his observations also pertain to fish schools and other swarms. *First*, the homeostat was a self-organizing machine that could reconfigure itself in response to external disturbances. *Second*, it mimicked the functions of living organisms with an electromechanical system: "This was the sense that Ashby could and did think of the homeostat as a brain, albeit a behaviorist brain that included no internal representations."[60] *Third*, by means of the homeostat it was possible to observe self-organization in action and to conduct experiments that could not be undertaken with a brain: "[T]he homeostat was a wonderful machine to think with. Contemplating the homeostat, one could see directly what self-organization might amount to, free from the biological complexity of the biological morphogenesis of brains."[61] Above all, however, the homeostat exhibited a fascinating sort of temporality. It was a machine that operated

57 Ibid., 103 (my emphasis).
58 Ibid.
59 Pickering, 'A Gallery of Monsters,' 237.
60 Ibid., 238.
61 Ibid.

in real time, reacting to events as they occurred, and thus it synchronized, in one and the same temporal process, the internal conditions of its four units with the variables of external disturbances:

> If a needle were pushed this way *now*, either the uniselectors would trip *now* or they would not. [...] [T]he homeostat obeyed the injunction 'never look back.' [...] The homeostat lived right there in the present and, one might say, it *looked the future in the face*. [...] The trick was *randomness*. The homeostat could not know what was coming at it from the future – the future was itself random as far at the homeostat was concerned – and it responded to that by reacting randomly to whatever came along. [...] It reconfigured and self-organized itself this way and that until it had learned to cope with the vicissitudes of real time. The intersection of two random series – external events unfolding and inner reconfigurations of the gadget – produced order. [...] More than a device to think with, the homeostat was a true *philosophical machine* – contemplating its workings and performance, solutions to very difficult conceptual puzzles become literally visible.[62]

That said, this form of coping with time did not lead the homeostat to oscillate regularly in response to its environmental system. Rather, its machine logic reacted according to particular internal parameters to an environmental system that likewise had rules of its own, and it only did so on a case-by-case basis. The two systems were not in harmony; on account of its design, the homeostat was merely able to offset any 'dissonance' – in an instantaneous and auto-regulatory manner – in that its units reacted synchronously and thus in a temporally coordinated way to such disturbances.

It is this process of synchronizing internal units for the purpose of adapting to external influences and events that links the homeostat to the manner in which swarms operate. Schools of fish and flocks of birds react to events in their environments on the basis of specific rules governing the interactions among their individuals, and these rules are reflected in their synchronized global movements. Like Ashby's machine, these collectives, which organize themselves in four dimensions (three dimensions of space plus time), remain in a dynamic equilibrium that ensures their continued existence over time. Of course, swarms could never be accused of drowsiness, for the interactions among their elements are not based on the strength of electrical currents but rather on the intensity of their movements. Whereas

62 Ibid. (emphasis original).

the homeostat sought and found rest and relaxation, the global structure of swarms is constantly transforming over time. Moreover, their internal reconfigurations do not depend on a selection of random values but rather on simple, predetermined rules. Nevertheless, both systems manage to remain 'viable' by reacting in real time to their respective environments. The additional question posed by swarms concerns the rapid transmission of information that makes synchronized swimming or flying possible in the first place. And although Ralph Gerard was unable to detect any 'wave of change' while observing schools of fish at Woods Hole, more recent fish-school research has identified just such waves of information, which serve to organize coordinated, decentralized, and synchronized movements and reactions.

3. Anchovy ex Machina

Falling into Formation

> Imagine a heavy hammer that waits for nothing else but to be gripped and swung. And imagine a strong, limber man who waits for nothing else but to grip and swing the hammer and to strike a heavy metal plate, which will then vibrate. This scene could take place in a dark shed in someone's backyard in the city, which is otherwise drowsy in the heat of a summer afternoon. At the same time, you happen to be walking down a nearby street and hear the hammer's strike, which sounds like a gong: that is the information at hand, and everything else can be deduced from it.[63]

As is well known, the modern history of information theory and a technically informed concept of information began in 1948 with the publication of Claude Shannon's article 'A Mathematical Theory of Communication.' Without dwelling on this history for too long, I would like here to summarize what constitutes this concept of information and how, together with cybernetic approaches, it provided a new way to describe the phenomenon of swarms. In his theory, Shannon removed all semantics from whatever 'messages' were to be transmitted, for he considered their meaning to be irrelevant to the technical problem of communication, which is to pass along information as accurately as possible to another location. On a purely syntactical level, the flow of a message should be made measurable; in this

63 Serres and Farouki, *Thesaurus der exakten Wissenschaften*, 411.

regard, Shannon borrowed ideas from Ralph V. L. Hartley, who in 1928 had argued that all psychological factors should be eliminated that might play a role in the process of communication in order "to establish a measure of information in terms of purely physical quantities."[64] According to Michel Serres and Nayla Farouki, these were conceived as basic quantities that could not be broken down any further: "A unit of information cannot be segmented any further and is limited to the most elementary sign that one can imagine: to yes or no, presence or absence. In short, it is the simplest element that can be transmitted."[65] With such a logarithm, information can thus be measured in terms of binary digits or 'bits.'

Though liberated from all meaning and substance, information still requires a physical substrate to be transmitted (like the hammer and the metal plate, for instance), even if it is independent of the latter. It also requires interpretation in order to 'arrive' somewhere and be of use.[66] Moreover, Shannon's model accounted for the fact that the transmission of information could be influenced by potential disturbances or interference. Communication is thus not only the reproduction of a selective process to ensure that a sequence of syntactically coded signs remains as intact as possible as it travels from its transmitter to its receiver. Communication must also contend with the *noise* that introduces a fundamental sense of uncertainty about whether the received signal really corresponds to that which was sent: "On the syntactical level, the transmission of information thus appears to be a potentially random process that can nevertheless be determined by means of statistical probability theory. Accordingly, a signal whose transmission is probable contains less informational content than a signal whose appearance on the other end is improbable."[67] Shannon's concept of information can be applied, for instance, to the manner in which all sorts of objects, sounds, texts, or images are coded as well to questions about how to increase the speed of communication (questions, that is, about the transmission of data, about how to deal with disruptions and interference effectively, and about how to correct or reduce errors during the transmission process). As soon as data are binary coded and converted into 'information sequences,' moreover, they not only become transmittable but can also be processed and manipulated by computers.[68] It was Norbert Wiener who first described the

64 Hartley, 'Transmission of Information.' For an introduction to this concept of information, see Münker, 'Information.'

65 Serres and Farouki, *Thesaurus der exakten Wissenschaften*, 412.

66 Ibid.

67 Münker, 'Information,' 97.

68 See Serres and Farouki, *Thesaurus der exakten Wissenschaften*, 413.

principle of selectivity that governs the communication processes between systems and their environments – a principle that applies to biological systems and living organisms as well as in the case of controlling technical apparatuses and machines. Wiener's use of Shannon's ideas culminated in his famous dictum about information standing in a category of its own: "Information is information, not matter or energy."[69]

This concept of information also makes it possible to speak about swarms and describe their dynamics in a different way. Rooted in communication technology and cybernetics, the concept allows the relations, transmission processes, and disruptive potential among swarming individuals to be regarded as phenomena that can be quantified both physically and in terms of probability theory. The substance of an entire swarming body or the material level of the swarming individuals and their aggregation were no longer matters of debate, and the analogies made to energy, which had often found their way into discourses concerned with collectives, began to lose their clout. Thinking about swarms as information machines actualized a biological perspective that could explain communication (or interaction) between individuals without recourse to psychological theories of contagion, sociological laws of imitation, immeasurable thought waves, or the idea of a 'super-sense' – all by way of a theory that was based on technically formalizable concepts. In that information theory and cybernetics understood swarms as biological systems whose swarming individuals organized themselves and synchronized their movements in response to dynamic environmental factors, it was possible to conceptualize swarms in more than just technical terms. The understanding of swarms was also transformed by the manner in which – a few decades later, with the rise in computing power – they were made operable as computerized applications for solving technical problems.

Sensory Integration Systems

If swarms can be described as philosophical machines, in Pickering's sense of the term, then it ought to be asked in which form, precisely, they can be described as mechanical arrangements and how, precisely, they organize themselves. The participants at the eighth Macy Conference focused their discussions on the significance of the sense of sight. It was the latter, they thought, that could explain the formation of typical swarm structures and the transmission of (motion-based) information within such collectives:

69 Wiener, *Cybernetics*, 132.

> In the schooling behavior of fish, the investigations [...] primarily have indicated that this behavior is related to the visual system of the fish, as such; that it is related to the kind of visual angle which the fish has, and to its directly determined responses to certain kinds of visual stimuli. There is a certain optimal position of the visual fixation on objects between Fish A and Fish B and Fish C, such that a change in the distance between them produces a distortion of image, and the fish then tend to maintain relative positions of relative optimal fixation.[70]

Movements are thus linked to visual stimulation, which is in turn reflected in movements. It was argued that schools of fish are sensorily determined, repetitive structures that maintain their dynamic equilibrium by means of continuous, visually stimulated feedback mechanisms and not, for instance, by means of their social preference for other fish. The arrangement depends on the distance at which each neighbor (one per eye) of a schooling individual can remain sharply in sight.[71]

Yet how is it possible that such a structure for transmitting information can result in the extremely rapid organizational achievements that are observed in schools of fish and flocks of birds? At the conference, Ralph Gerard related a story about how he once attempted to measure a simultaneous shift in direction enacted by a flock of birds by synchronizing their activity with his moving car:

> I was once able to check that in case of birds. A flight of birds was going along parallel to my car, so I could time them. I happened to be watching them as they veered away, and I would certainly have seen one bird go forward or drop back relative to the others if its timing was off. As I remember, I calculated there was less than five milliseconds possible time for cueing from one to another.[72]

How is it possible to explain the speed with which a flock is able to transfer information about a change of direction? Herbert Birch proposed that this was caused by a common stimulus from the environment. Each of the flocking birds would then react to this stimulus individually, but in the same way; he thought that simultaneity could be the function of a given

70 Birch, 'Communication Between Animals,' 461.

71 Ibid., 461–462. On this matter, see also Parr's article 'A Contribution to the Theoretical Analysis of the Schooling Behavior of Fishes.'

72 Birch, 'Communication Between Animals,' 468.

environmental change. Even if one had to assume that the birds' reaction would be statistically scattered, each individual would also be confronted at the same time by continuous feedback from its neighbors, and any delay in one bird would soon be levelled out by a sort of 'lag-correcting operation' (in Julian Bigelow's words). This hypothesis, however, did not seem very likely to Gerard, who thought that such beautiful synchronization – in response to a given signal – could not even be expected from a trained group of like organisms. (It was then asked, as related above, whether all of the participants at the conference would jump up within a millisecond of each other if someone happened to fire off a gun in the room.)

The physicist Donald M. MacKay refined this line of questioning by focusing on the amount of information that would be needed for such systems to function, and thus he created a link between biological and technical swarm systems: "In a nonlinear system of this sort the rate of change is the important thing. It is really a question of how many bits of information you need, and how fast."[73] Here, the concept of information employed in cybernetics and mathematical communication theory made it possible to think of a quantifiable 'currency' that was 'traded' during the processes of exchange and interaction taking place between the individuals within a swarm. What had previously been regarded as a biological matter of sensory stimulation involving the organs and brains of fish and birds was reconceived as a mathematically formalizable and (under certain conditions) even technically realizable concept. If the control logic of swarms could be described as a locally organized yet comprehensive (in that it structures the swarm as a whole) process of exchanging information, then it would be possible at the same time to describe the limits of this capacity for exchange. How much information does dynamic control require, and how fast does this information have to move? And above all: How much information – about the positions and movements of neighbors, about external environmental influences, and about the structure of the entire swarm, for instance – can a single swarming individual even possess?

Later, agent-based computer simulations would be designed on the suspicion that this capacity was highly limited and that it was precisely the ignorance of individuals about the swarm's overall structure that enables swarms to adapt effectively to changing situations and to synchronize their activity quickly and continuously. The interaction between biological swarm research and cybernetic concepts thus paved the way for the formulation of later ideas about dynamic models, even though no direct answer to MacKay's

73 Ibid., 469.

question would be provided at this or at the later Macy conferences. Or in Birch's words: "All we are doing today, probably, is opening up areas for investigation more than we are answering questions."[74] Enlightened by cybernetics and information theory, biological swarm research did, however, go on to produce some tangible results. At the end of the twentieth century, these efforts were summarized by the biologists Carl R. Schilt and Kenneth S. Norris under the biophysical concept known as 'sensory integration systems.'

Schilt and Norris followed the cybernetic paradigm in that they defined swarms as information infrastructures *in process*. In order to remain together over a long period of time within a fluid or ephemeral three-dimensional medium such as water or air, swarming individuals, according to Schilt and Norris, have to follow paths that run more or less parallel to one another. The preservation of this cohesion thus represents a synchronization problem in which movements are coordinated over time:

> The fundamental tenets of sensory integration systems are: 1. transduction of environmental stimuli external to the group via the sensory capacities of many individuals; 2. propagation of resulting social signals across the group, possibly with attenuation or amplification or other signal conditioning; 3. coordinated group response based on a summation of these social signals from various sources in various directions *at any moment*.[75]

Here, the act of synchronizing a multifaceted information transfer is explicitly cited as the basis for generating the structures of swarms. Beyond that, however, the authors also examined the possible ways that such structures enable signals to be manipulated: as an "interacting array of sensors and effectors," a swarm is not only able to receive and process more information from its environment; by combining knowledge and non-knowledge about its own spatial structure, it can also synchronize itself to this environmental information. Fundamental to this is, *first*, the interconnection of social signals, which makes up for the inability of certain members of a swarm to see such things as approaching predators: "Like a nerve cell in which more distal inputs are weighted less than those nearer the decision making 'trigger zone' of the neuron, an individual can weigh the alarm of the other members in its group by their proximity. Individual error by oversight or

74 Ibid., 468.

75 Schilt and Norris, 'Perspectives on Sensory Integration Systems,' 229 (my emphasis).

overreaction can be damped by the group"[76] – which sounds quite familiar to the lag-correcting operation mentioned by Julian Bigelow, only with a large number of simultaneous trigger mechanisms.

Second, sensory integration systems, which Schilt and Norris jokingly refer to as "anchovy ex machina," are systems that connect various senses to the processes of gathering and filtering the incoming data that flows through them into the swarming collective. In the case of fish, this involves, in addition to the sense of sight, special sensory organs for perceiving acoustic and hydrodynamic information (such as the lateral line, which can detect distant movements and changes in pressure). Unlike the case of homeostats, various sorts of information are synchronized through various points of entry in that they are integrated and converted into precise and coordinated movements:

> In sensory integration systems, individuals receive, process, and respond to stimuli from the environment. Their responses may influence (change) other near neighbors, which may in turn influence still others. The signal thereby generated may die out or may, by propagation and summation, change the greater group's behavior. Group members may also generate social signals (i.e. internally derived) that propagate through the group. We use the word 'integration' in the sense of combining or blinding into a unified response. The functional result of such a process is that the individuals in the group can respond in a coordinated manner to stimuli to which most of them have no direct access.[77]

Throughout this process, the creation of 'social signals' is of particular interest. The transmission of information from one neighboring fish to another leads to "discrete pulses of change" that propagate throughout the sensory integration system.[78] It so happens that these 'waves of behavior' spread like physical waves – an observation that had already been made during the 1950s and 1960s. Dimitri Radakov, for instance, described "waves of agitation" as "a rapidly shifting zone in which the fish react to the actions of their neighbors by changing their position [...]. The speed of propagation [...] is much higher than the maximum (spurt) speed of forward movement of individual specimens."[79] While in motion, the silver underbellies of school-

76 Ibid., 228.
77 Ibid., 229.
78 Ibid., 231.
79 Radakov, *Schooling in the Ecology of Fish*, 82.

ing fish produce a light reflection that is imitated by all the neighboring individuals, so that a so-called 'flash frontline' courses through the school. At the time, the speed with which schools can change direction exceeded the resolution of the film technology used to observe them. The medial process of analysis, that is, could not be synchronized with the object of study, and so the transmission of such waves could not be followed without gaps. A good fifty years would pass before hydroacoustic methods and their computer-graphical visualization processes made it possible to quantify waves of agitation geostatistically. In 2006, a team of French and Peruvian researchers wrote the following:

> The main process of information transfer we could observe was that of waves of agitation crossing large anchovy schools. The average speed of these waves (7.45 m/s) was much greater than the average 0.3 m/s school speeds measured during this experiment. The internal organization of each school modified dramatically after the waves of agitation had crossed them. Changes in school external morphology and internal structure were described and measured using geostatistics. Our results show that information transfer is a crucial process for the cohesion and plasticity of schools. As such, it allows efficient reactions of schools of pelagic fish to variations in their immediate environment in general.[80]

What is important in this process is that the same information can potentially be transferred without loss to all individuals simultaneously, regardless of how far they are from the wave's point of origin. Their direction and speed provide a precise indication of the effects triggered by environmental influences, and the research team was able to use variograms to demonstrate that, over the course of a wave of agitation, the internal structure of a school would change significantly and adopt a far higher level of regularity, homogeneity, and synchronicity.[81] Moreover, it was shown that sensory integration systems have to be capable of 'resetting' quickly in order to process new stimuli and that the speed of their reactions does fall as the number of processed stimuli happens to increase.

In the oscillation between the asynchronous reactions of schooling individuals to perceived environmental stimuli and their self-organized synchronization to the global maneuvers of the school, waves of agitation play a decisive role. They form the school into a more efficient collective

80 Gerlotto et al., 'Waves of Agitation,' 1405.

81 Ibid., 1415.

structure by coordinating the school's movements in time and space. Out of schools, they thus create successful synchronization projects in which the temporal synchronization processes are reflected in the modification and coordination of spatial movements and structures, thereby becoming measurable and accessible. Swarm-time and swarm-space therefore become inseparable in the context of a "rhythmic cadence of signals, [...] related to locomotory movements, that keep the mutual monitoring system engaged and operating."[82]

Here, fish schools are conceived as *information machines* in Michel Serres's sense of the term. Reduced to their functions and chains of operations, they are manageable as organizational models for coordinating multi-modal processes. What is more, the inseparable nature of swarm-time and swarm-space, which is responsible for this coordination, also belongs to a broader context in which the tendency to think about and model 'relational being' in terms of forces and impulses gives way to the concept of physical *fields*. In an article published in 1951, which Herbert Birch drew upon in his presentation at the eighth Macy Conference, Charles Breder thus proposed, as already noted, that the dynamics of schooling fish could be modeled according to the behavior of iron filings caught in a magnetic field. The waves of agitation in fish schools can thus be conceptualized as being analogous to the propagation of electromagnetic processes – as described by Maxwell's equation – from one area in space to another. For, as in the case of such physical fields, the following statement is also true of schools: "That which takes place at a particular place at a particular time is clearly determined by occurrences in the immediate spatial and temporal vicinity."[83]

The nature of swarms as self-organizing synchronization projects makes them attractive biological models not only for technical applications to do such things as coordinate the flow of traffic, solve all sorts of logistical problem, or optimize human group behavior via computer simulations (this context will be examined in greater detail in my final chapter). Up to this point, however, swarms can be regarded as synchronization projects in two respects, first in their adaptive behavior in relation to their environment and second in the local synchronization processes that enable them to react globally to external influences and thus to survive as collective structures. The matter of their synchronization-based adaptive behavior links them to the history of cybernetic concepts and models. Their functionality is visualized in the irreducible relationality of swarm-space and swarm-time.

82 Schilt and Norris, 'Perspectives on Sensory Integration Systems,' 242.

83 Franke, ed., *Dtv-Lexikon der Physik*, vol. 3, 110.

Whereas Ashby's homeostat – that sleepy machine operating in the here and now – achieved stability by combining internal randomness with the accidental events of its environment, biological swarms modify their collective arrangement by means of mobile swarming individuals. They adjust their arrangement to their space and its event in a time-sensitive manner. Here, synchronization processes are converted into structural changes, the observation of which depends on an additional synchronized relation – namely that between the swarm as an object of observation or study and the media techniques that are applied to it. The synchronization of swarms can perhaps best summarized with the following chiasmus: their formation is produced by information, while they themselves generate information through their formation by making synchronization processes visible as a structure. Their synchronization is thus not only a temporal project; it is simultaneously a spatial project as well.

3. The Third Dimension of Science

Space Lattices and Crystalized Schools

It was not only the case that the ideas of the founding figures of cybernetics were shaped by the opaque organizational structures of biological collectives. In the other direction, biological swarm research was influenced by cybernetic ideas as well. Charles Breder, at least, was certainly informed by them. In an article in which he compared the arrangement of fish schools to a geometric space lattice, for instance, he refers to W. Ross Ashby's book *An Introduction to Cybernetics*. Like other models of fish schools produced during the 1970s, Breder's was less concerned with how aggregations come about than it was with the geometric form that resulted from this phenomenon.[84] Published in 1976, Breder's study 'Fish Schools as Operational Structures' presents a three-dimensional space-grid model – a model derived from that "third dimension of science" whose significance in the history of producing knowledge has been underscored by Soraya de Chadarevian and Nick Hopwood.[85] Drawing on earlier criticism of studies that had described the distribution of schooling individuals in purely stochastic

84 Breder, 'Fish Schools as Operational Structures'; see also Pitcher, 'The Three-Dimensional Structure of Schools in the Minnow'; and Weihs, 'A Hydrodynamical Analysis of Fish Turning Maneouvers.'
85 See De Chadarevian and Hopwood, eds., *Models: The Third Dimension of Science.*

terms, Breder likewise doubted that the distribution of living organisms with advanced capabilities for mobility and complex sensory systems could ever be truly random. In natural systems, he believed that it was far more likely to encounter an underlying degree of order, which for its part could be disrupted by any number of factors. Breder maintained that there were two possible approaches to analyzing fish schools – the empirical and the theoretical – and he described the former approach as follows: "One approach to the structure of a fish school, the empirical, can be made by measuring the distance, angle, or other parameter between a given fish and the other members of the school. The mathematical measurement can establish values that may serve as an index to the school's organization. One's imagination alone limits the selection of data."[86]

Building upon Albert Parr's work from the early 1950s, Breder had already conducted a number of experiments that demonstrated the limitations of empirical methods in the case studying fish schools, and his early attempts to model them may be regarded as a personal strategy for overcoming this obstacle.[87] Instead of having to rely on his imagination when selecting relevant data, Breder thought that it would be no less legitimate to base a model on three-dimensional geometric concepts and constructs. To do so, however, meant that he had to select one such model – out of an infinite number of possibilities – that might have "some conceivable application" to schools of fish.[88] This turned out to be easier than one might suppose: "The establishment of a geometrical model of a fish school is relatively simple, for whatever else a fish school may be, it is essentially a closely packed group of very similar individuals united by their uniformity of orientation," wrote Breder, adding that their activity is also defined by a high degree of synchronicity.[89] With this approach, the organization of individual fish into a congregation became a problem of packing and making the most ideal possible use of the space in question.

Owing to the need for each fish to have sufficient room for swimming, the basic 'unit' of his model consisted of a single fish enveloped in a sphere of water; a school is therefore the packed arrangement of these identical units. Breder began by transforming the traditional cubic lattice into a rhombic model, and he did so because the latter not only better resembles the

86 Breder, 'Fish Schools as Operational Structures,' 472.

87 See Breder, 'Studies on the Structure of the Fish School.'

88 Breder, 'Fish Schools as Operational Structures,' 472.

89 Ibid. Here Breder refers to Van Olst and Hunter, 'Some Aspects of the Organization of Fish Schools.'

empirically observed structures of schools but also because it can achieve a greater density: the number of neighboring points that were now equidistant from a central point increased from four to six. As regards the use of space, this hexagonal arrangement was optimal from a two-dimensional perspective. Oriented within a hexagonal grid, that is, two-dimensional circular units represented the optimal use of space for each circle.[90]

Imagined as a system with spheres instead of circles, the model could be expanded into three-dimensional system with multiple stations. If several of these stations were then layered on top of one another in such a way that the centers of the spheres in the upper layer would align with the center of an equilateral triangle that connected the centers of the supporting first layer of spheres, then the spherical units in the second layer would fit into the empty space between the three spheres of the first, thus creating an optimal three-dimensional use of space. In this model, each unit (aside from those on the outer edge) is thus in contact with twelve neighboring units (its 'nearest neighbors').[91]

As a point of reference for his geometric model, Breder relied on empirical data from Tony Pitcher's 1973 article 'The Three-Dimensional Structure of Schools in the Minnow.' When swimming in a strong current, the minnows analyzed in Pitcher's study arranged themselves in a manner quite similar to Breder's model.[92] With an additional modification, Breder was able to make his geometric concept even more realistic: "Schooling fishes should not be expected to space themselves exactly as spheres and they do not so in precise detail [...], but a basic resemblance exists. If the rigid sphere of geometry be mentally replaced by a soft rubber ball, the approximation comes closer to that of a fish embedded in a school of its fellows."[93] Effects of force on Breder's flexible spheres made his fish-school geometry malleable in that the neighboring areas could expand or contract in space without losing their basic hexagonal form. This geometric model of space made it possible to represent the connection between individual and collective movements in a conglomeration of formable fish-school units. The freedom of schooling individuals to move within the congregation

90 For an in-depth cultural-theoretical analysis of hexagonal geometry with reference to the structure of honeycomb, see Berz, 'Die Wabe.'

91 Breder, 'Fish Schools as Operational Structures,' 474–476.

92 See Pitcher, 'Three-Dimensional Structure of Schools in the Minnow.' For his part, Breder concentrated on the potentially different dynamics exhibited by schools maintaining their position in flowing water as opposed to those swimming ahead in still water (see Breder, 'Fish Schools as Operational Structures,' 476).

93 Ibid.

was limited by the position of their neighbors; every individual movement influenced the movements of the nearest neighbors and vice versa. Any change in the form of an individual unit therefore changed the global form of the school. Because the basic geometric structure of the model was maintained and because neighboring units never lost contact with one another, the result was something like a 'three-dimensional blob' that could, at least in theory, adopt almost any conceivable form.[94] Together with the restrictions placed on the individuals' movement, the schooling 'blob geometry' described by Breder could also be used to model the behavior of fish schools making sudden and acute changes in direction. The 'forbidden zones' and maximum turning angles allowed by the model's geometry had also been demonstrated with empirical observations: "The turns made by real fish schools, measured by motion picture analysis [...] indicate the absence of intrusion into the enclosed areas. This examination of the sharp turnings of fish schools would not have shown these features if they had been organized on some pattern other than that of the hexagonal lattice."[95]

Yet even a lattice of malleable rubber balls could not account for the constant changes of position made by the individual fish within a school, and neither could it integrate any deviant activity undertaken by one fish or another (such outlying behavior had been observed in previous studies).[96] Breder's model was essentially concerned with only one aspect of a school's entire repertoire of movement, namely the case of 'orderly' swimming and changes of direction. His three-dimensional hexagonal lattice defined the units' possibilities for movement by placing them in fixed relations with one another. As a structural model for individual schooling fish, Breder's space lattice was rigidly limited by its close confinements and densely packed space. It glued individual trajectories to relatively static neighboring clusters that failed to account for the ability of schools to create flexible global structures and to form entirely different patterns of movement – that is, to adapt dynamically – in response to environmental stimuli in real time. In Breder's model, fish schools were represented as a gridded, even crystalline space. In this regard, the model discussed in my next section was entirely different, for it made it possible to represent neighboring fish as instigators of highly dynamic processes.

94 Ibid., 482–483.

95 Ibid., 485.

96 See Hunter, 'Procedure for Analysis of Schooling Behavior.'

SelFish Behavior

> A frog hunts on land by vision. He escapes enemies mainly by seeing them. His eyes do not move, as do ours, to follow prey, attend suspicious events, or search for things of interest. If his body changes its position with respect to gravity or the whole visual world is rotated about him, then he shows compensatory eye movements.[97]

Frogs – it must be said – have not been treated very kindly in scientific research. Whether they are made to twitch in electricity experiments or whether their optic nerves are cut and switched around for the sake of studying perception – to name just two examples – they have often been used by scientists as "media materialities" (in Stefan Rieger's terms).[98] What is interesting for fish-school research, however, is that frogs were mentioned as an example in the frequently cited article 'Geometry of the Selfish Herd' by William D. Hamilton, who argued that the phenomenon of animal aggregation was caused by selfish behavior.[99] Earlier approaches to this question, which sought explanations in various social instincts or vitalistic forces, were thus turned on their head. In this model, the frog (as a figure of knowledge) became relevant to fish schools (as an object of knowledge) because Hamilton, like Breder a few years later, conceptualized the genesis of dense animal collectives in geometric terms and thereby, again, treated the periphery as a central consideration. Hamilton supplemented Breder's mathematical model with a perspective that aimed to answer both *how* and *why* aggregations formed. Hamilton's approach should not be seen as competing with Breder's ideas but rather as complementing them.

As is clear from the quotation cited at the beginning of this section, Hamilton's point of departure was not how the frog *sees* but rather how it *is seen*. For the frog, as Hamilton explains, staying in one place all day long is simply not an option, and it is this need to relocate that served as the basis of his model. The beginning of his text reads like a sinister fairy tale:

> Imagine a circular lily pond. Imagine that the pond shelters a colony of frogs and a water snake. The snake preys on the frogs but only does so at a certain time of day – up to this time it sleeps on the bottom of the pond. Shortly before the snake is due to wake up all the frogs climb out

97 Lettvin et al., 'What the Frog's Eye Tells the Frog's Brain,' 233.

98 See Rieger, 'Der Frosch – ein Medium?'

99 Hamilton, 'Geometry for the Selfish Herd.'

> onto the rim of the pond. This is because the snake prefers to catch frogs in the water. If it can't find any, however, it rears its head out of the water and surveys the disconsolate line sitting on the rim – it is supposed that fear of terrestrial predators prevents the frogs from going back from the rim – the snake surveys this line and snatches *the nearest one*.[100]

As it turns out, this situation triggers a truly dynamic chain reaction. For, given the condition that Hamilton's hypothetical frogs can move freely around the edge of the pond, they will not be content with the position that they have randomly taken upon leaving the water. Rather, the hypothetical frogs are well aware of the fact that the danger of being eaten by a snake is less if they situate themselves between two other frogs. Reducing the 'domain of danger' – that is, the length of half the distance between its neighbors on either side – is the goal of every frog. This danger zone becomes smaller as a frog moves closer to its neighbors. Naturally, however, any frog's neighbors will also attempt to optimize their own positions: "[O]ne can imagine," according to Hamilton, "a confused toing-and-froing in which the desirable narrow gaps are as elusive as the croquet hoops in Alice's game in Wonderland."[101]

This model was played through with a hundred hypothetical frogs, which, at the beginning, were randomly distributed in 10° segments along the edge of a pool. In each 'round' of jumping, a frog would remain in place only if the gap that it occupied was smaller than the gaps on either side of it. Otherwise, it would jump into the smaller of these gaps and pass its neighbor's position by one-third of the width of the gap. After just a few rounds, a few large agglomerations were formed, of which only the largest continued to grow by the end of the experiment. Without any help whatsoever from physical forces of attraction, the selfish desire to avoid predators alone led to aggregations – and this in an edgeless universe in which the one-dimensional rim of the pond formed a closed circle.[102]

In the three-dimensional world of fish-schools and other swarms, such a result would seem even more likely, given that aggregations of that sort perceive threats from the outside and not from the middle (thus making it even more advantageous to bond tightly together than in the case of frogs). Hamilton thus assigned an ecologically sensible function to the formation of schools: because there are few opportunities in the open water for fish to

100 Ibid., 295 (emphasis original).
101 Ibid., 296.
102 Ibid., 297.

conceal themselves from predators, they have to hide among themselves. Moreover, he believed that the constant and dynamic changes of position undertaken by individual schooling fish and the cohesion of the entire system were motivated by the instinct of every fish to avoid any position on the edge of the school. In two-dimensional space, the aggregation of individuals and their respective domains of danger should give rise to an arrangement of polygons that resembles a Voronoi tessellation[103] – the larger the individual polygon in comparison with those adjacent to it, the greater the danger of becoming prey. A simple local rule – seek cover by moving closer to one's nearest neighbors – causes this polygon structure to become more compact.

Hamilton's model, however, contained a problem with respect to the edge: as the polygon structure became denser, the domains of danger relating to the individuals on the periphery continued to extend out to infinity, which is unrealistic. One consequence of this was that the motivation for aggregating would weaken, especially in the case of smaller collectives; even for larger aggregations, as Richard James and his colleagues have pointed out, "peripheral animals may still play an important role in the aggregation process."[104] More recent models have therefore modified Hamilton's approach by including an empirically informed 'limited domain of danger.' In this case, it is not the entire space that is partitioned into Voronoi polygons; rather, a defined number of individuals is each assigned a circular domain of danger that is limited by the maximum 'range' of a potential predator. The result is that peripheral individuals, too, are modeled as having a more realistic risk of predation.[105]

The relevance of Hamilton's work to later research lay in his model's perspective, which sought to describe, in mathematical and geometric terms, the creation of global patterns as a result of individual and local behavioral processes. He thus introduced an approach that would be pursued further in the *dispositif* of agent-based simulation models, which will be discussed in greater detail in the next section and in the next chapter. Hamilton's model of the selfish herd also attributed constitutive significance to the edge of fish schools. In his view, the periphery is not the outmost protective

103 In a Voronoi tessellation or diagram, a plane is partitioned into regions that are defined by a predetermined set of points. Each region is defined by a point at its exact center and encompasses an area of space that consists of all the points that are closer to its center than to that of all the neighboring regions. The boundary lines of a Voronoi diagram are formed by all of the points that lie at an equal distance from more than one region's center.
104 James et al., 'Geometry for Mutualistic and Selfish Herds,' 108.
105 Ibid., 109.

position of a social and altruistic collective but rather a transitional area or permeable boundary that every schooling individual selfishly endeavors to avoid, and for this reason it is also the driving force behind the dynamics of animal aggregations.

From the 1950s to the 1980s, as I hope to have shown, models were used in fish-school research to define a variety of phenomena on a formal level and on the basis of just a few governing rules. Among other things, they were employed to describe the basic factors in the creation of conglomerations, global structures, and the clearly delineated edges of schools. Scientists designed geometrically presentable and calculable spatial models based on the predetermined characteristics of the *relations* among a given school's individual elements. On the one hand, this led to types of modeling that, by relying on mathematical and physical laws, tended to represent the structures of fish schools as geometric forms. On the other hand, scientists focusing on the *interactions* between individual schooling fish opened up the field of fish-school research to new ideas from information theory. With the cybernetic vocabulary that, during the late 1940s and early 1950s, had been used specifically to understand communication and interaction in animal collectives, it was possible to model fish schools in a second way. They were now defined as 'networks of integrated sensors' (in Schilt and Norris's terms) – that is, as information-processing arrangements. In 'social media' of this sort, 'social instincts' were no longer regarded as a defining factor in the global dynamics of schools. Rather, the 'social' itself was defined as a function of various information inputs, time-lags, and lag-correcting operations. On this level, schools were integrated with 'schooling' in that their no less fundamental 'temporality' was built into the models in question. Fish schools could thus be described as *synchronization projects* in which swarm-space and swarm-time converge in an informal space-time structured by the exchange of information.

4. Ahead of Their Time: Schooling Simulations in Japan

In previous sections of this book, Yanchao Li's visualizations of acoustic analyses, Carl Schilt and Kenneth Norris's idea of sensory integration systems, and Hamilton's notion of selfish herds were each discussed as historical precursors to the dynamic and animated computer models that would later be used in biological swarm research. Two other approaches, however, can likewise be situated along this historiographical trajectory, and they happen to be paradigmatic examples of how difficult it can be to draw a sharp line between (mathematical) models and computer simulations. Developed in the context of Japanese fisheries research, both approaches tested agent-based models that could be run through various scenarios – that is, temporalized – with the support of digital computers. Yet both projects had to manage without advanced visualization software, and thus they offered little added value over what had been done before. It is probably for this reason, too, that their epistemic potential went largely unnoticed at the time by the broader community of fish-school researchers. Although the two computer models described the formation of various types of schools in terms of just a few basic physical variables (attraction, repulsion, the alignment of speed), the dynamic visualization software with which it would have been possible to 'test' these scenarios did not yet exist.

In a study published only in Japanese (and this might explain its limited influence), Sumiko Sakai used a PDP-12 computer to model the movements of individual schooling fish in relation to one another as well as the behavior of entire schools that should result if these inter-individual relations were multiplied.[106] Far ahead of her time in media-historical terms, Sakai formulated a functioning simulation model according to the principles that would later be known as 'agent-based methods.' For the inter-individual relations, she defined – as did Charles Breder in the 1950s – forces of attraction and repulsion, the combination of which would result in relative changes in direction and adjustments on the part the individual schooling fish. A change in these forces would not only lead to entirely different individual courses of movement; by modulating these forces, it was also possible to cause structural changes on the global level (schools could even be broken up and reunited).

It should be noted that in Sakai's model, like others before it, psychological or behavioral-biological influences were removed from the equation. The 'natural behavior' of fish schools manifested itself simply as a function of physical, quantified variables. Here, schooling behavior was represented as the behavior of a system that was not oriented toward 'natural' or biological factors but rather toward clearly definable and analogously applied parameters. As a whole, schools were in turn modeled as a technical system with multiple elements, each with its own defined set of characteristics. Models of this sort enabled biological swarm research to take a far more general research direction as regards its ability to describe multiplicities composed of homogeneous elements. As for what sort of 'nature' these systems actually represented, this would become a matter of secondary importance. Sakai's

106 Sakai, 'A Model for Group Structure and Its Behavior.'

model became a point of reference for later studies, especially in the context of Japanese fisheries research.

In 1976, for instance, Tadashi Inagaki and his colleagues built upon Sakai's study in an analysis of the long-term cohesion of fish schools. Two strategies presented themselves for such an analysis:

> For the purpose to investigate the mechanism, the sure and longtime recording method of tracking of fish school is needed, which will enable us to analyze the fish schooling. But such bio-telemetry system does not yet be developed. The other method to estimate the relationship between mutual force and schooling form varying the combination of forces which are considered to exist in a school.[107]

The authors developed a mathematical model with five variables: "mutual attractive or repulsive force, mean swimming force, random force, force exerted by the change of circumstances and frictional force of swimming motion."[108] Only certain combinations of these individual parameters would allow schools to remain cohesive over long periods of time.

In 1978, moreover, Ko Matuda and Nobuo Sannomiya focused on the possibility of computer simulations by refining Sakai's model to represent the potential reactions of fish schools to nearby fishing nets. In their article on the topic, which appeared in 1980 as 'Computer Simulation of Fish Behavior in Relation to Fishing Gear,' they justified their approach as follows:

> Schooling behavior to a fishing gear has been studied by making use of such techniques as underwater visual observations, underwater cameras, hydroacoustic measurements, under water television. However, all of these observation techniques are subject to restrictions caused by illumination, underwater visibility, underwater transparency and sea conditions. In addition, these techniques give only a partial information under a specific condition, and then may be difficult and laborious to describe the general behavior of fish school to fishing gear under various conditions. Under the same circumstances as our cases, a computer simulation technique has been used as an effective means in many fields of science. However, there are few studies on the computer simulation in the fisheries science. This study is directed to a development of a new

107 Inagaki et al., 'Studies on the Schooling Behavior of Fish,' 265 (*sic*).
108 Ibid.

> method of approach in the fishing techniques and tactics in addition to the traditional methods.[109]

This quotation reads like a summary of the chapter at hand, and it also points to the new epistemic direction represented by the processes of computer simulation, which will be the subject of my next and final chapter. As far as underlying parameters are concerned, Matuda and Sannomiya's model was far more complex than those of Sakai and Inagaki:

> The fundamental equations of motion are described by regarding the motion of a fish as that of a particle. The equations of motion contain the fundamental elements of fish behavior such as mass, drag coefficient and external forces acting on the fishes. As external forces, the following six forces are introduced: propulsive force, interactive force, schooling force, repulsive force against wall, directional force and random force. The computer simulation is carried out by solving a system of the nonlinear difference equations by means of TSS (Time-Sharing System). The position of each individual is plotted as a result of the computer experiment in order to check the propriety of the model. The moving patterns of fishes obtained by the simulation are quite similar to actual behavior of fish school.[110]

Here, the schooling of fish was explicitly formalized according to the movements of far more general 'particles,' which were governed by defined rules and properties. With a variety of 'computer experiments' run on the FACOM-M200 computer at the University of Kyoto, the researchers then simulated the behavior of individual fish or groups of individuals in various situations to see what would happen if they encountered obstacles such as a 'wall' (a simulated fishing net). Because of limiting processing power and the complexity of the experiments, they were carried out in two instead of three dimensions. It was shown, for instance, that several schooling individuals were noticeably faster than the others at finding the passage from one partitioned area of a virtual aquarium to another – a feature that had not been written into the program but rather arose from the interactions themselves. Yet in the case of this simulation model, too, there was no way to animate the schooling behavior with visualization processes. Instead, Matuda and Sannomiya represented the various speeds of the individual

109 Matuda and Sannomiya, 'Computer Simulation of Fish Behavior in Relation to Fishing Gear,' 689 (*sic*).
110 Ibid.

fish with longer or shorter directional arrows in the graphics accompanying their article.

Despite these preliminary studies, the work that is commonly credited as the first simulation study devoted to the schooling behavior of fish is Ichiro Aoki's 'A Simulation Study on the Schooling Mechanism in Fish,' which was published in 1982.[111] Articles written during the 1990s cite Aoki's work over and over again. Upon its publication, however, it hardly attracted any attention at all (as with the previous studies, this was likely because Aoki did not have access to dynamic visualization software). Although he acknowledged the early computer- simulations studies conducted in the context of Japanese fisheries research, his work differed from those by concentrating on the intransparent mechanisms of self-organization that govern schooling fish. That is, Aoki focused on the transitional area between the multiple, synchronous, relative behavior of individual fish and the global behavior of the school. He was interested not only in how schools form but also in how they execute and synchronize collective movements: "In other words, given the group exists, under what conditions will it react as a whole?"[112] I should note up front that, despite the promise and novelty of Aoki's approach, his article ends on a rather disillusioned note: "The simulation method is limited in ability to clarify actual complex biological systems. However, results provide a useful guide in the conduct of further biological research."[113] His skepticism concerning the methods of computer simulation is of course understandable, given that agent-based mathematical modeling had yet to take off.

Much like Matuda and Sannomiya's, Aoki's starting point was a numerical, probability-theoretical model, though now with the following specifications: Time (t) was quantized into discrete intervals, and all movements were made independent of the previous step; the 'hypothetical' organisms moved in two dimensions; movement itself consisted of speed and direction, two interdependent stochastic variables characterized by a probability distribution. The latter variables were determined at every interval by means of a Monte-Carlo simulation, which involved the repeated generation of random numbers. With this process, Aoki intended to integrate the fundamental stochastic structure of real-life phenomena into his simulation model and thus to generate a sort of artificial reality whose homomorphic image of the world had been the stuff, as Peter Galison has pointed out, of John

111 Aoki, 'A Simulation Study on the Schooling Mechanism in Fish.'
112 Parrish and Viscido, 'Traffic Rules in Fish Schools,' 57.
113 Aoki, 'A Simulation Study,' 1088.

von Neumann's and Stanislaw Ulam's dreams.[114] In short, the process was meant to breath 'life' into the system. In order to keep the simulation model simple, interactions between individual particles were restricted to the directional variable.

In the initial state of the model, individuals were distributed at random within a quadratic (and thus aquarium-like) area. In this virtual environment, predefined parameters influenced the interactive behavior of the schooling individuals, which responded according to the known schooling parameters of mutual attraction, avoidance, and parallel-orientation movements. For each form of movement, Aoki defined particular ranges of influence around each swarming individual. These variables were put into relation with one another according to weighing factors based on empirical data, and thus various interactions between individuals were defined by such things as the nature of swimming forward, the range of a real fish's visual field, and the tendency of fish to orient themselves more often according to the angle of neighboring fish ahead of them as opposed to that of the fish to their side. If a hypothetical fish had no neighbor in the 'approach' area, then it would move toward the nearest neighbor within the range which circumscribed the 'search' distance. If no neighbor could be found there, then the direction of the fish's movement would be determined at random.[115]

The structure of the simulation model as a whole could be represented with a flowchart. The model itself could be played through with various parameters in place, with each simulation containing more than two thousand steps. Everything was carried out on a FACOM-M160 computer, and the results were represented graphically with a line-printer and an *x-y* plotter. For some simulations, Aoki made use of an Apple II microcomputer attached to a television screen, but even this media apparatus could only display static 'movement patterns.' The output generated by a 'standard run' showed the movement behavior and the direction of schooling individuals at various intervals. This simulation scenario not only confirmed accepted theories about the formation of schools: "We found that group movement in unity occurred despite each individual lacking knowledge of the movement of the entire school, and in the absence of a consistent leader."[116] Aoki was also able to scale it up from eight to thirty-two virtual fish in order to demonstrate the respective relevance of the three major parameters in question. If, for instance, the parameters of attraction and avoidance were set to zero and

114 See Galison, 'Computer Simulations and the Trading Zone,' 144.

115 See Aoki, 'A Simulation Study,' 1081–1082.

116 Ibid., 1085.

only parallel-orientation movement remained in effect, the structure of the simulated school would immediately dissolve. One advantage of Aoki's simulation method was the ease with which he could add, exclude, and change the parameters governing individual movements. Applied in such a way, however, his computer simulation of schooling fish did little more than test and confirm longstanding theories about the formation of schools by using established empirical values to define their most characteristic and 'realistic' zones of attraction and avoidance (it is perhaps no surprise, then, to find references in Aoki's article to the nearly contemporaneous empirical studies conducted by Partridge).[117] The results of this computer simulation were in fact highly similar to those produced a few years before by Partridge's data tablets, which enabled observational data to be represented graphically. For the sake of reducing complexity, Aoki, like Partridge, could do no more than present his results in just two dimensions plus time.

In these early computer-simulation models, the 'nature' of fish schools as biological objects was fully transformed into the mathematical, physical, and technical parameters of collectives known as 'particle systems,' which were composed of homogeneous 'agents' with specific characteristics. Here, schools no longer appeared as a particular biological social form but rather as a relational and decentered ordering principle of multi-particle systems. The organizational knowledge of schools, which was still insufficiently understood, was thus removed from the exclusive context of biological collectives and applied to other multiplicities, for instance to the virtual agents in computer programs. As already mentioned, however, this new application would not catch on for some years to come. In the early 1980s, the knowledge produced by biological researchers about the dynamics of swarms had in large part not been transferred to other disciplines. Before this could happen, swarms would first have to undergo yet another 'media-becoming.' This, it turned out, would involve the animation of swarming principles with visualization processes that could enable swarms to appear, yet again, as new and different entities written in their own medium.

117 Ibid., 1081.

V. Transformations

Abstract

Concerned with the general concept of *transformations*, this chapter focuses on the overlappings of 'fish and chips.' Against the backdrop of an epistemology of computer simulation, it describes how a biologization of computer science coincided with a computerization of (swarm) biology. Biological studies, beginning around the year 1980, were increasingly informed by digital media. As a *retreat from nature*, they employed computer-supported data processing, agent-based computer simulation models, and sophisticated computer graphic imagery. For this epistemology, interference and noise received a *constitutive function* for setting the parameters and tuning the dynamic models themselves. Conversely, biological knowledge about swarms made its way into computational programming routines and likewise informed fields like agent-based modelling and collective robotics.

Keywords: boids, computer graphic imagery, swarm intelligence, agent-based modelling, artificial life, history and epistemology of simulation

1. Fish and Chips

A book investigating the role of fish schools at the intersection of biology and computer science can hardly refrain from using the subtitle 'Fish and Chips.'[1] Although fish-school research has done much, as we have seen, to assist the fishing industry, I am less interested in British comfort food here than I am in the operative and performative function of agent-based

1 I confess that this wordplay is not entirely new. For a similar formulation, see Simon Schaffer's essay 'Fish and Ships: Models in the Age of Reason.' According to an advertisement by its publisher, moreover, Gary William Flake's book *The Computational Beauty of Nature: Computer Explorations of Fractals, Chaos, Complex Systems, and Adaptations*, whose cover art features a circuit board and flying fish, is "affectionately known as 'The Fish and Chips Book'." Quoted from http://mathforum.org/library/view/16466.html (accessed 19 May 2018).

Vehlken, S., *Zootechnologies. A Media History of Swarm Research.*
Taylor & Francis Group, 2019
DOI 10.5117/9789462986206_CH05

computer simulations of schools. Whereas the previous two chapters dealt with the 'fishy business' of experimental and empirical studies, their optical and acoustic methods of observation, and their physico-mathematical modeling – over the course of which the 'naturalness' of fish schools was gradually subtracted and replaced by sets of functional principles – the aim of the present chapter is to elucidate the final stages of this subtraction. In addition to fulfilling this aim, however, I will also have to focus on a historiographical and epistemological recursion that marked a transformation in our knowledge of swarm collectives: against the backdrop of an epistemology of computer simulation, a biologization of computer science coincided with a computerization of (swarm) biology.

In what follows, I intend to examine the liminal area between fish and chips, within which biology and computer science overlap. The technology of computer simulation expands the realm of addressable problems by increasing the applicability of quantitative analyses. Simulations treat multiple variables simultaneously and make them manageable in time. In real time, that is, they directly address the behavior of complex *systems* without needing to be based on any specific reference to empirical data. They are thus especially applicable to sets of problems in which such data are lacking (because they defy the methods of traditional experimentation). Computer simulations can therefore be regarded as extensions of mathematical models that generate knowledge from an inverse perspective: only by running through or processing a simulation scenario with particular results does it become possible to recognize similarities in a system's behavior that are based on certain configurations of parameters whose own similarity to empirical data is then recognized after the fact – or not. This computer-supported manner of producing knowledge operates by running through multiple iterations of simulations, each with a different set of parameters. In principle, computer experiments are based on trial and error.

The basis function of this knowledge is 'seeing in time.' With their ability to temporalize phenomena, computer simulations can animate mathematical models – that is, they can fill them with 'life' in real time. Like the global systems of swarms, which seem to have a peculiar life of their own, simulations of biological processes and animations of models also raise new questions about life and the living. They do more, however, than simply expand existing epistemic strategies or improve upon numerical calculation methods with the processing power of computers. Computer simulations experiment with theories in such a way that they can be attributed a unique epistemic status of their own, one in which pragmatic operationality supersedes any firm basis in theory. In short, "performance beats theoretical accuracy."[2] Unlike the case of theories, it is not important whether computer simulations are true or false; what matters is whether they are useful.[3] Freed from any specific materialization and yet always designed with their own materiality in mind, computer simulations open up spaces of possibility; they allow scenarios to be played through and they make it possible to make recursive comparisons with data gained from observations and experiments. They also, however, make it possible to write "synthetic history."[4] Intermediary stages and intermediary spaces for epistemic things or model organisms – those common themes in the work of Rheinberger and Latour – are thus compressed together into the space-time of virtual scenarios. Or put another way, the use of computer simulations leads to a simultaneous explosion and implosion of epistemic things: an explosion because they can always be multiplied into new scenarios, and an implosion because they thereby lose their resistant character and become fluid and processable.[5] In light of computer simulations, too, Rheinberger's own question about whether it makes sense to distinguish between epistemic and technical things has to be posed anew. The division between these two types of things, which helped Rheinberger "to assess the game of innovation,"[6] blurs in computer-simulated scenarios. Epistemic and technical 'things' blend together when – as in the case of swarm research – the epistemic 'thing' (the swarm) is simultaneously the instrument of technical analysis as well: swarms have been and continue to be studied with swarm-intelligence systems that, for their part, have been inspired by biological principles of swarming. Technical developments or refinements of simulation software and hardware take place on the same level and are therefore also a manner of working on the epistemic thing.

Moreover, simulation systems, their programming, and their algorithms are now often differentially compared and validated with other simulations or with alternative sets of parameters. For they are used precisely in those cases – like that of swarms – in which empirical data are precarious

2 Küppers and Lenhard, 'The Controversial Status of Simulations,' 274.

3 Sergio Sismondo, 'Models, Simulations, and Their Objects,' 247.

4 Pias, 'Synthetic History,' 176.

5 Although Rheinberger has discussed the significance of "iterations" – that is, the goal-oriented re-application of intermediary results in open and calculable systems (of equations, texts, and experiments) – these are better understood as linear problem-solving methods. In contrast, computer-simulation methods and their scenarios can be defined as examples of *parallel processing* in which multiple re-applications of various variables are undertaken. See Hans-Jörg Rheinberger, *Iterationen*, 9–29.

6 Rheinberger, *Toward a History of Epistemic Things*, 31.

and difficult to acquire: "[S]chooling behavior remains largely an enigma, primarily because of the difficulty to obtain such data experimentally. As a result, simulations [...] continue to be based more on the presumptions of their authors than on actual data."[7] It is not exactly straightforward to compare the processes of swarm simulations with the processes of biological swarms and thus to test their representationality. Rather, simulations have to be verified more or less on their own terms:

> Since simulations are used to generate representations of systems for which data are sparse, the transformations they make use of need to be justified internally; that is, the transformations need to be considered well motivated based on their own internal form, and not solely on the basis of what they produce. Simulation requires an epistemology that will guide us in evaluating the trustworthiness of an approximation qua technique, in advance of being able to compare the results with the broad range of the phenomena we wish to study. In general, the inferential moves made in simulations are evaluated on a variety of fronts, and they can be justified based on considerations coming from theory, from empirical generalizations, from data, or from experience in modeling similar phenomena in other contexts.[8]

In addition to epistemological questions, however, computer simulations of swarms raise media-theoretical and media-historical questions as well. One essential matter is the replacement of analytical by numerical methods of calculation, which, by means of computers, make approximate solutions possible. This media-technological history of computer simulation will be examined more closely below. Above all, however, computer simulations enable the dynamic visualization of the phenomena under investigation. This new way of producing images has therefore also opened up access to a type of knowledge that operates entirely in the symbolic and yet, for this very reason, creates operative approaches to complex real-world phenomena. Accordingly, computer simulations of swarms can only be laid bare by means of a media-theoretical analysis of their dual epistemological function: On the one hand, they have been used as a technical tool in order to overcome the deficiencies of optical and acoustic analyses of swarms. On the other hand, they have broadened the modes of representation used in the (rather static) dynamic models of fish schools discussed at the end of

7 Viscido et al., 'The Effect of Population Size,' 361.

8 Winsberg, 'Simulations, Models, and Theories,' 447.

my last chapter. In short, under the conditions of computer graphics and digital images, computer simulations make use of every tool in the epistemic toolbox by combining aspects of experiments, theories, and models. They thereby create not only alternative worlds and scenarios but alternative temporalities as well.

Far more so than the types of models already examined, computer simulations are concerned above all with the *relations* within systems: relations that can only be conceived as time unfolds. It is at this point that the swarm (as an object of knowledge) and the episteme of simulation encounter one another. The relational being of swarms, in which the microscopic and the macroscopic intersect, can only be adequately represented by a technology that, blurring the distinction between epistemic and technical things, focuses on *epistemological relations*. In contrast to other media-technological processes, each of which is distorted in one way or another by the dynamics of swarms, computer simulations align the visual unclarity and organizational opacity of swarms with their own epistemological unclarity. It cannot be stressed enough, however, that the epistemological unclarity of computer simulations is precisely programmed and 'codified' in advance. Information processes, which presumably take place in the movements of swarming individuals and in those of the swarm collective but defy accurate analysis, can be made to appear in simulation scenarios as a processed and procedural form (in Fritz Heider's sense of the term).[9] That which cannot be adequately recorded *in vivo* or *in vitro* can be represented *in silico*.

The transmission of information in swarms cannot be separated from the level of the form in which it is manifested. It cannot be separated from the local and global forms and patterns of movement that take place in three dimensions plus time. The architect Stan Allen, who, together with Jeffrey Kipnis in the late 1990s, theorized about architectonic concepts beyond the dichotomy of objects and spaces (Kipnis, for his part, referred explicitly to schools of fish), summarized this idea succinctly: "Form matters, but not so much the form of things, but the form between things."[10] As regards the dynamics of swarms, of course, it would be more accurate to speak of a constant process of *formation* and *deformation*.[11] Computer simulations synthesize the dynamic formations of biological swarms by means of

9 See Heider, *Ding und Medium*.

10 Stan Allen, 'From Objects to Fields,' 24.

11 About the dynamics of fish schools, Jeffrey Kipnis writes that they are "always *in* form, but always *changing* form." Quoted from his essay '(Architecture) After Geometry – An Anthology of Mysteries,' 43 (the emphasis is original).

artificial model parameters and they make these processes comprehensible by visualizing them in computer-graphical sequences. When biological swarms are studied by means of computer simulations that are themselves based on highly similar rules, it is necessary to consider, in addition to correspondences on the levels of relationality and performativity, a historical index as well. For, as Claus Pias has stressed, "Data do not exist without data carriers. There can be no images without screens. All information is linked to material technologies and historically variable processes."[12] When the epistemic methods of computer simulation combine with the information in swarms, mathematical models come together with computer graphics to form an amalgam that foils the neat separation of image, text, and number and enables a new form of knowledge.

It is only in the digital image, "which does not exist," that a media theory of swarms can truly come into its own. Not until the rise of computer-graphical visualizations and their underlying algorithms at the end of the 1980s did a synthesis of the four-dimensional relations within swarm collectives appear on the horizon of knowledge.[13] Swarms pose a certain set of problems that can be addressed by the epistemic strategies of computer simulation, the general dissemination of which in variety of scientific disciplines could be referred to as a media culture of intransparency. In such applications, computer graphics make it possible to compare different global structures. By means of animated digital processes of visualization, alterable parameters in

12 Pias, 'Das digitale Bild gibt es nicht.'

13 The term *visualization* is used here in the sense proposed by Hans-Jörg Rheinberger, who has employed it to problematize the reference of representations in scientific practices. If someone is working with a representation technique – with an electron microscope, for instance – he or she cannot, while looking at its images, view the objects under investigation without also viewing the representation technique itself (otherwise the technique would not be needed at all). According to Rheinberger, who here draws upon the work of Bachelard, "scientific reality" only appears in representations, which can only be compared with other representations and not with the represented itself. Any reference to the object becomes objectless, and thus Rheinberger prefers the term *visualization* instead of *image* and *depiction*: "In the sciences, we typically use the term visualization to denote a process that relies on graphical images instead of on verbal descriptions or formulas." In the case of computer simulations, this separation of image production and code is of course invalid. That said, the objectless reference to the represented can be seen quite clearly in the example of swarms. When discussing image-generating processes, I will therefore speak of both *visualizations* and *presentations*. The graphical *presentation* in computer simulations gains its epistemic freedom from the fact that operates without a direct '*re*,' without recourse to any real process. Processes and scenarios in computer simulations verify themselves internally; in comparison with data 'from the real world,' moreover, they are confronted with *constructed* data, with the outputs of media-technological methods. More on this will be said below; for now, see Rheinberger, 'Objekt und Repräsentation,' 57.

the code of agent-based simulation models and the sporadic empirical data gathered from open-water or aquarium-based fish-school research can be combined with one another and tested in "virtual behavioral experiments." It is only through this process of animating swarm models that it can be determined whether the chosen combination of parameters has produced an outcome that resembles the behavior of a biological school of fish. In swarm research, the area of overlap between 'fish and chips' generates a specific form of computer simulation and is at the same time generated by a particular form of computer simulation that can be called, as it is in the relevant scholarly literature, either "agent-based modeling" (ABM) or "agent-based (computer) modeling and simulation."[14] Above all, there are three reasons that make this sort of computer simulation applicable to swarms, and each of these reasons involves, in one way or another, knowledge about biological swarms:

First, they take into account the fundamental aspect of *non-knowledge*: Agent-based models function on a bottom-up modeling paradigm that has advantages over other forms of computer simulation, such as system-dynamic or discrete-event simulations. Whereas the latter have to make top-down presuppositions about the constituents of a system and their relations to one another, agent-based models operate in a decentralized manner and without any such preconceived definitions of a system's global behavior. Rather, the behavior of the system under consideration emerges only from parameters that are implemented simply and locally (namely on the level of the individual agents or particles). For this reason, as Andrei Borshev and Alexei Filippov have noted, agent-based models are better suited for designing "*models in the absence of the knowledge about the global interdependencies*: you may know nothing or very little about how things affect each other at the aggregate level, or what is the global sequence of operations, etc., but if you have some perception of how the individual

14 Other common acronyms include ABS ("agent-based systems"), IBM ("individual-based models"), and MAS ("multi-agent systems"), the latter of which appears most frequently in work on robotics. Here I will stick as much as possible to the abbreviation ABM. With reference to swarms, James Kennedy and Russell C. Eberhart have pointed out that the use of the term *agent* can be somewhat problematic on the grounds that the individuals comprising a swarm lack such individual qualities as autonomy and specialization. From a media-historical perspective, however, the use of the term *agent* is fitting, and one of the topics discussed in this chapter will be the reciprocal effects, which can also be technologically induced, between systems with more-or-less autonomous agents. Swarming individuals can be regarded as a special case of agent system in which agents with broadly homogeneous characteristics interact and in which this interaction limits their autonomy to some degree. See Kennedy and Eberhart, *Swarm Intelligence*, xix; and Schieritz and Milling, 'Modeling the Forest or Modeling the Trees.'

participants of the process behave, you can construct the AB model and then obtain the global behavior."[15]

Second, they are distinguished by their degree of *autonomy* in the sense of Evelyn Fox Keller's definition of models. Keller has characterized cellular automata (CA) as the paradigmatic example for computer simulations that turn out to be effective as independent research instruments with their own epistemic strategy. In this regard, too, she also underscores the role and relevance of visualizations: "In actual practice, the presentation – and, I argue, the persuasiveness – of CA models of biological systems depends on translating formal similitude into visual similitude. In other words, a good part of the appeal of CA models [...] derives from the exhibition of computational results in forms that exhibit a compelling visual resemblance to the processes they are said to represent."[16]

Third and finally, agent-based models do much – and they do so in an interdisciplinary manner – to blur the lines between objects and their context. This act of blurring, moreover, becomes especially explicit in the media history of swarm research, as when a fragmentary biological knowledge of swarms inspires programmers designing animation systems with ABM methods, which in turn inspire biologists to use similar agent-based models – now computer-supported – in their research on swarms.

In the space of possibility between biology and computer science that has been opened up by computer simulation, swarms are conceptualized by coming to terms with (and ultimately dissolving) the distinction between epistemic and technical things. This happens in an exemplary manner in their oscillation between two poles, in which they are alternately understood, in Michel Serres's terms, as "object, still, sign, already; sign, still, object, already."[17] The level of visualizing data has become essential for the production of knowledge, and the reversibility, flexibility, and adaptability of the *immaterial culture* of computer simulation (at least as far as these features pertain to simulation software) evade the descriptive practices and terminologies that have been so productive for the *material culture* of laboratory research throughout the history of science. Moreover, my focus on the performativity of these graphical methods in biological fish-school simulations – and on descriptions of complex global dynamics by means of a simple and local set of instructions plus large-scale, simultaneous

15 Borshchev and Filippov, 'From System Dynamics and Discrete Event to Practical Agent-Based Modeling.'

16 Keller, *Making Sense of Life*, 272.

17 Serres, *Statues*, 109. Quoted from Rheinberger, *Toward a History of Epistemic Things*, 29.

interactions conducted on the basis of this set of rules – is also of some relevance to the media theory of animation: for, at around this same time, certain applications were developed in the field of graphic design for creating realistic-looking 'fuzzy objects' with as little programming effort as possible, and these applications derive from highly similar modeling parameters.

Of course, it would be possible to approach the convergence of fish and chips and the related developments in CGI and ABM in an entirely different way.[18] However, I would like to limit my examination here to a specific set of questions: How have swarm research and ABM mutually informed one another? What has been the role of digital visualization processes in this interaction? And what role, in general, have all of these factors played in the epistemology of computer simulation? To answer these questions, it will be necessary to provide – among other things – a detailed history of agent-based processes, which are now being applied in a wide range of scientific disciplines. Here my focus will be on computer models that have been relevant to swarm research. My discussion of cellular automata, for instance, will touch upon their many applications in the history of media and technology but will not cover all of their aspects. The same can be said of my treatment of field of artificial life (AL), the context of which gave rise to the animations by the graphic designer Craig Reynolds. It would take me too far away from the topic of fish and chips to address the fundamental ontological questions that AL raises about life and "how it could be."[19] That story, moreover, has already been told from a variety of perspectives.[20]

After this introduction, the chapter at hand contains three main sections. In the first, CGI methods that have been inspired by biological swarm research will be connected with the epistemology of ABM. Here, an excursion will be made to discuss the relevance of this connection to the representation and conception of "swarm-intelligent" multiplicities, the visual performance of which has done much to popularize the concept of the swarm in recent years. The second part then examines the connection of fish and chips from

18 See Garnier et al., 'The Biological Principles of Swarm Intelligence,' 10; and, more generally, Camazine et al., *Self-Organization in Biological Systems*.

19 See Langton, 'Life at the Edge of Chaos.' More recently, Eva Horn, in her essay 'Das Leben ein Schwarm,' has sought a connection between the question of 'life' and swarms by examining their representation in popular thrillers such as Frank Schätzing's *The Swarm*, Michael Crichton's *Prey*, and Stanislaw Lem's *The Invincible*.

20 See, for instance, Langton, *Artificial Life: An Overview*; Levy, *Artificial Life: A Report from the Frontier Where Computers Meet Biology*; Emmeche, *The Garden in the Machine*; Hayles, *How We Became Posthuman*; Forbes, *Imitation of Life: How Biology is Inspiring Computing*; Helmreich, *Silicon Second Nature*; and Metzger, 'Genesis in Silico: Zur digitalen Biosynthese.'

the other direction; that is, it looks at the use of computer-supported models in the field of biological swarm research. Finally, the third part follows the ideas of the software developer and computer scientist Frederick Brooks to discuss the specific transformation of programming paradigms ("writing, building, growing") that was brought about by software environments and tools based on 'swarm logic.'

2. Agent Games

Playing with Fire

As dubious objects with unclear edges – as 'bodies without surfaces,' in Leonardo da Vinci's words – swarms combine an aesthetic with an epistemological borderline experience. The problem of representing "fuzzy objects"[21] such as clouds, smoke, dust, or fire not only occupied the painters and architects of the Renaissance; a few centuries later, it would also cause difficulties for the developers of digital graphic and animation design. In the early 1980s – that is, when Japanese fish-school researchers were first beginning to experiment with computer simulations and when Brian Partridge and his research group were busy quantifying and analyzing film sequences of schools on an image-by-image basis – so-called *particle systems* were developed in the field of graphic design.[22] By defining the behavior of numerous virtual elements, particle systems make it possible to visualize dynamic relational objects like fire, water, dust, or clouds.[23]

And one of the very first animations of this sort was in fact created to represent a world 'where no man has gone before.' The graphic designer William T. Reeves introduced the term particle system in an article published in 1983. Here, his approach differed pointedly from the other methods being used during the early stages of computer-generated special effects.[24] Also, his essay derived from a project that had been commissioned by the famous

21 Reeves, 'Particle Systems,' 91.

22 In this context, the term *particle* designates mathematical abstractions without mass or volume.

23 See ibid.

24 Admittedly, Reeves himself made a point to mention that it was not a new idea to model objects as "collections of particles." Earlier attempts to do so, however, were far less detailed and did not take random processes into account. Regarding their scientific applications, they were used, for example, in the first computer-supported model of fluid dynamics. See ibid., 92; and Harlow and Meixner, *The Particle-and-Force Computing Method for Fluid Dynamics*.

special-effects company Lucas Arts. The objective of the project was to create an animation sequence for the film *Star Trek II: The Wrath of Khan* (1982). Known by the name 'Genesis Demo,'[25] it is the first CGI animation ever to appear in a movie. In the scene, Kirk, Scottie, and Spock watch a computer simulation of planet being 'terraformed' (the planet is gradually engulfed in a dancing, particle-system-generated flame). The sequence represents, according to Christopher Kelty and Hannah Landecker, "the science-fiction dream of a complete reorganization of matter into a new reality, mirrored inside the software by the reorganization of a formal biological theory into a graphical representation of this fantasy."[26]

Reeves defined particle systems as stochastic processes controlled by a few global parameters.[27] At the time, the design of a particle system differed from the that of other computer-graphical representations in three essential respects. *First*, its objects were not represented by a set of primitive surface elements defining their boundary; rather, they were represented as clouds of particles defining their volume. In Reeves's system, every generated particle is defined by the following seven attributes: "(1) initial position, (2) initial velocity (both speed and direction), (3) initial size, (4) initial color, (5) initial transparency, (6) shape, (7) lifetime."[28] This, too, represented a breach from traditional methods of defining spaces; it was a departure from rigid systems of coordinates that fix every point in space. As has already become clear in the examples of the blob and of diverse synchronization projects, researchers were attempting to design systems that create their own space and geometry. *Second*, particle systems are not static entities. Their particles change their form and move over time. And *third*, objects represented by a particle system are not deterministic, because their shape and form are not entirely specified. Instead, the system uses stochastic processes to create and alter the shape and appearance of its objects.[29]

All of this created significant advantages over traditional, surface-oriented approaches. For one, because a particle – conceived as a point in three-dimensional space – is a much simpler geometric form, a relatively small amount of computing capacity could deal with a far greater number of such basic formal elements and thus create more complex images. On account of the simplicity of the system, according to Reeves, it was also fairly easy to

25 Reeves, 'Particle Systems,' 97.
26 Kelty and Landecker, 'A Theory of Animation: Cells, L-Systems, and Film,' 54.
27 See Reeves, 'Particle Systems,' 107.
28 Ibid., 94.
29 See ibid., 92.

"motion-blur." Moreover, a particle system can change its level of resolution or definition by zooming in on objects to suit specific viewing parameters. Perhaps the main advantage, however, was the procedural definition of the model itself, which was controlled by random numbers and allowed detailed surface images to be produced with minimal human effort: "[O]btaining a highly detailed model does not necessarily require a great deal of human design time as is often the case with existing surface-based systems." Here, the complex nature of computer simulation was decoupled from the complexities of the programming process in that a certain degree of control was transferred from the designer to the simulation environment itself (that is, to the 'behavioral intelligence' of the program).[30] Not least, it was now possible to model a 'living' system that changes its own form over time – a process that surface-based methods could not achieve without considerable effort.[31] Reeves's own concise definition of his system is as follows: "A particle system is a collection of many minute particles that together represent a fuzzy object. Over a period of time, particles are generated into a system, move and change from within the system, and die from the system."[32] A mixture of 'biological' and 'technological' elements was thus part of the program from the beginning.

To create the flickering flame for the Genesis Demo, Reeves combined two hierarchical levels of particle systems. First, the top-level system was generated at the impact point of terraforming bomb. In turn, this particle system initiated a random concentric distribution of additional particle systems around the surface of the simulated planet, whereby the number of new particle systems was based on the circumference of its concentric ring and a predetermined density parameter.[33] At the same time, the second-level system was modeled to appear like explosions to generate the effect of dancing flames. The particles were not identical but were rather programmed to differentiate themselves at random:

> Their average color and the rate at which colors changed were inherited from the parent particle system, but varied stochastically. The initial mean velocity, generation circle radius, ejection angle, mean particle size, mean lifetime, mean particle generation rate, and mean particle transparency parameters were also based on their parent's parameters,

30 Ibid., 92.
31 See ibid., 91–92.
32 Ibid., 92.
33 Ibid., 98.

> but varied stochastically. Varying the mean velocity parameter caused the explosions to be on different heights.[34]

Several phases of the Genesis Demo involved an extremely high number of particles: The impact was generated with one large particle system and twenty smaller ones, together using around 25,000 particles in all, whereas the ring of fire spreading across the planet required two hundred particle systems and 85,000 particles in total. The scene ends with a 'wall of fire' engulfing the 'camera,' an effect that ultimately required four hundred particle systems and approximately 750,000 individual particles to be achieved.[35]

Terms such as *generation, parent,* or *lifetime* are the biological analogies with which the virtual particles are brought to 'life' – even though, in the animation itself, they were only used to represent lifeless material elements. Reeves's system was not, however, based on mimetic processes. Rather, the metaphorical 'liveliness' of his simulation arose specifically from the program's 'investment' in processes of movement that depended on randomness and were *not* defined in advance, two features that could be said to be typical of such animation and simulation processes. With their (self) control logic, moreover, these 'enlivened' particle systems could be used as a medium for simulated a variety of fuzzy objects. Or, as Reeves put it, "The most important aspect of particle systems is that they move: good dynamics are quite often the key to making objects look real."[36]

Despite all the affinity for representing dynamic processes and the potential for "temporal anti-aliasing" (that is, the reduction of undesired or jagged effects in the movements of animated objects) that Reeves saw in his particle system, the latter was still not suitable for representing swarms.[37] This is because his particles did not mutually influence one another over time and because it was not yet possible to model the effects of external random influences on a particle system that was already running its course: "[A]ll stochastic decisions concerning a particle are performed when it is generated. After that, its motion is deterministic."[38] What was needed were *interactive particle systems,* which could visualize the dynamics of biological collectives in digital computer-simulation models. A few years

34 Ibid., 98–99.
35 See ibid., 100.
36 Ibid., 107.
37 See ibid.
38 Ibid., 103.

later, this was achieved on the software level and, shortly thereafter, it was implemented on hardware that used parallel computers (such as W. Daniel Hillis's Connection Machines) to carry out the specific logic of the program.[39]

The Boid King

In the middle of the 1980s, the graphic designer Craig Reynolds developed just such an interactive particle system, but he gave it an entirely different name (one that was influenced far more by biological systems). His animation model for swarms, which he described in an article titled 'Flocks, Herds, and Schools: A Distributed Behavioral Model,' not only sounds like a text from the context of behavioral biology. In fact, it would be cited as a foundational work in almost every later article on computer-supported biological swarm research. Admittedly, Reynolds was not interested in designing realistic behavioral variables based on actual biological swarms – and, as we have seen, the available data about such variables would have been too meager in any case. His aim was simply to make the performance of his "bird-oid objects" ("boids," for short) *appear* to be true to life.[40] In his case, too, his first challenge was to achieve temporal anti-aliasing. Any unrealistic movements made by his virtual swarms would be unacceptable because, like Reeves, Reynolds was SFX designer for the movies. His model ended up being used to simulate colonies of bats in the film *Batman Returns* (1992) and to create an extended stampede sequence in *The Lion King* (1994). Yet, as Reynolds states in his introduction, "The aggregate motion of a flock of birds, a herd of land animals, or a school of fish is a beautiful and familiar part of the natural world. But this type of complex motion is rarely seen in computer animation."[41] Thus his modeling strategy was to follow, as closely

39 See Hillis, *The Connection Machine*; and Sims, 'Particle Animation and Rendering Using Data Parallel Computation.'

40 See Parrish and Viscido, 'Traffic Rules of Fish Schools,' 66; and Reynolds, 'Flocks, Herds, and Schools.' According to Reynolds, boids could be used just as well to simulate fish or other moving collectives. He acknowledged that his work was not the first in the field of CGI to simulate swarms. Researchers at Ohio State University, for instance, had already developed an animation technique to generate a film scene with simulated flocking birds. Their approach, however, differed from Reynold's model (see ibid., 26). A short time later, moreover, the ornithologists Frank Heppner and Ulf Grenader developed (apparently without knowledge of Reynold's work) a similar simulation model, but it was more like a particle system than an interactive particle system. Also, the authors confessed: "In essence, the model worked, but it was not altogether clear why." See Heppner and Grenader, 'A Stochastic Nonlinear Model for Coordinated Bird Flocks.'

41 Reynolds, 'Flocks, Herds, and Schools,' 25.

as he could, the known basic rules that had been established in biological swarm research. In his article, Reynolds not only refers to a number of studies concerned with the behavioral biology of birds; he also cites the fish-school research conducted by Shaw and Partridge in the 1970s, Wayne Potts's chorus-line hypothesis, and even Edmund Selous's theories about thought transference in birds. As with Reeves, one of Reynolds's first steps was to yield control to the layout of his system – to the relationality that his programming allowed:

> The simulated flock is an elaboration of a particle system, with the simulated birds being the particles. The aggregate motion of the simulated flock is created by a distributed behavioral model like that at work in a natural flock; the birds choose their own course. [...] The aggregate motion of the simulated flock is the result of the dense interaction of the relatively simple behaviors of the individual simulated birds.[42]

Thus, not only are analogical relations between biology and computer graphics used to generate the desired fuzzy objects. In principle, Reynolds's model functions (or claims to function) *exactly* like natural swarms – not necessary on the detailed level involving the individual distances, acceleration abilities, and interactive potential of particular species of fish or birds but certainly on the level of the processes that enable the global patterns of movement of entire swarms to arise from a multitude of individual movements.

At the basis of Reynolds's agent-based simulation was the old question about leadership in swarms. How can such complex dynamics take place without any strict central control? The answer that Reynolds settled on for his model: Dynamics of this sort are *only* possible without such central control! And this answer arose from the pragmatic need to increase the efficiency of CGI programming. To create a realistic animation of a swarm, according to Reynolds, it would be a Sisyphean and error-prone task to program, separately, the path of every single boid within a large number of particles. In doing so, it would be nearly impossible to guarantee that, in every frame, the animated birds would maintain their flock motion and avoid colliding. Moreover, such programming would be inflexible, because a change in a single flight path would also affect those of every other swarming individual: "It is not impossible to script flock motion, but a better approach is needed for efficient, robust, and believable animation of flocks and related

42 Ibid., 25.

group motions."[43] By following the assumption that global swarm dynamics ultimately derive from the interaction between simply defined individuals, Reynolds was able to clean up his simulation model considerably. He took as his example a flock of birds: all that would be needed for every boid, he thought, would be a control structure governing the behavioral features that enable participation in a flock, and this could be enhanced with a simulation of "portions of the bird's perceptual mechanisms and aspects of the physics of aerodynamic flight." The model would then be complete, for everything else would be created during the run-time of the simulation: "If this simulated bird model has the correct flock-member behavior, all that should be required to create a simulated flock is to create some instances of the simulated bird model and allow them to interact."[44] The point is that, ultimately, the specification of "correct flock-member behavior" could also be retroactively modified on the basis of the interactions and behavior of the flock as a whole – whereby "correct" means only that the performance of the simulation model was persuasive (as Reynolds mentions, it is impossible to measure the success and validity of simulations objectively). In the end, all that mattered was that many viewers found his simulated and computer-graphically visualized flocks "delightful to watch."[45]

The behavior of these simplified and universal (or better: *principle-based*) swarming individuals was of course far less complex than that of their inspirations from real life, given that Reynolds only modeled a small portion of the behavioral repertoire of biological swarms. But this, he wrote, was a difference "of degree, not of kind."[46] The model did not directly simulate, for instance, the senses used by real swarm individuals (sight, hearing, and the lateral line in the case of fish): "Rather, the perception model tries to make available to the behavior model approximately the same information that is available to a real animal as the end result of its perceptual and cognitive processes."[47] Here one sees how the cybernetic concept of information had made such a transfer from animals to digital machines conceivable. Thus it was possible to define a biological system and a computer model on the basis of their systemic behavior over time and by means of multiple relational exchange processes between neighboring 'units.' Regarding the dynamic relationality of swarms, it was no longer a matter of figuring out *what* was

43 See ibid., 25.
44 Ibid.
45 Ibid., 26.
46 Ibid.
47 Ibid., 29.

taking place in any substantial sense; rather, the focus had shifted entirely to processes of exchange and to manners of organization and control in order to answer the question of *how* the system behaves as a whole. Or, as the computer scientist Christopher Langton, has written:

> It is important to distinguish the ontological status of the various levels of behavior in such systems. At the level of the individual behaviors we have a clear difference in kind: Boids are *not* birds, they are not even remotely like birds, they have no cohesive physical structure, but rather exist as information structures – processes – within a computer. But – and this is the critical "But" – at the level of behaviors, *flocking Boids and flocking birds are two instances of the same phenomenon:* flocking.[48]

Boids behave in a more complex manner than Reeves's particles. Whereas, in particle systems, the individual elements do not interact with one another, the functionality of Reynolds's model *depends* on the interaction of the boids. Reynolds defined every point of his system as a subsystem of its own, each with a local system of coordinates and a geometric frame of reference. He thereby created a geometric orientation for each boid. He defined the behavioral repertoire and the status of a boid as *instances* of an *object* in the sense of object-oriented programming (OOP). For every instance – that is, for every boid – its respective internal conditions were combined with established behavioral rules by means of a standardized programming process known simply as a *method*. The result, in Reynolds's words, is an "actor, [...] essentially a virtual computer that communicates with other virtual computers by *passing messages*." As he noted, it had already been proposed that such an "actor model" might be well-suited to simulate biological behavior and, inversely, flocks and schools had been cited as examples of robust self-organizing systems in the scholarly literature on parallel and distributed computer systems.[49] Reynolds himself used the programming language Lisp, though he produced his code and animations with a sequential Symbolics 3600 workstation and not with a parallel computer.[50]

In contrast to the stochastic diffusion process of particle systems, the units in Reynolds's model were able to arrange themselves in relation to one another on the basis of a simple algorithm with three defined "traffic rules":

48 Langton, 'Artificial Life,' 32 (emphasis original).

49 Reynolds, 'Flocks, Herds, and Schools,' 26 (emphasis original). Reynolds refers here to Kleinrock, 'Distributed Systems.'

50 See Reynolds, 'Flocks, Herds, and Schools,' 32.

"(1) Collision Avoidance: avoid collisions with nearby flockmates; (2) Velocity Matching: attempt to match velocity with nearby flockmates; (3) Flock Centering: attempt to stay close to nearby flockmates."[51] Once the process was underway, the animator became something like a "meta-animator." Rather than dealing directly with the motions of his animation, he instead designed the behavioral parameters that created this activity during the run-time of the program. To some extent, then, the outcome was out of his hands: "One of the charming aspects of the work reported here is not knowing how a simulation is going to proceed from the specified behaviors and initial conditions. [...] On the other hand, this charm starts to wear thin as deadlines approach and unexpected annoyances pop up. This author has spent a lot of time recently trying to get uncooperative flocks to move as intended [...]."[52] In testing various values for these parameters, it became clear that a true-to-life simulation of swarm activity could only be achieved when each of the boids oriented itself toward the center of the flock:

> Before the current implementation of localized flock centering behavior was implemented, the flocks used a central force model. This leads to unusual effects such as causing all members of a widely scattered flock to simultaneously converge toward the flock's centroid. An interesting result of the experiments reported in this paper is that the aggregate motion that we intuitively recognize as "flocking" (or schooling or herding) *depends* upon a limited, localized view of the world.[53]

Swarms are thus immersed in the "heart of a principle of invisibility" that Michel Foucault had identified as the basis for economic collective thinking.[54] Reynolds's model makes it clear that the complex, dynamic movements and control mechanisms of a multi-agent system are created by the highly limited knowledge and the highly reduced behavioral repertoire of its agents. Whereas too much knowledge – too much information about the condition of the swarm – is counterproductive, it is in fact extensive non-knowledge that happens to be productive in dynamic collectives.

Reynolds thus believed that a fundamental limitation of this sort must also exist in the case of biological swarm individuals – something that, in the jargon of computer science would be called a 'constant time algorithm.' The latter

51 Ibid., 28.

52 Ibid., 27.

53 Ibid., 29–30 (emphasis original).

54 Foucault, *The Birth of Biopolitics*, 279.

limitation would decouple the amount of 'thinking' or 'processing' that a bird, for instance, has to do from the number of birds in its flock: "Otherwise we would expect to see a sharp upper bound on the size of natural flocks when the individual birds became overloaded by the complexity of their navigation task. This has not been observed in nature."[55] Here, too, a pragmatic approach to computational complexity led to a sort of feedback loop with biological swarms.

In addition, Reynolds also had to determine how the size of a neighborhood should influence the movements of a boid – and how strongly this should be weighted. This decision, too, was informed by biological research. Citing Brian Partridge's quantitative studies, Reynolds programmed his model in such a way that a boid would be influenced more strongly by its nearest neighbors than by any distant members of the simulated flock (following Partridge, he made the level of influence inversely proportional to the square of the distance). With these specifications in place, the oriented boids forged their own paths by constantly comparing their own activity to that of their neighbors. The result was collective movements that came close to those of biological swarms – virtual swarms that, without the addition of further modeling parameters, could independently avoid obstacles or suddenly change their direction.[56] That said, Reynolds's model differed from biological swarms in one crucial point. It did not have a constant time algorithm; every boid knew the status of the *entire* flock, even though they ignored the majority of this information and only took into account the activity of their nearest neighbors. This led to a sort of computational bottleneck, because the complexity of the model increased as the size of the virtual flock grew: "Doubling the number of boids quadruples the amount of time taken." This was a problem that Reynolds hoped to solve with distributive processing: "If we used a separate processor for each boid, then even the naive implementation of the flocking algorithm would be $O(N)$, or linear with respect to the population."[57] Unlike the case of biological swarms, however, the complexity would always increase with the addition of new individuals, and so in future models Reynolds intended to incorporate a constant time algorithm that was insensitive to the total population.

Because of their simplicity and flexibility, computer-graphical boid collectives were soon put to use in special effects. In this form, swarms returned to the big screen not merely to represent visual threats or deformations but rather as part of the organizational principle of the animation itself.

55 Reynolds, 'Flocks, Herds, and Schools,' 28.

56 See ibid., 29–31.

57 Ibid., 32.

As simulations, they marked a culmination point in dealing with swarms as vague phenomena, in that this very vagueness became the model itself – the condition of possibility. In order to simulate realistic swarms on the computer, 'experiments' were conducted with distributed-behavior parameters that, *after the fact*, appeared to resemble the basic rules of biological behavior. It is no surprise, then, that Reynolds mentions biological research as a potential application of his model:

> One serious application would be to aid in the scientific investigation of flocks, herds, and schools. These scientists must work almost exclusively in the observational mode; experiments with natural flocks and schools are difficult to perform and are likely to disturb the behaviors under study. It might be possible, using a more carefully crafted model of the realistic behavior of a certain species of bird, to perform controlled and repeatable experiments with 'simulated natural flocks.' A theory of flock organization can be unambiguously tested by implementing a distributed behavioral model and simply comparing the aggregate motion of the simulated flock with the natural one.[58]

And in fact, if Steven Levy's account can be believed, Reynolds soon received a phone call from an interested biologist wanting to learn more about the model's control algorithm.[59] Shortly thereafter, the distributed behavioral model of his boid system was reimported into the very sort of biological swarm research that had inspired Reynolds's design in the first place. Structurally, Reynolds's model hardly differed from that of its Japanese precursors a decade before. Yet the availability of more powerful computers and graphics hardware now made it possible to produce *dynamic* models that took into account the dimension of time. Swarms could now be modeled and observed in virtual laboratories as four-dimensional collectives, and hardly any later work in the field of biological swarm research fails to cite his article (more on this later).

Even though Reynolds's model may in part be "biologically improbable,"[60] and even though he was indebted to the "natural sciences of behavior, evolution, and zoology [...] for doing the hard work, the Real Science, on which this computer graphics approximation is based,"[61] his dynamic computer-

58 Ibid.

59 See Levy, *Artificial Life*, 80.

60 Parrish and Viscido, 'Traffic Rules of Fish Schools,' 66.

61 Reynolds, 'Flocks, Herds, and Schools,' 33.

graphical visualizations of swarm simulations pointed to an epistemic strategy beyond that of the aforementioned 'technological morass' in which biological swarm research was then mired. They presented a real-time, procedural approach to knowledge in which the central perspective and geometric code that traditionally failed to represent diffuse bodies without surfaces was replaced by a code that enabled swarming individuals to locate and organize themselves independently. Central perspective was abandoned for a topological system that created its own space and could be implemented in computer experiments which, in a process of recursion, could repeatedly redefine and reshape the specifications and modulations of the simulation program itself. For, in order to study the behavior of the boid system over time, it was necessary to work with its graphical and digital presentation.[62]

Artifishial Life

> Imagine a virtual marine world inhabited by a variety of realistic fishes. In the presence of underwater currents, the fishes employ their muscles and fins to gracefully swim around obstacles and among moving aquatic plants and other fishes. They autonomously explore their dynamic world in search of food. Large, hungry predatory fishes hunt for smaller prey fishes. Prey fishes swim around happily until they see a predator, at which point they take evasive action. When a predator appears in the distance, species of prey form schools to improve their chances of escape. When a predator approaches the school, the fishes scatter in terror. A chase ensues in which the predator selects victims and consumes them until satiated.[63]

What reads here like a description of the best of all virtual aquarium worlds – if one has the good fortune, at least, to live as a predator in this environment – is in fact the introduction to a third landmark project in graphic design that happens to be of interest to the media history of swarm research. Much like Reynolds's boid-based simulation model, this project was also concerned with the efficient use of self-organized behavior and self-learning agents: "The key to achieving this level of complexity with minimal intervention by the animator," according to the authors, "is to create fully functional artificial animals."[64] Again, the aim was to create, without explicitly preprogramming it, as much control knowledge as possible

62 See Keller, 'Models, Simulations, and "Computer Experiments".'
63 Tu and Terzopoulos, 'Artificial Fishes: Physics, Locomotion, Perception, Behavior,' 43.
64 Ibid.

from the bottom up with a set of behavioral parameters. Entitled 'Artificial Fishes,' the study in question by Xiaoyuan Tu and Demetri Terzopoulos rode the momentum that artificial life was building around 1990 as the 'new thing' in the field of computational intelligence. Their artificial fish-agents approximated the realistic behavioral repertoire of fish far more closely than Reynolds's boids did that of birds.[65]

Based at the University of Toronto, the researchers modeled every individual fish in their simulated aquarium as a lifelike autonomous agent that took both biomechanical and hydromechanical factors into account.[66] Each was given a mutable body that moved by means of simulated internal muscles; moreover, each fish was also equipped with 'eyes' (that is, 'virtual on-board sensors') and a 'brain.' The latter consisted of various areas for controlling sequences of motion, perception, the behavioral repertoire, and learning. It was thus possible to simulate controlled muscle and fin movements, and the fish were able to move around in simulated water while taking into account its hydrodynamic 'realities.' What is more, they were able to optimize these movements autodidactically over the course of their 'lifetime' by evaluating the efficiency of their combined muscle movements in relation to a fitness function, which monitored such things as their speed of forward motion.

With recourse to the information processed by the modeled sensory organs in the 'brain,' a series of 'fishy' behavioral traits were created, including "collision avoidance, foraging, preying, schooling, and mating."[67] The outside of the fish agents was generated by means of digital photographs of real fish, which were run through a NURBS model (NURBS stands for "Non-Uniform Rational B-Splines").[68] This, together with a "motor system" model consisting of twenty-three lumped masses interconnected with ninety-one viscoelastic elements, made it possible to simulate realistic bodily movements. The latter influenced the modeled hydrodynamics of the simulated environment, which in turn influenced the behavior of the simulated fish – an interplay that yielded a particular set of motion:

> As the body flexes, it displaces virtual fluid, producing thrust-inducing reaction forces that propel the fish forward. The mechanics are governed

65 See ibid.; and Terzopoulos et al., 'Artificial Fishes.'

66 See Terzopoulos, 'Artificial Life for Computer Graphics'; and Maes, ed., *Designing Autonomous Agents*.

67 Terzopoulos, 'Artificial Life for Computer Graphics,' 41.

68 NURBS are models for generating smooth curves or surfaces on the basis of defined vertex points. See, for instance, Rogers, *An Introduction to NURBS*.

> by systems of Lagrangian equations of motion (69 equations per fish) driven by hydrodynamic forces. [...] The model achieves a good compromise between realism and computational efficiency, while permitting the design of motor controllers using data gleaned from the literature on fish biomechanics.[69]

The internal 'character' of a fish was based on a set of 'habit parameters,' which determined such things as its sex and its preference for darkness. A so-called 'intention generator' would then combine these habits with the incoming stream of sensory data to generate the 'goals' or behavioral dispositions of the fish (this generated disposition would be saved in short-term memory to ensure the coherence of the model). This combination gave rise to dynamic behavior that was never explicitly programmed, such as hunting and feeding on prey. The intention generator was also responsible for filtering the incoming information from the environment, and thus it focused an agent's behavior: "At every simulation time step, [it] activates behavior routines that input the filtered sensory information and compute the appropriate motor control parameters to carry the fish one step closer to fulfilling the current intention."[70] With this process, a parameter such as 'avoid collision' could be associated with larger or smaller regions of sensitivity. The former would lead to "timid" behavior (because the fish in question would take evasive action far in advance), while the latter would generate "courageous" or "curious" behavior (the fish would not attempt to dodge a collision until the last moment).[71] To keep matters simple, Tu and Terzopoulos programmed predators in such a way that they would not prey on other predators, and they also limited their behavioral parameters: "*escape*, *school*, and *mate* intentions are disabled."[72] Schooling, too, was generated procedurally by combining various types of behavior (such as the tendency to seek proximity to other fish and the propensity for a fish to adjust its speed and direction to its nearest neighbors').

Though from an entirely different direction, the behavioral science of artificial fish led to an approach that was similar to that taken by certain practitioners of ethology in the 1930s and beyond. Psychological factors and attributes related to animals were determined on the basis of potential movements, the integration of environmental factors via sensory organs, and

69 Terzopoulos, 'Artificial Life for Computer Graphics,' 41.

70 Tu and Terzopoulos, 'Artificial Fishes,' 44.

71 See ibid., 47.

72 Ibid. (emphasis original).

a few basic needs (such as eating, procreating, and avoiding danger), which an animator could set on a defined scale for each agent. Vaguely defined motivations or instincts were thus transformed into quantifications of physical parameters that could be represented in the computer-simulation model. In this case, as in a great deal of ethological swarm research, the 'behavioral science' of computer simulations became a science of movements. Now, however, such movements could be reenacted and modified by means of computer-graphically animated underwater sequences.

As with all the simulation systems and visualizations discussed so far, Tu and Terzopoulos's model of 'artificial fishes' had to deal with the technical limitations of its time, which in their case involved certain hardware specifications that are only of anecdotal interest today. At any rate, the research team was at first able to conduct a simulation with ten fish, fifteen food particles, and five static objects at a rate of four frames per second. The model was implemented on a Silicon Graphics R4400 workstation, and a greater number of fish or a more elaborate environment would have required considerably more computing power. With their system, they were able to produce short animated films with playful titles such as *Go Fish!* (1993) and *The Undersea World of Jack Cousto* (1994). Unlike Reynolds, however, who thought that his model could be used in biological research, Tu and Terzopoulos positioned themselves in the field of artificial life, their hope being that computer models would soon be able simulate such things as the spawning and fertilization behaviors of fish. That said, they also mentioned that their computer-supported environment might be applicable as a testing ground for developing systems of cooperating robots. After all, virtual robots (which they ultimately considered their artificial fish to be) would be much simpler, faster, and cheaper to design and manipulate than physical prototypes.[73]

Cellular Automata

Of course, artificial fish and lifelike autonomous agents did not simply pop out of the sea or drop from the sky at random; rather, they were part of the development of certain modeling and programming paradigms in which they were gradually able to establish their autonomy. So much is clear from both the development of distributed animation approaches (particle systems) for the simulation of flocking birds and the computer-generated behavior of artificial fish. The latter belong to an older tradition

73 See Terzopoulos et al., 'Artificial Fishes,' 350.

of computer-technical simulations that involves modeling activity on the basis of local, neighborhood-based organization. So-called cellular automata (CA), which were popularized in the 1940s and 50s by the work of John von Neumann and Stanislaw Ulam, have two crucial characteristics: First, they serve as a blueprint for developmental environments that make it possible to explore, by means of computer experiments, the complex macrostructures that can emerge from simple and local rules for interaction.[74] Second, they undermine anthropocentrism (in typical cybernetic fashion) and in certain respects even blur the distinction between computers and biological organisms, in that the foundation of their 'bio-logic' is not defined by mechanical components or chemical bonds but rather by *information*. On account of their basis in computer programming, moreover, cellular automata also possess certain characteristics that make them useful (and interesting) for modeling dynamic systems: compared to systems of differential equations, for instance, they have the advantage that computer simulations do not produce any rounding errors. In dynamic systems in particular, errors of this sort can quickly get out of hand. At the same time, it is easy to incorporate stochastic elements into their rules for interaction in order to model disruptive influences. Moreover, cellular automata are characterized by their dynamics in time and space. Mathematically, they are defined by the following factors:

> 1. Cellular space, i.e., the size of the configuration, the number of dimensions (a line, a flat surface, a cube, etc.), and its geometry (rectangular, hexagonal, etc.); 2. Boundary conditions, i.e., the behavior of those cells with an insufficient number of neighbors; 3. Neighborhood, i.e., a cell's radius of influence (e.g., the von Neumann neighborhood of four surrounding cells or the Moore neighborhood of eight surrounding cells); 4. The number of a cell's possible states [...]; and 5. The rules that govern the evolution of states.[75]

74 Von Neumann, *Theory of Self-Reproducing Automata*, 20. Von Neumann also described these characteristics in his article 'Probabilistic Logics and the Synthesis of Reliable Organisms from Unreliable Components.' For further discussion of cellular automata, see Ulam, 'On Some Mathematical Problems Connected with Patterns of Growth of Figures'; Wolfram, *A New Kind of Science*; Forbes, *Imitation of Life*; Schiff, *Cellular Automata: A Discrete View of the World*; Charbonneau, *Natural Complexity: A Modeling Handbook*; and Minati and Pessa, *Collective Beings*. Regarding the application of cellular automata to various academic fields, see Haefner, *Modeling Biological Systems*; Klüver, *The Dynamics and Evolution of Social Systems*; Batty, *Cities and Complexity*; and Adamatzky and Martínez, eds., *Designing Beauty: The Art of Cellular Automata*.

75 Pias, *Computer Game Worlds*, 271.

The particular epistemic potential of cellular automata comes to light when their dynamics can be represented with computer graphics. The Norwegian researchers Rune Vabø and Leif Nøttestad accomplished exactly this in the 1990s, and were the first to apply cellular automata to fish-school research. "The precise biological values are of minor importance in this model," they wrote, for their goal was to formulate the most general possible statements about the self-organizing schooling behavior of fish and their reactions to approaching predators. Although they borrowed some of their model's parameters from the patchy available data about schools of herring, they had more fundamental factors in mind when designing their animation.[76] In the episteme of computer simulation, limiting matters to just a few carefully chosen parameters is often the most revealing approach, for any increase in their number will exponentially increase the number of reciprocal effects and thus obscure the informative value of the simulation model.[77] Furthermore:

> The conceptual CA model introduced in this paper is based on a philosophy of allowing individual fish to perform separate actions on the basis of simple behavioural strategies. [...] The model includes stochastic elements which assume that individual herring do not have perfect information about their surroundings. [...] The cooperative dynamics of the school should occur as a result of all the local actions taken by each individual. In our model, fish are represented as objects moving between cells or fixed points in a two-dimensional grid equivalent to the open sea, *which provide a useful visual representation to study school behavior.* [...] By visualizing the lattice with the positions of individual herring over sequential time steps, the dynamics of this CA could resemble some of the dynamical schooling structures seen in nature. It may thus be possible to see realistic schooling behavior from the changing structure of the CA configuration.[78]

Here, fish schools were not only situated within a two-dimensional lattice but, analogous to the visualizations of accumulated target strengths in the voxel graphics of multi-beam sonar systems, it was also possible to

76 Vabø and Nøttestad, 'An Individual Based Model of Fish School Reactions,' 155.

77 That said, if the relevance of the variables in question is falsely assessed, this can lead to unreliable and misleading reductions. See Pias, *Details zählen*, 17: "Every reduction contains the risk that the simulation will no longer be 'correct', and every addition increases the chance that the simulation will now concern a qualitatively new sphere of phenomena."

78 Vabø and Nøttestad, 'An Individual Based Model of Fish School Reactions,' 155–156 (my emphasis).

populate each cell with up to nine schooling fish.[79] This cell density was color-coded, and for every time step a contour map was produced that indicated the position changes of all the individuals. From a random initial configuration, various scenarios developed through the application and adjustment of parameters such as the grid size; the number of predators (and their 'perception length'); and the number, perception length, density, and so-called 'panic distance' of the simulated herring. An increase in the 'stochastic factor,' for instance, would lead to congregations that were more amorphous and less dense, while the presence of fewer disruptions would cause larger collectives to split into many smaller and denser schools. Changes to the perception length of the fish and to their initial distribution would also lead to clearly different results and, in simulations of predator attacks, schools exhibited evasive maneuvers that resembled the 'split, join, and vacuole' formations observed in open-water research.

Vabø and Nøttestad were nevertheless sure to mention the limitations of their simulation. It is not only that biological fish can swim far more freely than virtual fish confined to two-dimensional cells; the interactions between simulated fish and their neighbors were also unambiguously (and therefore unrealistically) established according to the parameters of the computer model. On top of this, there was also the issue of the standardized rhythm of the cellular automata, which was synchronized in advance with centrally controlled updates. Questions concerning synchronization processes, such as those discussed at the Macy Conferences, were thus beyond the scope of their study. Regarding the autonomy of the individual fish, the model was therefore not very sophisticated. Vabø and Nøttestad also admitted to facing certain temptations: "When constructing a model of a particular biological or ecological system, one is tempted to relax the strict definition of CA so as to match as well as possible the design of the system under study. The trade-off is always between the realism and the tractability of the model."[80] In other words, it is not with realistic but rather with unrealistic models that important knowledge can be gained about the behavior of distributed dynamic systems. There is, then, a productive epistemological divide between natural and simulated systems, and that of the latter can be applied to "various fields of biological science, including studies on a wide spectrum of schooling fish, flocking mechanisms and herd behaviour in mammals."[81]

79 Ibid., 156.
80 Ibid., 169.
81 Ibid.

The use of cellular automata as environments for computer simulations pertains to the multiplication and fluidification – the simultaneous explosion and implosion – of epistemic things that I discussed at the beginning of this chapter. Their materiality and structurality have a colonialist character; instead of encircling one identifiable epistemic thing, models based on cellular automata bridge together various disciplines, various objects of study, and they combine biological and computer-technical approaches. They do so not by leveling out differences in structure and content; rather, by making these differences differentially negotiable, by making epistemic divides operational, and by making dynamic visualizations processable, cellular automata create epistemic aggregations. The latter are not only defined in exact mathematical terms but are also represented in dynamic sequences of images. In the 'bathroom-tile' structure of cellular automata, we have a case of grids that, unlike the tiled floors of the aquaria used by Radakov and others, generate actual dynamics. These grids no longer simply make analytic observations and measurements possible; rather, they synthesize them into collective patterns of motion. In doing so, moreover, they are not merely graphical *representations* that make it possible to observe artificial swarms; at the same time, they are also unambiguously described programming processes that are implemented on a logical structure and even correspond – at least when used in conjunction with parallel computers like Hillis's connection machines – to a material structure which itself corresponds to that of the graphical surfaces. Appearance and reality therefore become one.

Object Orientation

Freedom, to paraphrase Rosa Luxemburg, is always the freedom of those who calculate differently. The development of autonomous agents – their 'emancipation' in the form of particular computer-programming software and simulation models – took place within a discourse that brought together biological and computer-scientific approaches. Cellular automata were only one technical form in this historical trajectory and, in light of the autonomy of later agent-based models, they were a relatively stiff corset for the simulation of distributed dynamic systems. More broadly, the development of agent-based models can be understood as a process that was embedded in a transformation of the guiding principles of programming – from structured programming to evolutionary designs, as seen in object-oriented programming languages.[82] Both agent-based modelling and

82 See Pflüger, 'Writing, Building, Growing,' 300–301.

object-oriented programming can thus be assigned to the same paradigm, one that Frederick Brooks subsumed under the concept of "growing" (in its dual sense of 'increase' and 'cultivate').[83] To a certain extent, control and 'intelligence' were delegated to a self-regulating system.

The basic idea of an object-oriented programming style is to regard every sort of programming development as a modeling problem. "From this perspective," as Wolfgang Kreuzer has noted, "programs describe the simulation of a nexus of ideas and they model relevant excerpts of reality with groups of interacting objects."[84] In their very constitution, then, the programming environments in which agent-based computer-simulation models are built originate in the same spirit as the resulting applications: before simulations were ever run on, in, or with these program tools, the tools themselves originated in a mode of distributive experimentation and testing. Simply put, they are themselves always already products of a simulating and synthesizing arrangement and thus already contain the very dynamics that they will later make available as an application or finished product for modeling other contexts and excerpts of reality. Object-oriented programming (OOP) follows a twofold bottom-up approach: On the one hand, it propagates an incremental procedure in which complex systems are developed by way of aggregating and networking the fragments of models. On the other hand, they function as a framework for "open model development," in which the configuration of an anticipated product (or of a system to be simulated) is no longer significant to the development process; rather, the limited knowledge of the software developer is taken into account, for it is with this that the functionality of complex systems is incrementally approximated.[85]

In the case of OOP, data and operations are integrated into an 'object,' and this object is then 'encapsulated' so that the data contained within it cannot be inadvertently altered by the 'methods' of other objects. Objects can thus be regarded as abstract *agents* whose data structures are determined by defining *attributes* and whose behavior is determined by *methods*. The latter are divided into 'classes,' that is, into abstract generic concepts for an object. Through the *inheritance* of attributes and methods, new classes can be derived from one that already exists. By means of specifying attributes and methods, various *instances* of an object can be produced from one

83 Brooks, 'No Silver Bullet,' 10–19.

84 Kreutzer, 'Grundkonzepte und Werkzeugsysteme objektorientierter Systementwicklung,' 213.

85 Pflüger, 'Writing, Building, Growing,' 297.

object class. With this principle of polymorphism, OOP is something like a warehouse of spare parts, with which it is possible to model complex objects and their systemic relations out of a multitude of simple basic components. This happens, for instance, when messages are exchanged between the objects (which can be understood as operating subprograms). In this case, the encapsulation of the objects defines which messages can be received and answered, the corresponding *method* is carried out, and the object then 'behaves' in a particular way. With this principle, it is possible and simple to execute both changes and subsequent modifications.

The early stages of structural software design were characterized at first by a struggle against limited computing recourses (Frederick Brooks and Jörg Pflüger refer to this phase as "writing") and then by the challenge of *building* a status description in the program that is preceded by the analytical specification of the problem. Here, the level of data and that of operations are strictly separated (and this corresponds to thinking about things in terms of cause and effect). Or, in the words of Herbert Simon, "[t]he general paradigm is: Given a blueprint, to find the corresponding recipe."[86] According to Pflüger, this perspective must first experience reality in order to appropriate it by redoubling it into a status description and a process description:

> With limited means, it attempts to reproduce what it has endured. The construction of its structural thought is an act of reconstruction, and this constitutes its understanding as an essentially analytical activity. It is indebted to the Enlightenment and operates teleologically by regarding nature as purposeful and objectives as prescribed. Its task has been accomplished when [...] process description and status description come into alignment.[87]

In the case of intransparent dynamic systems – when the goal, that is, is to reconstruct a world that defies legibility – it is often impossible to determine such status and process descriptions in advance because no (or insufficient) knowledge happens to exist about the system in question. In many areas of knowledge, according to Pflüger, there has thus been "an observable epistemic shift away from the scientific rigor of concepts toward the purposeful understanding of forms."[88] In the *dispositif* of self-organization

86 Simon, *The Sciences of the Artificial*, 211.

87 Pflüger, 'Writing, Building, Growing,' 314.

88 Ibid., 313.

and its open systems, there does not need to be a preexisting blueprint or overall plan to create structures and patterns; all that is needed are "relations within a field or 'excitable medium'"[89] – a medium such as a modular OOP or an agent-based simulation environment. With this bottom-up approach, the dual modes of description – process description here, status description there – can no longer be separated: "It is an iterative reality construction that, instead of analyzing, reflexively modernizes."[90] Over the course of the 1990s, program libraries and toolkits such as SWARM (from the Santa Fe Institute), RePAST ("Recursive Porous Agent Simulation Toolkit," developed at the University of Chicago), Ascape (which refers to Epstein's and Axtell's earlier "Sugarscape" model, and originated from the Brookings Institution), or MASON ("Multi-Agent Simulation On Neighborhoods," developed at George Mason University) made it possible to acquire the "warehouses of spare parts" mentioned above, and simulation environments such as Starlogo, Netlogo, and AgentSheets were designed explicitly for agent-based modeling.

This interminable "progress of compositional synthesis" apparently managed to cope without any principles or laws: "It dispenses with the regulative idea of expediency of the prediscovered [...] and instead trusts the rules of learning processes, according to which useful orders are established [...] by following the principle 'just squish things around until you like the total effect.' [...] Reality has to be conceived as a consequence of transitions."[91] This sounds like a rather sarcastic critique of the naive belief in self-organizational principles, a belief that resonates quite blatantly with a market-liberal ideology. It is no coincidence that the reformation of process structures was first adopted and disseminated in the field of management (swarm intelligence would be an example of this as well). That which might be sensible as part of an epistemic shift in computer-programming processes and its focus on real-world phenomena can have outright catastrophic effects in economic and socio-political spheres. Unlike computer environments, the 'real' environment often responds more sensitively to trial-and-error methods; moreover, as far as testing the effects of social processes is concerned, rapid prototyping is probably not the wisest approach.

As an ideological function, according to Pflüger, computer technology thus maintains the "background metaphor of the 'writability of the world' in the world."[92] One wonders, however, whether there might in fact be two

89 Ibid., 314.
90 Ibid., 315.
91 Ibid.
92 Ibid.

steps here that should be kept apart: first, a medial level of design-oriented, computer-supported writing processes, upon which "designers explore [...] the problem through a series of attempts to create solutions. There is no meaningful division to be found between analysis and synthesis [...] but rather a simultaneous learning about the nature of the problem and the range of possible solutions."[93] This indeed is an epistemically transformed and 'more open' approach to 'the world' made possible by computer technology and its structurally reformed methods and possibilities to create effective simulations. Yet, on a second level, it remains to be seen whether scenario-based methods of this sort might ultimately just serve to establish the 'real world' more efficiently and 'purposefully' without any shift whatsoever in the structural effects of power. On the media-technological level, self-organization by no means entails that society can automatically be reconfigured to have fewer hierarchical structures, despite the fact that self-organization (along with the concept of swarm intelligence) has become a ubiquitous buzzword in today's discourse.

Beginning in the early 1990s, that which appeared in the realm of computer technology as a paradigmatic 'design-oriented' framework that could realign the internal formation of computer programming with its relation to the world resulted in an epistemic strategy highly similar to that of agent-based computer-simulation processes, which were intended to shed light on the dynamic behavior of mobile collectives. And this perhaps underscores why the media history of such approaches based on graphic design is so relevant to understanding the broader epistemology of computer simulation. Here, the limits of what can be calculated or predicted represent a shift toward 'natural' or biological principles, even though the objects of OOP act with less autonomy than those in agent-based models. Agent-based models, moreover, are less interested in the self-organization of the 'program maker' and the organization of the programming process than they are in the specific orientation of such logic toward objects of knowledge. On account of their programming paradigms and their general design, however, object-oriented programming languages are especially suitable for programming agent-based simulation models, with which the dynamic activity of such things as animal collectives can in turn be simulated. It was within an object-oriented programming paradigm, of all things, that the non-object of swarms was addressed and made addressable – a paradigm that is part of the same poststructural episteme as agent-based processes.

93 Lawson How Designers Think, 44.

The KISS Principle

Robert Axelrod knows his stuff. As a longstanding consultant for the U.S. Department of Defense and the RAND Corporation, he understood that, in the jargon of the American army, something called the 'KISS principle' would have a less awkward ring to it than one might suppose. For, simply enough, it stands for "Keep it simple, stupid."[94] Of all things, it was this military slogan that, for Axelrod, initiated the freedoms of autonomous agents and agent-based models:

> The KISS Principle is vital [...]. When a surprising result occurs, it is very helpful to be confident that we can understand everything that went into the model. Although the topic being investigated may be complicated, the assumptions underlying the agent-based model should be simple. The complexity of agent-based models should be in the simulation's results, not in the assumptions of the model.[95]

Like the reformation of the programming process in OOP, the use of agent-based computer simulations shifted the modes of understanding and describing dynamic systems by generating macroscopic effects from the interaction of simple local rules in a population of autonomous heterogeneous agents within a relevant environment. Joshua M. Epstein and Robert L. Axtell have summarized this transformation as follows: "[ABM] may change the way we think about explanations [...]. What constitutes an explanation of an observed [...] phenomenon? Perhaps one day people will interpret the question, 'Can you explain it?' as asking 'Can you grow it?'"[96] A famous example of this approach are Thomas C. Schelling's studies of the racial dynamics in large American cities.[97] With a pen-and-paper model of an agent population consisting of two different groups, he explained that mild preferences for neighbors of the same race could soon lead to complete segregation on the macro level. The formation of ghettos did not automatically have to be a sign of racism but could simply be caused by a large number of low-threshold individual motives. *Keep it simple, stupid*: a simple model with a few different preference parameters could be carried out in a flexible manner and lead to counterintuitive results. In Axelrod's

94 Axelrod, *The Complexity of Cooperation*, 5.

95 Ibid., 5.

96 Epstein and Axtell, *Growing Artificial Societies*, 20.

97 See Schelling, 'Models of Segregation'; and idem, 'Dynamic Models of Segregation.'

opinion, ABM could thus be described as a "third way of doing science."[98] Like deduction, it starts with a set of explicit assumptions. ABM provides an orderly framework for formally implementing various hypotheses in a dynamic simulation model (which is usually based on an underlying mathematical model that is static).[99] Unlike deduction, however, ABM does not prove theorems: "Instead, an agent-based model generates simulated data that can be analyzed inductively. Unlike typical induction, however, the simulated data come from a set of rules rather than direct measurement of the real world. Whereas the purpose of induction is to find patterns in data and that of deduction is to find consequences of assumptions, the purpose of agent-based modeling is to aid intuition."[100]

As the physicist Eric Bonabeau has explained, agent-based systems therefore possess a number of advantageous features, and during the 1990s (and thus with the increased availability of powerful computers) they came to be used in more and more areas of application:

> Individual behavior is nonlinear and can be characterized by thresholds, if-then-rules, or nonlinear coupling. Describing discontinuity in individual behavior is difficult with differential equations. [...] Agent interactions are heterogeneous and can generate network effects. Aggregate flow equations usually assume global homogeneous mixing, but the topology of the interaction network can lead to significant deviations from predicted aggregate behavior. Averages will not work. Aggregate differential equations tend to smooth out fluctuations, not ABM, which is important because under certain conditions, fluctuations can be amplified: the system is linearly stable but unstable to larger perturbations.[101]

To this it can be added that it is more natural or accurate to model the behavior of entities on the basis of local behavioral rules than it is to do so with equations which prescribe the dynamics of density distributions on the global level. Moreover, the flexibility of ABM consists in the fact that additional agents can be added at any time, their parameters and the relationships between the agents can be adjusted, and it is possible to tinker with aggregate agents, subgroups of agents, and individual agents within a

98 Axelrod, *The Complexity of Cooperation*, 3–4.

99 See Foster, 'A Two-Way Street to Science's Future'; and Grüne-Yanoff, 'Artificial Worlds and Agent-Based Simulation,' 619.

100 Axelrod, *The Complexity of Cooperation*, 3–4.

101 Bonabeau, 'Agent-Based Modeling,' 7281.

single model.[102] Agent-based models can also be easily combined with other techniques of computer simulation to describe such things as traffic jams, stock markets, and other types of organization.[103]

Yet what, exactly, is an agent according to ABM?[104] In their tutorial on this technique, Charles M. Macal and Michael J. North felt the need to summarize its most important features, because most of the authors who had written about ABM tended to focus on various aspects at the expense of others. Bonabeau, for instance, treated all sorts of independent components as agents regardless of whether they are capable of complex adaptive activity or merely primitive reactions. John L. Casti, on the contrary, claimed that the term should only apply to components that behave adaptively, learn from experiences in their environment, and adjust their behavior to it – components, in other words, governed by rules that enable them to change their rules.[105] For his part, Nicholas R. Jennings focused on the role of autonomy, that is, on the active ability of agents to make independent decisions instead of passively reacting to a given system.[106]

Such terminological distinctions aside, Macal and North list a number of features that play an important role from the pragmatic perspective of the model's designer or "simulator":

> An agent is identifiable, a discrete individual with a set of characteristics and rules governing its behaviors and decision-making capability. Agents are self-contained. The discreteness requirement implies that an agent has a boundary and one can easily determine whether something is part of an agent, is not part of an agent, or is a shared characteristic. An agent is situated, living in an environment with which it interacts along with other agents. Agents have protocols for interaction with other agents, such as for communication, and the capability to respond to the environment. Agents have the ability to recognize and distinguish the traits of other agents. An agent may be goal-directed, having goals to achieve (not necessarily objectives to maximize) with respect to its behavior. This allows an agent to compare the outcome of its behavior

102 See ibid.

103 See Helbing, *Social Self-Organization Agent-Based Simulations*, 29.

104 Here I have limited my definitions to those used in the field of agent-based computer simulation. Of course, there have been other concepts of the 'agent' in the history of science and theory, such as that applied in Bruno Latour's 'actor-network theory,' but these will have to be left out of the discussion.

105 See Casti, *Would-Be-Worlds*.

106 See Jennings, 'On Agent-Based Software Engineering.'

> relative to its goals. An agent is autonomous and self-directed. An agent can function independently in its environment and in its dealings with other agents, at least over a limited range of situations that are of interest. An agent is flexible, having the ability to learn and adapt its behaviors based on experience. This requires some form of memory. An agent may have rules that modify its rules of behavior.[107]

In light of these basic features, it is no surprise that the authors explicitly cite the adaptive behavior of swarms and other biological animal collectives (such as ants) as sources of inspiration for the development of an agent-based "mindset" within a *dispositif* of self-organization.[108] For, unlike particle systems, the individual agents here are heterogeneous and dynamic with respect to their attributes and behavioral rules. The implementation of ABM, moreover, overlaps considerably with the programming paradigm of OOP – above all to the extent that the process-based perspective of other simulation models is abandoned in both cases. As a foundation, the object-orientation paradigm is useful because it allows an agent to be defined as a self-orienting object with the ability to make autonomous decisions that are dependent on the situation that the agent happens to be in: "The O-O paradigm is natural for agent modeling, with its use of object classes as agent templates and object methods to represent agent behaviors. O-O modeling takes a data-driven rather than process-driven perspective."[109] There are thus five steps to designing an agent-based model: *First*, the types of agents and other objects within the simulation model have to be defined together with their respective attributes. *Second*, it is necessary to model the environment and how it might affect the agents and their possible interactions. *Third*, so-called "agent methods" have to be specified; these determine how an agent's attributes will update in response to its interactions with other agents and with the environment. *Fourth*, the methods have to be added to determine when and how agents will interact over the course of the simulation. *Fifth* and finally, the agent model has to be implemented in computational software:

> Developing an agent-based simulation is part of the more general model software development process. The development timeline typically has several highly interleaved stages. The *concept development and articulation*

107 Macal and North, 'Tutorial on Agent-Based Modeling and Simulation,' 74.
108 Ibid., 75.
109 Ibid., 78.

> *stage* defines the project goals. The *requirements definition stage* makes the goals specific. The *design stage* defines the model structure and function. The *implementation stage* builds the model using the design. The *operationalization stage* puts the model into use. In practice, successful ABM projects typically iterate over these stages several times with more detailed models resulting from each iteration.[110]

Through the implementation of ABM, the modeling paradigm therefore shifted toward self-organizing, object-oriented simulations. This yielded a way of understanding that is procedural: by repeatedly running through the dynamic model and observing the varying attributes and methods of the agents between each iteration, the system behavior of the simulation is modulated and adjusted. What is interesting about this process is that both theoretical considerations as well as 'experimental' (or empirical) data can be drawn upon when designing an agent-based model, and both sides can complement one another at various stages of the model's development:

> One may begin with a normative model in which agents attempt to optimize and use this model as a starting point for developing a simpler and more heuristic model of behavior. One may also begin with a behavioral model if applicable behavioral theory is available [...], based on empirical studies. Alternatively, a number of formal logic frameworks have been developed in order to reason about agents [...].[111]

Remember: "Performance beats theoretical accuracy." For, in the *dispositif* of self-organization, it is only by running through the visualized performance of an agent-based system's global behavior that theories can be developed about nonlinear processes and complex activity. Only during a simulation's run-time is it possible to test the relevance of empirical data and theoretical ideas to the behavior of the system, and only then is it possible to formulate a sound theory of self-organization. For this reason, however, the manner in which the 'performances' of agent-based models are accessed has also become a matter of debate – at issue, that is, are the dynamic visualizations that make it possible to deal in an epistemic and 'intuitive' manner with the behavior of complex systems in real time. In computer simulation's orders of knowledge, theory and experimentation come together in an unfamiliar

110 Ibid., 79 (emphasis original).

111 Ibid.

and unprecedented dynamic. Here, the line between epistemic and technical things is blurred, for instance, when self-organization or swarm behavior is in turn studied with the help of self-organizing or 'swarm-intelligent' computer simulations.

Simulation and Similarity

The mathematician (and swarm researcher) David J. T. Sumpter has pointed out that the *performance* of dynamic systems always derives from its underlying parameters:

> If we are to build a useful theory of self-organization of animal groups it is not enough to say that certain things 'look' similar. [...] [T]he aspect that links different systems together is similarity in the mathematical models we use to describe their behavior. [...] Describing a system as self-organized tells us little about how it actually works, while providing a slight sense of mysticism. From a practical point of view it is better to say that the behaviour of a system arises from a particular combination of, for example, positive feedback, response thresholds and negative feedback. Such description allows for more detailed between system comparisons, not only between different types of collective animal behaviour but across all complex systems.[112]

Thus, it is possible to use more than just visually supported programming languages – such as the Unified Modeling Language (UML) – in order to access, in an 'intuitive' manner, the often-multifaceted interactions and attributes in agent-based models. From the output side of things, too, it is of little help to receive the results of simulations in the form of discrete values, mere diagrams, or statistics. Therefore, that which has to be easily accessible and adjustable on the programming side depends on interfaces that resemble dynamic flowcharts upon which potential interactions and attributes can be processed graphically. It must be possible to understand the output side intuitively as well: in the end, it is the graphic presentation of interrelated model parameters that provides information about such things as automobile traffic, the movement of panicked crowds, or the formations of fish schools and flocks of birds. And it is only by comparing various simulated scenarios on diverse levels of observation (by zooming in on something within a visualization model, for instance) that it is possible

112 Sumpter, 'The Principles of Collective Animal Behaviour,' 11 and 19.

to evaluate and identify the combinations of parameters that best represent real-life processes.'[113]

The relevance of visualization methods to ABM-based knowledge processes lends the latter an explicitly media-aesthetic tinge. When imaging methods make it possible to evaluate the utility of simulation runs and thus to select 'interesting cases' (and this includes outliers or extreme cases as well), or when they make it possible for researchers to deal with such models 'intuitively,' then they not only pose questions about the validation of data gained in simulations. Beyond that, they also necessitate an examination of the media-technological transformations that take place between input data, simulation models, visualization tools, and output data: How have non-scientific fields such as graphic design and animation influenced the development of scientific visualizations? For their part, how are graphic methods technically linked to hardware developments? And not least, such 'dynamic data-imaging sequences' certainly unsettle the concept of 'representation.' They form a realm of scientific image sequences that has to be mapped out on its own. To cite one example, Hans-Jörg Rheinberger developed his concept of visualization in the context of distinguishing between epistemic and technical things – a concept that has also been applied to computer simulations (by Gabriele Gramelsberger, for instance).[114] 'Scientific reality' (in Bachelard's terms) appears in representations that can only be compared to *other* representations, and not with the represented itself. The relation to the object of reference becomes objectless, and therefore Rheinberger preferred to speak of 'visualizations' instead of 'images' and 'depictions.' Visualizations, he thought, are always based on a separation of graphic-imaging means and their descriptions and formulas. In the case of computer simulations, this separation of image-production and code is of course irrelevant. The graphic *presentation* in computer simulations gains its epistemic freedom precisely from the fact that it gets by without any direct '*re*-,' that is, without any recourse to a real process. As mentioned above, processes and scenarios in computer simulations verify themselves internally and, though compared with data 'from the real world,' they are confronted with *constructed* data, that is, with the inputs of processes that are always already media-technological. In the foreground here is thus an approach that attempts to describe images of data on the basis of the technical methods of their *production*.

113 On the problem of perspective in visualizations of data, see Schubbach, '"... A Display (Not a Representation) ...".'

114 Rheinberger, 'Objekt und Repräsentation,' 55–61; and Gramelsberger, *Computerexperimente*.

In the scientific production process, users of interactive computer-simulation environments are consequently liberated from the 'technological morass' of (often impossible) manual data processing and can deal with the development of dynamic scenarios in a more productive manner because visualizations provide greater insight into the inner connections that exist between data.[115] At the same time, computer-technical visualizations and animations of model parameters also entail a differential way of understanding, for which no more is needed than a (mathematically simplified) *similarity* in the system's global behavior. Often enough, the primary goal of agent-based modeling is to preserve a degree of openness and simplicity in a model's basic structure and parameters. Only in such a way will there be enough leeway for modulating the model, and only in such a way is it possible to model a system's relevant factors with sufficient clarity. What is crucial is the ability to identify the reciprocal effects between various factors that have led to particular types of behavior on the global level. The true advantage of computer simulations is that they make it possible to visualize processes. The latter become ascertainable in real time and thus allow assertions to be made about the solution behavior of the underling mathematical model under specific conditions. By means of visualizations, moreover, this modeling of natural-scientifically interesting system processes can be compared with observed systems or with other visualized processes that cannot be observed *in situ*.[116]

What is produced over the course of this are dynamic data images that maintain complex interrelations with the programmed specifications of the system. Every visualization of an agent is tied to its specification in the simulation model, and every formation of global patterns and dynamic orders (which can only be comprehended visually and in time) is an effect of these inter-individual relations.[117] The visualization of data, as the philosopher Arno Schubbach has noted, therefore has to be understood as a complex and irreducible coupling between data, their algorithmic representation, and visible computer images. This sort of visualization is not to be confused with a "direct look at the data," for the latter have already been structured in a particular way by their implementation in a simulation model. This determines what can be visualized in the first place and thus entails the "inevitable danger that a structure has been created that the display only seems to make obvious."[118]

115 See Schubbach, '"... A Display (Not a Representation) ...",' 17.

116 See Gramelsberger, *Semiotik und Simulation*, 96.

117 See Adelmann et al., *Datenbilder*.

118 Schubbach, '"... A Display (Not a Representation) ...",' 16.

Moreover, the referentiality of these processes to real objects or processes is not transparent. Visualizations of a swarm simulation are based on combinations of parameters that only have an indexical relation to reality to the extent that they can be compared to (experimental and patchy) data sets that were themselves produced with the help of media-technological methods of visual or acoustic observation and the analytic instruments associated with them. In this context, data are always something constructed rather than given. In dynamic data-image sequences of this sort, distributions, reactions, and movements become legible, and the formation of specific patterns can be seen as epistemic aggregations in a single glance. On the one hand, their perception is no longer encrypted in columns of numbers whose development might be difficult if not impossible to ascertain. On the other hand, however, the visualizations are themselves the basis for modulating and adjusting the stock of data: "This means that data are visualized in order to be visually controlled and interpreted, whereupon the data are in turn altered on account of these visual findings."[119] The interference between 'real space' and the space of perception is eliminated. As Evelyn Fox Keller has noted, the transference of simulated processes into the processes that they are meant to describe in real life only succeeds to the extent that formal similarities are translated into visual similarities: "[A] good part of the appeal derives from the exhibition of computational results in forms that exhibit a compelling visual resemblance to the processes they are said to represent."[120] Without visualizations there can be no iterations of simulations, and without these iterations (and their differential mode of understanding) there is no way to evaluate simulation models. Only under these conditions is it possible to deal with the disruptive nature, intransparency, and 'non-objectness' of swarms. The distributed 'intelligence' of swarms can therefore be described as an *animated* intelligence in two respects: as a motion-based intelligence established by interrelations within the swarm and as a form of 'intelligence' to which the field of animation design has made a crucial contribution.

Massive Attack

Agent-based modeling, however, has become increasingly relevant in areas beyond the scientific study of complex systems and collectives. In the field of graphic design, too, it has been influential in a number of particle and

119 Gramelsberger, *Semiotik und Simulation*, 88–89.
120 Keller, *Making Sense of Life*, 272.

swarm models. Under the enabling condition of immensely more powerful computers, such models have taken the behavior of 'lifelike autonomous agents' far beyond that of the 'boids' or 'artifishial life' from the early 1990s. Because it has since become possible with computer graphics to represent the 'natural' behavior of swarms in a convincing manner, sequences of this sort have been used more and more in special effects. They thus contributed their share to the prominence of swarms and their (presumed or factual) intelligence in the popular discourse around the year 2000. The (re)presentation of swarm intelligence in agent-based simulation and animation models on television and in films is undoubtedly related to the epidemic spread of the swarm concept as a metaphor throughout so many areas of society.

Unlike the scientific use of agent-based modeling, however, this visualization context is focused above all on the *believability* of the models, that is, on making the simulated phenomena seem as true to life as possible. Realistic assumptions about the precise contexts of their formation, on the contrary, typically play a subordinate role. In the case of scientific agent-based models, of course, the opposite is true. They are often less concerned with believability than they are in implementing highly abstract graphic visualizations and in limiting matters to just a small set of characteristic features. In short, "scientific simulations are more focused on getting the processes rather than the visual representation right."[121] That said, although such dichotomies are made to seem rather straightforward in scholarly literature, they are in fact not so clearly delineated from a media-historical perspective.

As far as swarms are concerned, the field of animation design has revived the traditional dichotomy between fascination and uncanniness on a new media-technological level. Whereas it may be true that the uncanniness of swarms is depicted and their flexible, adaptive, and rapid form of organization are celebrated on the *plot level* of science fiction, action, or fantasy films, agent-based systems are implemented on the *production level* as an entirely pragmatic and operative modeling tool. Simply on account of the computing power at its disposal, animation design can assist in clarifying the intransparent modes of organization seen in swarms and thus further integrate science fact and science fiction.

Yet in the field of special effects, too, things often turn out in unexpected ways. In life, of course, it is best to be ready for anything, and this applies just as well to the software-based 'life' of artificial agents. The programmer Stephen Regelous experienced this first-hand in the late 1990s. As

121 Helbing, *Social Self-Organization Agent-Based Simulations*, 28.

the chief software developer at the company WETA Digital, Regelous was commissioned at the time by the director Peter Jackson to develop CGI software for the *Lord of the Rings* trilogy. One of his main challenges was to produce lifelike visualizations of thronging masses coming together in Middle Earth's epic battles ("mass ornaments," one could say).[122] The result was a CGI program called MASSIVE (an acronym for "Multiple Agent Simulation System in Virtual Environment") that could generate realistic crowd sequences. Like Reynolds's boid system, this program made it possible to circumvent a fundamental problem, namely that it would have been far too time-consuming to program such crowd scenes by hand in such a way as to keep the multiple individual movements of all the elements under control. What is more, the results were far more adequate than those of previous programs, which were based on just a handful of 'cloned' agent models with predefined features and courses of action. The latter programs could admittedly simulate mass movements, but the behavior of their agents was too restricted to animate the bustle of mythic battlefields in a realistic manner.

In the case of MASSIVE, however, the animators created agents composed of a defined set of parameters, individual sequences of motion, various possibilities for action, and a simulated 'brain' for making independent decisions. According to Regelous, his aim was "to take the processes of nature and apply them to generate computer imagery."[123] To this end, he experimented with Lindenmayer systems,[124] he imported natural sequences of human and animal movement on the basis of motion-capture footage, and he evaluated the evasive maneuvers of people walking down busy streets. The final result was agents in which anywhere between 150 and 350 different, second-long sequences of motion could be implemented. When they came into contact with one another, these agents could engage in complex actions such as "attacking" or "searching for combatants." Such actions, in turn, were limited by the respective bodily form of the agents, their clothing, certain simulated laws of physics, and even by the weather conditions.

Depending on the complexity of an agent's role, its 'brain' was made of anywhere between one hundred and eight thousand so-called "behavioral nodes." These defined the spectrum of their sensory perceptions, controlled their movements, described their potential for aggression in 'combat mode,'

122 See Kracauer, 'The Mass Ornament.'

123 Macavinta, 'Digital Actors in *Rings* Can Think.'

124 For a discussion of Lindenmayer systems, see Kelty and Landecker, 'A Theory of Animation,' 46–58.

and determined possible courses of action in response to encounters with other agents or the simulated environment. The latter reactions were not negotiated by means of dualistic, yes-or-no decisions but rather by a form of fuzzy logic that resulted in gradual and relational conditions such as 'a little dangerous,' 'far away,' or 'very loud.' With the help of a special editing tool, which was represented in the software as a sort of elaborate flowchart, the animators could combine specific behavioral nodes and thereby create individual sets of characteristics to simulate various behavioral patterns and levels of aggression, fear, or energy.[125]

Defined in this way, the agents were implemented in a virtual scenario and the animation began. With recourse to their sets of parameters, they were able to find their ways independently through the artificial environment, avoid collisions with other agents, or start fights with one another. What is noteworthy is that, form this point on, the animators no longer had any influence over the agents' behavior. Modifications could only be made after the fact, and so in this case, too, the programming method was based to some degree on trial and error: "When an animator places agents into a simulation," as the journalist Courtney Macavinta wrote in *Wired Magazine*, "they're released to do what they will. It's not crowd control but anarchy. That's because each agent makes decisions from its point of view."[126]

It remains to be seen whether autonomous agents should be discussed in the same context as street protesters who form so-called 'black blocs' and have thus been treated as examples of 'human swarm intelligence' and smart mobs. In any case, it was the 'anarchic behavior' of this artificial life – anarchic because it was not entirely predetermined by the program but rather unfolded as the agents would interact – that astounded Regelous and his team. He had the following to say about some early test runs of the program and their unexpected results: "It's possible to rig fights, but it hasn't been done. In the first test fight we had 1000 silver guys and 1000 golden guys. We set off the simulation, and in the distance you could see several guys running for the hills."[127] Not only did these artificial life forms contain surprises; they also exhibited something like a will to survive that compelled some of them to desert the battle rather than dying, as their creator intended, a heroic death.

Over the course of these developments, it is interesting that there loomed in the background a profound (and rather esoteric) philosophy of "life as it

125 See Mecklenburg, 'Digitale Ork-Massen.'

126 Macavinta, 'Digital Actors in *The Ring*,' n.p.

127 Ibid.

could be," a line of thinking that was formulated as part of the artificial-life movement at the end of the 1980s and which went on to find both advocates and critics throughout the 1990s.[128] Surfaces were polished and, to some extent, the primacy of fictionality was accepted. The artificial life of multi-agent systems was used to produce realistic simulations of natural or biological processes in research and in film sequences, and biological principles of organization were applied to solve technical problems or to model social dynamics. Through the use of computer simulations in a broad variety of scientific disciplines, fictitious scenarios and the dynamic processes of computer models became the foundation for generating knowledge; at the same time, however, the status of artificial life was also changing within the fields of computer science, animation, and graphic design. For instance, artificial-life principles such as swarm dynamics were no longer simply discussed philosophically but at long last became an element of software packages for nearly every promising graphical development environment. Moreover, the discussions that had been instigated about the equal rights of artificial life and about what really constitutes 'life' at all seemed to have run their course.[129] Given that today's computer games are populated by highly complex agents interacting in the most diverse possible manner within highly complex artificial environments, the real world and its ontology are no longer a matter of debate. Rather, there is now a pragmatic and pleasant exchange between computer-generated and real worlds, and the pleasure involved with this exchange derives from new possibilities of narration that are 'liberated' and perhaps even anarchic.[130]

3. Written in Their Own Medium

Fish and chips come together in an epistemology of computer simulation or, to be more precise, in an epistemology of agent-based simulation approaches and their processes of visualization and presentation. Whereas the previous sections focused on the question of how biological knowledge has inspired and informed the development of computer hardware, software, and graphical applications, I will now turn to examine the other side of

128 See, for instance, Langton, *Artificial Life*; and Shanken, 'Life as We Know It and/or Life as It Could Be.'

129 If these discussions can be said to have lived on at all, it was in the discourse concerning computer viruses. See Mayer and Weingart, eds., *Virus!*

130 See, for instance, Moorstedt, 'Düstere Entscheidungen.'

the equation: What were the reciprocal effects that this biologization of computer technology exerted on the computerization of biology? Following up on my discussion in the previous chapter of the early fish-school simulations developed in Japan, I will now focus primarily on simulation models that have opened up new research perspectives since the early 1990s. Beginning around this time, too, and in the wake of animation models from the field of graphic design, biological fish-school research came to be influenced more and more by computer simulation models that used dynamic visualizations to generate new knowledge about the dynamic self-organization of schools.

The epistemological perspective behind these developments followed a recursive model of knowledge. This was not just a case of combining and recombining old and new knowledge bases (a process discussed by Ana Ofak and Philipp von Hilgers as a historiographical operation beyond an accumulative, progress-centric historiography and thus beyond a discourse-analytical historiography of epochal caesurae).[131] Here, the model appears as an example of the "formal and conceptual fine-tuning that recursion has received in mathematics and computer science" – it is well-suited for "allowing the aspect of self-referentiality to emerge."[132] In the media-technological chiasmus of fish and chips achieved by ABM, software-swarms inspired by the behavior of biological fish schools have been used to study the behavior of biological collectives. The results of this have differed somewhat from some of the aspects of recursion underscored by Ofak and Hilgers: "Recursion therefore involves the assumption that today's culture of science is open – that it refers back to unfinished pasts and refers ahead to future constellations. And this it does in the sense of what recursive functions have already established in the field of calculability and programmability, namely the ability to produce that which is still unknown by reproducing that which is known."[133]

Here, in light of the concept of recursion as it used in computer science and in light of the case studies discussed below, it is rather the case that the recursive intertwinement of biological and computer-simulated swarms has involved a strategy of knowledge that does *not* refer to what is known in order to reveal what is unknown. In the process of recursion between biological swarm principles and computer-based simulation environments, two aspects of opaque self-regulatory processes are brought closer together

131 See Ofak and Von Hilgers, 'Einleitung,' esp. 7–18.

132 Ibid., 13.

133 Ibid., 14.

on the basis of just partially known or imprecisely defined parameters. In terms of computer science, recursion is defined as the reapplication of a processing instruction to a variable that is itself already an output of this instruction: "The value of the variable changes with every run through the loop, and the effect of this repetition is not the production of identity but rather a predefined form of variation. [...] Recursion combines repetition and variation with the aim of generating something new."[134] It describes the ability of a program or program routine to run on itself, and this is more or less what has happened in the conflation of fish and chips.[135]

With this historiographical and epistemological figure in mind, I will now turn to the computerization of swarm research. As in the second section of this chapter, my first order of business will be to discuss particle systems and agent-based models, but here the focus will be on their use in scientific research. I will then examine recent efforts to generate more conclusive empirical data with the help of automatic algorithms for image analysis and discuss how such data have been used to improve existing simulation models. Finally, it will be shown how an additional level of feedback between natural and artificial fish schools has been created by means of robotic fish equipped with sensors.

Self-Propelled Particles

Similar to the optical and acoustic methods of analyzing fish schools, in the case of simulation models there are two conceivable approaches. On one hand, schools can be modeled 'from the inside out' in an agent-based manner; on the other hand, it is possible to define global motion equations and thus to model schools 'from the outside in.' The latter option has been used to produce mathematical descriptions of a number of physical processes, and biologists have typically employed it to describe processes at the population level.[136] As a *continuous* or *Eulerian* model of the activity of large aggregations

134 Ernst, 'Der Appell der Medien,' 185. Here the author refers to Winkler, 'Rekursion: Über Programmierbarkeit, Wiederholung, Verdichtung und Schema,' 235.

135 This figure of *recursion* is to be distinguished from the *iterative* processes of scenario-based variation that are executed in the agent-based models themselves. In the latter case, various results or combinations of parameters are repeatedly implemented with recourse to the results of previous simulation runs in one and the same program environment. For further discussion of this distinction, see Rheinberger's book *Iterationen*. Regarding the significance of recursive processes to image-generating methods in computer graphics – and in particular the method of "raytracing" – see Kittler's essay 'Computer Graphics: A Semi-Technical Introduction,' 37–39.

136 See, for instance, Grünbaum and Okubo, 'Modelling Social Animal Aggregation.'

(of bacteria or plankton, for instance), it produces abstractions on the basis of population densities. Models of this sort describe motion in terms of distribution and concentration processes, and they formalize the latter into partial or integro-differential equations. Processes can thus be derived in light of both their temporal and spatial dimensions, and thus it is possible to describe such things as fluid motion as a reaction, for instance, to the arrival of predators within a population of plankton. Partial differential equations, however, fall short in producing realistic representations of swarms in motion. Around the end of the 1990s, as Alexander Mogilner and Leah Edelstein-Keshet have noted, attempts were made "to model locust swarm migration [...] based on biologically reasonable hypotheses. The conclusions are mostly negative, pointing to the difficulties of describing a cohesive, compact swarm with traditional models."[137] Integro-differential equations, on the contrary, can describe interactions across distances and can therefore represent such things as the range of an animal's sense of hearing and vision.[138] The use of such tools, however, becomes complicated when additional inter-individual interactions and influences from the environment are included as relevant mechanisms of organization.

For this reason, models have also been developed in the scientific context that take an individual-based approach. With their minimalistic variants, these *Lagrangian* models rely on the features of mathematical tools to describe processes of statistical physics, such as the movement of particles in gases and fluids or that of metal particles in magnetic fields. In this regard, the swarm researchers Iain Couzin and Jens Krause have stated the following: "While particles may be subject to physical forces, animal behavior can conceptually be considered to result from individuals responding to 'social forces,' for example, the positions and orientations of neighbors, internal motivations (e.g., degree of hunger), and external stimuli (such as the position of obstacles)."[139] Among the most conceptually simple models of coordination within biological aggregations are thus systems that combine the inclination of swarming individuals to orient themselves in parallel with the resulting directionality, on the global level, of a large population of *self-propelled particles.* Particles of this sort move at a constant speed (with some randomness involved) and orient themselves according to the average direction of other particles within a defined local neighborhood: "The only rule of the model is *at each time step a given particle driven with a constant*

137 Mogilner and Edelstein-Keshet, 'A Non-Local Model for a Swarm,' 535.

138 See ibid.

139 Couzin and Krause, 'Self-Organization and Collective Behavior in Vertebrates,' 4.

absolute velocity assumes the average direction of motion of the particles in its neighborhood of radius r with some random perturbation added."[140]

Because of their simplicity, these simulations can be analyzed with methods from non-equilibrium physics. In such a system by Tamás Vicsek and his colleagues, the change from one form of aggregation to another within a collective was modeled in analogy with physical phase transitions and simulated on a Connection Machine 5 parallel computer.[141] Yet the particles in this model neither avoid collisions nor react with mutual attraction to their nearest neighbors, as fish and birds do in their schools and flocks. Self-propelled particles can, however, be used to simulate other types of aggregations: "The present model, with some modifications, is already capable of reproducing the main observed features of the motion (collective rotation and flocking) of bacteria."[142] By changing the parameters of *density* and *noise*, it was possible to create typical forms of collective motion. In physical terms, these changes were implemented as the effects of changes in temperature, just as temperature changes will cause iron atoms to spin differently in a ferromagnet: the higher the temperature, the greater are the effects of random disruption on the system.[143] Unlike the Ising model of ferromagnetism in statistical mechanics, which Charles Breder had applied to fish schools as early as the 1950s, self-propelled particle systems are dynamic to the extent that their particles continue to move ahead between temporal intervals: "The rule corresponding to the ferromagnetic interaction tending to align the spins in the same direction, in the case of equilibrium models, is replaced by the rule of aligning the *direction of motion* of particles in our model of cooperative motion. The level of random perturbations we apply are in analogy with the temperature."[144]

Simulations based on self-propelled particles are too differentiated, however, to depict the dynamics of swarm behavior in a detailed manner, and the minimalism of their models come at the expense of "biological realism."[145] That said, they are capable of reproducing the dynamics of large numbers of particles and thus they provide, on the basis of simple rules, a different perspective on collective organizational phenomena from that offered by the first agent-based models. The latter, much like early

140 Vicsek et al., 'Novel Type of Phase Transition in a System of Self-Driven Particles,' 1226 (emphasis original).

141 See Vicsek et al., 'Spontaneously Ordered Motion of Self-Propelled Particles,' 1376.

142 Vicsek et al., 'Novel Type of Phase Transition,' 1226.

143 See Czirok and Vicsek, 'Collective Behavior of Interacting Self-Propelled Particles.'

144 Vicsek et al., 'Novel Type of Phase Transition,' 1226.

145 Couzin et al., 'Collective Memory and Spatial Sorting in Animal Groups,' 2.

observations of fish schools in aquaria, based their calculations on just a few swarming individuals (no more than fifty 'particles' by the mid-1990s), whereas self-propelled particle systems could work with more than ten thousand at a time.

Traffic Rules in Fish Schools

The fascination of swarms derives from the fact that they exhibit complex patterns of collective behavior even though they consist of relatively simple elements. This complexity arises from the nonlinear interactive processes in which the behavior of individuals is coupled to that of their neighbors and this, in turn, is connected to the ever-changing structure of the entire system. Simulations of fish schools consequently aim to describe this relationality in the simplest possible terms, which they then graphically depict over a defined period of time. They therefore function in the opposite direction of visual methods of observation, which involve tracking fish schools over time, plotting their trajectories, and only then drawing conclusions about their basic relational parameters. Whereas film and video seek to control a school's movement in time by stopping it, simulation processes create this very motion in time by implementing a set of (adjustable) basic rules. Whereas the former methods produce analyses of the past, the latter create projections of the future.

Agent-based models of fish schools are quantitative methods for simulating a set of possible 'traffic rules' within a school. As I showed in the third chapter, empirical methods for studying swarms remained mired in a technological morass well into the 1990s: "Three-dimensional tracking techniques have not yet advanced to the stage where it is feasible to observe large schools (*i.e.*, over 10), in three dimensions, over long times (*i.e.*, for more than seconds)."[146] With simulations, it was now possible to study the dynamics that arise through the variation of different hypothetical rules of interaction. These rules could be modeled by applying a series of forces that influence the velocity and direction of each swarming individual in relation to other individuals and in relation to environmental conditions: "Typical force components include locomotory (*e.g.*, biomechanical forces such as drag), aggregative (*e.g.*, long-range attraction, short-range repulsion), arrayal (*e.g.*, velocity matching), and random (*e.g.*, individual stochasticity)."[147] For the sake of simplicity, however, detailed biomechanical influences were ignored in most

146 Parrish et al., 'Self-Organized Fish Schools,' 297.
147 Ibid., 298.

models, which instead focused on the connection between local decisions to move in one way or another and the resulting motion on the global level.

In order to keep the models to a manageable size, the early simulations of fish schools tended to concentrate on a limited set of algorithms for analyzing three main categories of behavior: behavioral matching, positional preference, and numerical preference. Behavioral matching (when individual agents attempt to match their behavior to that of nearby agents) is modeled by defining a zone of parallel orientation or by programming the agents to adjust their speed to correspond to that of their neighbors or to a predetermined value. Positional preference is modeled by assuming that each fish tends to keep a preferred distance from its nearest neighbors – by creating such things as 'assigned distance zones' (which involve repulsion, parallel orientation, and attraction) or by setting other positional parameters to influence how agents react to one another (parameters involving such things as bearing angles and collision times). The third category, numerical preference, refers to the number of neighbors that a fish pays attention to, which could be determined by an *a priori* value, a conditional value, or situational factors (the number of fish that could be seen within a certain distance, for instance). Furthermore, the magnitudes of such effects could be weighted differently depending on the respective positions of the fish. That said, any given model explored just a small subset of all the possible variations, though they would ideally include such variations as "initial position and velocity; the strength and type of stochastic components; spatial distribution of repulsion, parallel orientation, and attraction; and degree of variation between individuals in a group."[148]

The beginning of the 1990s thus saw a steep rise in the use of agent-based models in fish-school research, and these models were conceptually related to those designed by Aoki (especially regarding the implementation of basic behavioral parameters and zones) and were at least in part inspired by the ideas and graphical visualizations of Reynolds, Terzopoulos, and others. It would be excessive to discuss each of these models in detail, and so instead I will focus on just a few developmental stages that have been especially relevant to the epistemology of agent-based computer simulation.[149] In 1992, the behavioral biologists Andreas Huth and Christian Wissel developed a two-dimensional simulation model that mimicked Aoki's concentration on

148 Ibid., 299. The next page of this article contains a tabular overview of fish-school simulation studies conducted from 1982 to 1992, along with their parameters and output variables.
149 For a survey of the use of agent-based models in fish-school research, see Parrish and Viscido, 'Traffic Rules of Fish Schools.'

the movement-based cohesion of already formed schools. Their approach thus excluded the phenomenon of a school's initial formation by defining a maximum area of attraction in which the individuals will orient their direction to that of their nearest neighbors. Moreover, polarized movement was programmed as an explicit parameter, and potential external influences were disregarded.[150] Their model was thus even more focused than Aoki's on a fish school's internal mechanisms of self-organization, and the authors stressed that they had attempted to construct the simplest possible model: "Only simple models promote a comprehension of the results. In other words, we are not interested in modelling every detail of the fish behaviour, but only the behaviours which are decisive for school organization."[151] *Simplicity* was one of their guiding principles; like others, they believed that simple simulation models were the precondition for producing insightful results about the factors driving the organization of schools: "The aim of modeling is often not to attempt to include all the known properties of a system, but rather to capture the essence of the biological organizing principles. One of the principle aims of self-organization theory is to find the simplest explanation for complex collective phenomena."[152]

In search of this explanation, Huth and Wissel tested various scenarios in which neighboring fish influence one another. They distinguished between a "decision model," in which each fish decides, on the basis of certain weighted factors, to which of its neighboring fish it will adjust, and an "averaging model," in which each fish takes into account the positions, speeds, and directions of multiple neighbors and orients itself according to their averages. It was only the latter model that yielded a coherent schooling structure. Two years later, the authors not only expanded this model to three dimensions, which allowed them to analyze the behavior of twenty to one hundred 'computational individuals,' but they also attempted to verify the model's parameters by comparing its internal variations to the sparse empirical data that had been collected by researchers such as Partridge.[153] They also examined how their model would work when two different schools converged, and they tested its scalability as well: "According to this we have no doubt that our model is also valid for schools of a thousand or more fish, which exist in nature, too."[154] Not least, they stressed that the factors which

150 Huth and Wissel, 'The Simulation of the Movement of Fish Schools,' 367.

151 Ibid.

152 Couzin and Krause, 'Self-Organization and Collective Behavior,' 5.

153 Huth and Wissel, 'The Simulation of Fish Schools in Comparison with Experimental Data.'

154 Ibid., 144.

they investigated were *universal* features of the organization of dynamic schools: "Our model shows that the self-organization of fish schools can be understood on the basis of some simple behaviour rules. It seems that special physiological details have no essential importance for the school organization."[155]

In the same year, the ecologists Hauke Reuter and Broder Breckling presented a simulation model that was intended to evaluate the movement of an individual fish in response to the influence of all the visible fish around it (not just the nearest neighbors), and their model also included external disruptions such as obstacles within the simulated environment.[156] This raised the questions of how many neighbors an individual could perceive and whether this number would change depending on environmental factors. In this regard, the authors underscored the relevance of a combination of field research, theoretical considerations, and laboratory studies to the behavioral-biological description of fish schools, whereby computer simulation played a special role of its own: "Even if at present important questions remain open, it is possible to exclude some behavioral patterns through theoretical consideration and simulation experiments."[157] In general, simulations of biological fish schools follow a *negative epistemic strategy*. By means of scenario-based variations, they exhibit improbable combinations of parameters that can be weeded out, and thus they enable an iterative approximation of the sets of rules and characteristics that define the self-organization of schools. In short: "[W]hile it is probably not possible to discern the exact rule(s) used in nature, it may be possible to rule out (or in) possibilities."[158]

At the University of Tokyo, Yoshinubu Inada reexamined the information-transfer processes between schooling individuals and their effects on the macroscopic behavior of schools by varying the directional orientations of local synchronization processes and running them through computer visualizations.[159] Because the fish in the front area of a school have relatively few or no neighbors with which to adjust their orientation, they exhibited more haphazard swimming behavior in Inada's model and generally swam at speeds that were slower than average. This led to a characteristic, unequal density distribution in the school, which was

155 Ibid.

156 Reuter and Breckling, 'Selforganisation [*sic*] of Fish Schools.'

157 Ibid., 157.

158 Parrish and Viscido, 'Traffic Rules on Fish Schools,' 73.

159 Inada, 'Steering Mechanism of Fish Schools.'

at its most dense in the front middle region (where the fish were most strongly influenced by the somewhat erratic swimming of the 'leading individuals' in front of them). Inada played through various orientation preferences, from a strong 'front priority' (the tendency of fish to follow the motion of those ahead of them) to a preference to follow neighbors swimming on either side. The former preference resulted in a sharply turning and rapidly synchronizing school structure, whereas the latter yielded more gradual and slower turning maneuvers because it took longer for information to be transferred to the individuals swimming in the back. The manner and direction of information transfer thus determined the global activity of the school; even in the case of entirely identical agents, their behavior varied depending on their position in the school and thus on their participation in the flow of information. In another study, Inada and his colleague Keiji Kawachi simulated the evasive maneuvers of a school responding to incoming predators and examined what sort of global structures would reform if the ability to exchange information were limited in various ways. These experiments revealed a connection between a low level of randomness in the motion of individual fish and the ability of the school as a whole to perform structured and flexible maneuvers.[160]

Other studies have investigated the influence of physically realistic agents (that is, agents with both shape and a mass), an approach to which graphical processing is even more relevant than usual. The typical punctiform shape of an agent was replaced by linear and ellipsoidal agent bodies, which in turn affected the ability of individuals to see their neighbors. Now it was no longer just the 'midpoint' of an agent that determined how individuals would modify their distance from one another but rather its entire length and form, and this was also reflected in the correspondingly modified shape of its 'zone of orientation.' Other simulations likewise modeled the mass of their agents, which led to a degree of sluggishness in the overall motion of the simulated schools. In general, as Parrish and Viscido observed, "agent shape (point, line, ellipse) had a significant effect on several group-level output variables, including polarity [...], the ratio of first to second-neighbor distances [...], group velocity, and group shape."[161] What is interesting is that, in models with 'realistic' ellipse-shaped agents, the global structure remained the most dense, the swimming velocity was more constant than

160 See Inada and Kawachi, 'Order and Flexibility in the Motion of Fish Schools.'
161 Parrish and Viscido, 'Traffic Rules of Fish Schools,' 61. Here Parrish and Viscido refer to Kunz and Hemelrijk, 'Artificial Fish Schools.'

usual, and the simulated schools were more receptive to disruptions. This was because the agents' elliptical zones of repulsion allowed for several possible reactions, thus making the agents' ability to synchronize somewhat more dynamic.[162] The *form* of the schooling individuals influenced the form of the information structure, and this in turn affected the dynamic shape of the school's formation.

In 2002, Iain Couzin and his colleagues took agent-based fish-school simulation to another level by creating three-dimensional simulations of large schools in which the model's parameters could be varied consistently and systematically. They were able to show that even relatively minor changes to just one parameter could lead to abrupt changes in a school's total structure. Thus, the transition between any of the four typical school formations – a diffuse 'swarm' formation, a torus formation, a dynamic-parallel group, and a highly parallel group – could only be brought about simply by varying the school's alignment factor. The transition between these states would happen rapidly because the intermediate types of formation were highly unstable. In biological schools, at least, the ability to change quickly from one structural type to another enables them to react successfully to changing environmental factors, for instance to external stimuli such as predators.[163] Furthermore, Couzin's models demonstrated something like structural memory. Over the course of modifying the alignment variable, it was seen that the system could jump directly from a dynamic-parallel formation to a swarming formation without having to pass through the other two types: "This demonstrates an important principle: that two completely different behavioral states can exists for identical individual behavioral rules, and that the transition between behavioral states depends on the previous history (structure) of the group, even though the individuals have no explicit knowledge of what that memory is. Thus, the system exhibits a form of 'collective memory.'"[164]

This principle of hysteresis made it clear that both the agents' present parameter settings as well as the system architecture preceding them could influence the organization of the school. With their three-dimensional methods of visualization, moreover, Couzin and his colleagues also analyzed the reactions of their simulated schools to external stimuli such as predatory attacks. Depending on their initial structure – and following the rules of

162 See Kunz and Hemelrijk, 'Artificial Fish Schools,' 252.

163 See Couzin and Krause, 'Self-Organization and Collective Behavior,' 24–27; and Couzin et al., 'Collective Memory and Spatial Sorting.'

164 Couzin and Krause, 'Self-Organization and Collective Behavior,' 29.

an "evade predator" subroutine – the simulated schools would adopt avoidance formations similar to the fountain effect, split effect, or vacuolation observed in nature whenever a simulated predator would approach their high-density areas. In this regard, too, the researchers tested the size of the orientation zone in which individuals could transmit information to others. If this space was too limited, no collective avoidance formations would result; if it was too large, the quality of the information that an individual could acquire from others would eventually be reduced. In the case of large schools, a decisive role was played by the way in which different neighboring clusters of individuals were structured: "When the population size far exceeded the number of influential neighbors, each fish took cues from a different, but overlapping, set of neighbors, which resulted in more mobile schools."[165] In addition, as the size of a simulated school continued to increase, scaling effects began to appear: "Our results showed that group properties such as polarity, group size and group speed are strongly influenced both by population size, and by the number of influential neighbors."[166] As in Reynolds's boid model, here too it became clear how relevant local knowledge was to the creation of dynamic schooling structures on the macro-level. Finally, the authors underscored the importance of their efforts to *systematize* the scenario-based combinations of parameters and their relation to one another, given that "most studies do not systematically vary all factors, and so the relative importance of each factor has remained a mystery."[167]

Another area of investigation that attracted attention was the productivity of disruptions or *noise* in the organization of schools. On the one hand, this involved describing the internal changes of position within a school's structure, and especially during turning maneuvers, for it was then that the individuals at the front of the school would usually change. For, in comparison to others, certain positions within a collective might have certain advantages when it comes to such things as finding food or avoiding predation (think of Hamilton's 'selfish herd'). Only an internal disruptive function, which causes schools to reconstitute themselves over and over again, could lead to dynamic formations capable of processing information from their environment so rapidly and adaptively. Only a degree of self-produced noise could enable schools to coordinate within "noisy environments": "The importance of fluctuations for attaining globally optimal

165 Parrish and Viscido, 'Traffic Rules of Fish Schools,' 63.
166 Ibid.
167 Ibid., 64.

states suggests that noise plays an important role in group organization."[168] To this, moreover, Couzin and Krause added the following observation: "Importantly, the sorting within the model depends on 'local rules of thumb,' that is, not on absolute parameters but rather on relative differences between individuals. Thus, an individual decreasing its zone of repulsion relative to near neighbors will tend to move toward the center of the group, even if it has no knowledge of where the center actually is."[169] In order to model such effects, the behavioral rules had to be made more flexible: "[C]ommon behaviors of schooling fish suggest that there may even be multiple rule sets that an individual can switch on and off as needed. In the next generation of agent-based schooling simulations, rule sets must allow a simultaneous exploration of individual movements provoking group-level pattern and fission-fusion of groups."[170] Thus it was hoped that it might soon be possible to study more closely the influence of individual behavior on the overall structure of schools in order to see, for instance, whether rapid positive feedback processes might enable the relatively deviant behavior of one or a few individuals to cause the entire school to change directions.

In biological swarm research, agent-based models and their visualizations thus bring together a number of concepts that are characteristic of computer simulation's particular order of knowledge. In scientific applications of agent-based models, their parameters are kept *simple* so that it might be possible to make *universal* statements about the factors behind the self-organization of dynamic collectives. At its best, this process involves *systematically varying* a model's parameters in relation to one another, and thus it represents a sort of trial-and-error approach to science. Combinations are played through, improbable scenarios are eliminated, and in this way, researchers are able to approximate a probable configuration. In all of this, image-generating processes play a decisive role, for this epistemic strategy is only possible on account of the comprehensible combination of *information* and *formation* that such processes allow. Questions concerning the *transitions* between different swarm structures, their *structural memory*, their *scalability*, and the effects of internal and external *disruptions* can only be addressed in dynamic, four-dimensional presentations – in real-time artificial life: "Unlike any set of laboratory

168 Parrish and Edelstein-Keshet, 'Complexity, Pattern, and Evolutionary Trade-Offs in Animal Aggregation.' See also Moreira et al., 'Efficient System-Wide Coordination in Noisy Environments,' 12085–12090.

169 Couzin and Krause, 'Self-Organization and Collective Behavior,' 43.

170 Parrish and Viscido, 'Traffic Rules of Fish Schools,' 74.

or field experiments, the computational approach has the potential to examine systematically the multiple adaptive peaks in the landscape of three-dimensional aggregation, pointing the way towards which rules are necessary, even fundamental."[171]

Unlike the case of graphic design, however, for biological swarm research it is insufficient to construct a computer-simulation model that produces behavior resembling that of natural swarms and yet introduces unrealistic parameters. This is a matter of evaluating influential factors that actually exist. Under the conditions of improved methods for analyzing images, researchers have thus begun to use their agent-based models to integrate new empirical data or to test and improve the models themselves on the basis of new data sets. The pride of Iain Couzin's former lab at Princeton University, where he and his team began to study the collective behavior of various life forms in the early 2000s, was consequently no longer the sort of research that one might imagine taking place in a typical biological laboratory. Of course, he had access there to aquaria for observing schools of fish, devices for investigating the physiology of locusts, and all the other technical apparatuses needed for such experimental systems. The laboratory's flagship project, however, took place in a normal office and consisted of a number of simple black boxes: commonplace PC towers with their blue interior lighting. It was no coincidence that these computers also looked as though they were still being used at a LAN party; they were equipped with powerful graphics processors that had been designed especially for CPU-intensive computer games. This was because the operations of Couzin's computer simulations, which involved an increasing number of individuals or computational agents, would have overtaxed the capacity of traditional computers:

> These computations can be become prohibitively slow if one is interested in simulating very large number of animals as seen in nature, which for example can be millions. In addition, the spatio-temporal variability in the environment, feedback between individuals and the environment together with how individuals evolve on evolutionary time scales make it virtually impossible to use traditional methods of CPU computing. These challenges emerge while analyzing the data from the experimental videos as well. To address these issues, CouzinLab develops simulation tools that combine models of swarming with an evolutionary game-theoretic

171 Ibid., 76.

> framework on massively parallel architecture on graphical processing units (GPU) developed by the NVidia Corporation.[172]

In cooperation with this producer of graphics cards, Couzin's team soon overcame these limitations by using, as co-processors, graphical processing units that each had several hundred processor cores. By means of the programming language CUDA ("Compute Unified Device Architecture"), it was possible to connect these to form a "personal super-computer" whose massively parallel processing architecture was ideal for modeling and simulating complex swarm dynamics.[173]

Although computer-graphical visualizations of agent-based models had been the pillars of biological swarm research since the mid-1980s, it was now the case that the (materially conditioned) computing speeds of traditional Von Neumann architectures were making it difficult for researchers to deal with simulations in a practical manner. GPGPU architectures (GPGPU stands for "General-Purpose Computing on Graphical Processing Units") provided a remedy in this regard, especially since being made into consumer versions that could be used both for traditional 3D-rendering operations as well as for general computing processes.

Meanwhile – and this is the interesting part from a media-theoretical perspective – the brute-force calculations of large collectives of interacting agents were now based on the properties of the massively parallel-processing hardware structure of graphics chips. Their architecture allowed various types of data to be organized in parallel streams which, though processed together by a single core, remained autonomous alongside one another. This architecture was thus ideally suited for the needs of modeling individual-based collectives. Furthermore, it also organized the data contained in a single stream into different "textures":

> Textures are bi-dimensional arrays of 4-dimensional components of float values. For each character, state-preserving attributes like position, velocity, mass, and size are stored compactly in place of pixels into 2D textures [...]. Textures are used not only for storing state-preserving information but also to store environment-related information [...]. During

172 Quoted from the homepage of Iain Couzin's former lab at Princeton, a website which is no longer active: http://webscript.princeton.edu/~icouzin/website/high-performance-computing-for-massively-parallel-simulations-of-animal-group-behavior/ (accessed 25 January 2011).
173 See Erra et al., 'An Efficient GPU Implementation for Large Scale Individual-Based Simulation of Collective Behavior.'

> simulation, textures are used as a rendering target to maintain output related to every single behavior.[174]

Here, interactive dynamics were created by so-called 'fragment shaders,' which connected the status-preserving information of individual data cells to information concerning the number of neighbors, their positions, and their orientation. This combination gave rise to the 'steering behavior' for each individual cell, which resulted in new positions and orientations in the system as a whole. This approach was further refined by Couzin and others, who presented the simulated 'world' as a three-dimensional grid that was analogous to GPU memory. Every individual cell had an identification number, as did every individual agent. Through a series of sequential operations, the simulation would update itself in steps by relating the different attributes of its agents (such as position, direction, and speed) to one another.[175]

The recursive intertwinement of agent-based computer-simulation processes and biological swarm research on the level of software development marks an epistemic rift whose consequences surpass the analytic capabilities of laboratory instruments. To understand them, what is needed is rather a technically informed and historical-epistemologically oriented media theory. The use of graphics chips as hardware for calculating swarm dynamics had little to do with the material culture that features so prominently in the history of science. Rather, this process was based on the *immaterial culture* of agent-based computer simulations, which harnesses the advantages of distributed structures not only on the software level (by means of object-oriented programming, for instance) but also deploys them in graphics chips and makes them useful for biological research on dynamic collective processes. As in the case of the agent-based models from the mid-1980s, here an inverted perspective pointed to the *zootechnicity* of such swarms and collectives. Of course, there was nothing 'natural' about Nvidia's G8x chips, but a computer-supported view of animals as *system animals* revealed connections between these realms that hinge on their abstract and formal knowledge of control and relations. And at the same time, these connections – along with quantitative leaps in computing power – led to qualitative leaps as well:

> And this has been an absolute revolution in terms of scientific computing for us. So we're investing heavily in our efforts to try and program all

174 De Chiara et al., 'An Architecture for Distributed Behavioral Models with GPUs,' 2.
175 See Erra et al., 'An Efficient GPU Implementation,' 54.

> of our simulations on these video game cards. And to give you sort of a rough impression, [...] if you can get 300 or 500 times as fast, if what used to take a month now takes you an afternoon, that changes the way we work. And also, because we can harness this vast computational power, we can start asking questions about evolution, we can start simulating these groups of reasonable size with the reasonable resolution in how they interact in space over such long-time scales that we can now start, you know, having a sort of virtual process of evolution to understand how and why collective behavior has evolved.[176]

It falls to a historical-epistemological media theory to describe such overlapping textures between hardware, software, and wetware and to trace their interrelations, in which computer-graphical visualizations have played such a prominent role as an epistemic tool. These constellations, which nullify the differential force of a concept like the epistemic thing, mark the beginning of a more comprehensive epistemology of computer simulation – with new concepts and categories of its own that remain to be discovered.

Robofish: Empiricism Strikes Back

Agent-based models and their visualizations were at first implemented in swarm research as a new epistemic tool in order to test the probable factors that are involved in the self-organization of swarms and that, just a few years before, could not be adequately described by observational media and empirical methods of data generation. Unlike the interactive methods of image analysis employed around 1980, which I discussed above with the examples of data tablets and the GALATEA system, recently developed automatic image-analysis algorithms have made it possible to reconstruct digital video and photograph sequences of large swarms. Computer-simulation models and the computer-graphical tracking of image sequences thus have a reciprocal relationship, given that various research groups have compared their simulation models with data produced by empirical research. Whereas Huth and Wissel still sought to validate their simulation model with reference to Partridge's findings, in more recent studies the output of simulation systems is made to encounter observational and experimental data on the same media level: the level of digital image sequences, where both sets of data can be evaluated, compared, or contrasted with one and the same program. Researchers have set aside the methods of

176 Quoted from Allen, 'Interview with Iain Couzin.'

interactive image-synthesis in favor of making direct comparisons between automatically collected measurement data and the information produced by their own simulation models. This has consequently resulted in yet another recursive interrelation. After being tracked, filtered, recorded, and stored, digital images of what is 'real' are then synthesized into trajectories, which is essentially a process of collecting data from image sequences. This is therefore an inversion of the process used to visualize computer simulations. In the latter case, image sequences are generated from 'artificial' data, and these data are then adjusted with the help of the visualizations in question. Empirical data have now been thrust into this process, so that measuring methods and simulation models mutually evaluate one another in a differential mode of knowledge. Under such conditions, it is difficult to speak of *validation* in the traditional sense.

In 2004, for instance, a research team at Princeton's Ecology Research Center made observations of fish schools in a river current. They filmed digital two-dimensional video sequences and analyzed the latter with a program called Tracker, which had been developed by NASA for tracking particles. In the natural environment of the river, however, the video equipment produced unclear images that impeded the automatic analysis of the data:

> [T]he background was too irregular and the contrast between the fish and streambed not sufficiently sharp. [...] Tracker does, however, have the ability to do image arithmetic. This was useful, in that the streambed background [...] could be subtracted away from the images, leaving only the fish. [...] Images resulting from subtraction were fuzzy due to the movement of the water in the stream. [...] Nevertheless, this method still facilitated the identification of fish in the images.[177]

Observations in the field turned out to be too troublesome for even the most advanced tracking programs, and so Steven Viscido and his colleagues decided to revert to the traditional method of observing small schools of fish in an aquarium.[178] They filmed their little schools stereoscopically in a one-cubic-meter tank for a period of thirty minutes, and they later analyzed the recorded sequences with Adobe Premier and NIH Image, an image-analysis tool developed by the National Institute of Health. With a self-written program called Tracker3D, the trajectories of the schooling

177 Tien et al., 'Dynamics of Fish Shoals,' 558.
178 Viscido et al., 'Individual Behavior and Emergent Properties of Fish Schools.'

individuals were then reassembled (at which point manual corrections were made) and the frame rate was filtered to a frequency of 5 Hz in order to eliminate high-frequency noise.[179] In this case, too, data were *produced.* A comparison of this information to ABM scenarios once again revealed the relevance of alignment factors in group responses, that is, the issue of how many neighbors a given individual can exchange information with. Beyond that, it was also shown that the deviant behavior of just a few individuals could influence the entire school, in effect making these fish unwitting 'leaders' of the collective.[180] And of course, it was also possible to compare all sorts of different parameter combinations, the results of which do not need to be rehearsed here.

As was already the case with Craig Reynolds's boid simulations, which pertained just as well to fish and birds, new strategies for obtaining empirical data were applicable both to fish-school research as well to the study of flocking birds. Both cases involve tracking highly dynamic particle systems, each with its own internal disruptive functions and each inhabiting an environment with its own set of disruptions. In what its authors called a "benchmark study in collective animal behaviour," a group of Italian researchers working in the field of complexity studies developed a novel image-analysis algorithm that dispelled the limitations of traditional avian-flock research (an insufficient number of individuals, loose formations, etc.). With this algorithm it was not only possible to automatically evaluate and measure digital photographs of the famous flocks of starlings that circle above Rome in greater and greater numbers. The researchers stressed that their interest lay not in the specific features of starling flocks but in the more general characteristics of collective behavior: "[T]he same techniques can easily be exported to other cases, most notably to fish schools, insect swarms, and even to flying mammals, such as bats. We hope that our methods may give rise to a new generation of empirical data."[181] Using stereoscopic methods (which later made it possible to produce a three-dimensional reconstruction of the data), they were able to analyze flocks consisting of up to 2,700 birds, and they were thus the first to look closely at such dynamic activity in the open air. Regarding questions concerning the organization and synchronization of motion, their study was thus far more interesting than the recordings of just a few individuals that had been made from the 1970s to the 1990s. Even in its ability to scale up the scope of its investigations

179 See ibid., 241.
180 See ibid., 248.
181 Cavagna et al., 'The STARFLAG Handbook on Collective Animal Behaviour,' 235.

over time, the ornithological branch of swarm research has closely resembled the fish-school research discussed in Chapter III.[182]

With continuous-mode photography, the researchers generated eight-second sequences with ten pictures per second and were then able reconstruct, with their software and the help of a third camera, the individual movements of 80-88 percent of the birds being observed. Even under the most modern technical conditions, efforts to separate signals from noise led only to approximate values; moreover, only around half of the recorded sequences turned out to be usable for analysis. For, as another researcher put it, "optical resolution is the main bottleneck."[183] Often there was a lack of contrast in the images, which could be caused by an unclear background, by birds flying out of the focal range of the cameras, or by there being so many birds clustered in one image that the software was unable to identify all of them. By means of a segmentation process, an algorithm would (attempt to) recognize all the objects within a pair of images and remove the background as much as possible (on account of the digital generation or compression of images, a certain degree of noise was unavoidable). So-called 'blobs' of overlapping objects had to be separated by means of a "blob-splitting algorithm," and the individuals within them had to be counted as accurately as possible:

> The effectiveness of the whole segmentation process, and in particular of the blob-splitting algorithm, can effectively tell us whether or not the group under study is too dense to be reconstructed. If, after careful optimization of the parameters, the segmentation produces huge super-blobs of hundreds of animals, then it is very likely that, even after applying the blob-splitter, the animals' positions will be so noisy that it will be very hard to continue with the analysis. In these cases, the only thing one can do is try to improve resolution, both digital and optical (more pixels and better lenses). On the other hand, if blobs contain few animals (up to ten, or a few more), then the blob-splitter can produce excellent results.[184]

By means of a 'matching' process, the segmented objects in one image were then coordinated with the corresponding objects in the second picture, whereby the images taken by a third camera could be consulted to help identify the matching pairs. Finally, a "3D-reconstrution algorithm" was

182 See, for instance, Heppner, 'Avian Flight Formations'; Major and Dill, 'The 3D Structure of Airborne Bird Flocks'; and Pomeroy and Heppner, 'Structure of Turning in Airborne Rock Dove.'
183 Ballerini et al., 'Empirical Investigation of Starling Flocks,' 205.
184 Cavagna et al., 'The STARFLAG Handbook,' 229.

used to consolidate all of this information, a process that involved a few distortive effects of its own on account of the function of the birds flying along the outer edge of the congregations:

> Consider a school of fishes circling clockwise around an empty core, a pattern known as milling. Now imagine that one wishes to compute the spatial distribution of nearest neighbours. The (many) individuals on the external border lack neighbours to their left, whereas the (few) individual on the internal border (the core) lack neighbours to their right. If individuals on the border are included in the analysis, one obtains a distribution indicating that fishes have, on average, fewer neighbours to their left. This result is not, of course, a general consequence of the local interaction rules among the fish: in fact, the completely opposite result would be obtained for a school milling counter-clockwise! This is a typical case where disregarding border effects results in the conflation of two levels of analysis that should remain separate; specifically, the morphological level (the mill-like shape of the school), and the behavioural level (individual interactions and nearest neighbour distribution).[185]

Statistical methods were needed, moreover, to identify marginal individuals with accuracy. Here the edge or border, which early fish-school studies such as those by Albert Parr and William Hamilton had used to define the cohesiveness of schools, turned out to be a disruptive factor in the researchers' technical-visual analysis.

That said, Cavagna and his colleagues' three-dimensional reconstructions provided new and detailed knowledge about such things as the directional changes that take place during a flock's dynamic maneuvers and about various density distributions within collectives. It must be kept in mind, however, that all of this knowledge was gained on the basis of still images. Even with this system, it was impossible to track the individual trajectories of a flock's members. Once again, the latter had to be animated with digital techniques for generating images. Yet despite such persistent problems, the system was able to demonstrate that flocks, which appear from the outside to maintain a stable dynamic equilibrium, are in fact always reconfiguring themselves and that their members, by constantly changing positions, create dynamic global movements, certain recurring shapes, and particular densities and internal structures. For the first time, according to the authors, it was now possible to improve existing simulation models by comparing

185 Ibid., 238.

and adjusting them to statistically reliable empirical data from 'urban field studies': "Some of our results [...] can be used as input parameters for existing models. Most of the results, however, should be used to refine and extend the models, to verify and assess their assumptions, and to identify the most appropriate theoretical frameworks."[186] Above all, they showed that the measurements of the areas of interaction and influence between neighboring individuals had to be revised. Whereas previous agent-based models had distinguished geometric zones for various modes of interaction, the empirical studies revealed a topological orientation: it was not the nearest neighbors that exerted influence over the individual flocking birds but rather a number of *perceived* neighbors. What mattered was not the metric distance between neighbors but rather a fixed number of neighbors, regardless of the distance between them. This topological orientation had not been reflected in prior simulation models.[187]

Finally, the coupling of computer-simulation models and new empirical research methods was taken a step further by projects that integrated biological and computer-generated specimens in a common experimental environment. In one international collaboration, for instance, an artificial fish that could be controlled by a computer was used to test the influence of individual schooling fish on a school's global dynamics and decision-making processes. Given the name 'Robofish,' this swimming robot was placed within a natural school of sticklebacks in an aquarium and controlled with electromagnets beneath the aquarium's floor.[188] Thus it was possible to investigate, by means of a completely controllable schooling individual, the extent to which somewhat deviant swimming behavior might gradually cause certain fish to become the 'leaders' of a school – a question that Konrad Lorenz had famously attempted to answer by damaging the brains of minnows (an experiment discussed in the first section of Chapter II). A fish swimming in an 'idiosyncratic' manner might be perceived by its neighbors as possessing relevant information and thus be followed, though only up to a certain point. Larger schools, for instance, could only be redirected by a critical mass of individuals, while the behavior of individuals would typically adjust to that of their neighbors and thus lead to optimal global behavior in relation to external factors. The complete controllability of the robofish made

186 Ballerini et al., 'Empirical Investigation of Starling Flocks,' 211.

187 See Ballerini et al., 'Interaction Ruling Animal Collective Behavior Depends on Topological Rather than Metric Distance'; and, more generally, Dell et al., 'Automated Image-Based Tracking and Its Application in Ecology.'

188 Faria et al., 'A Novel Method for Investigating the Collective Behaviour of Fish.' See also Landgraf et al., 'RoboFish.'

it possible to identify and quantify such threshold values in the decisions made by schooling collectives. It generated data that, in combination with computer models, further enriched our knowledge of fish schools. And further projects such as the European ASSISIbf research consortium show that such approaches have become more and more popular and feasible.[189] The rather peculiar acronym of this consortium refers to Saint Francis of Assisi, who, as legend has it, preached to birds; it stands for "Animal and Robot Societies Self-Organize and Integrate by Social Interaction (bees and fish)."

A team working at Iain Couzin's lab in Princeton opted to try a similar approach. This involved projecting computer-generated clusters of zooplankton into an aquarium containing a biological species (bluegill sunfish) that feeds upon real plankton. The virtual plankton could be fully controlled and equipped with various sorts of dynamics – from individualized and aggregate to highly polarized and neighbor-oriented behavior – and it was possible to evaluate the effects that this artificial prey had on its predators. It turned out that mobile and coordinated swimming schools of simulated plankton were attacked the least. This observation supported the thesis that school formation serves as a form of defense against predators and should therefore be regarded as a significant factor in evolutionary biology.[190]

In response to the question posed at the beginning of this chapter, namely that of how the recursive development of biologically inspired computer-simulation models has interacted with the computerization of biological research, it is now possible to make at least three assertions. *First*, a whole series of agent-based models has been used in biological swarm research since the early 1990s, or perhaps it would be better to say that a great deal of research projects in that field have welcomed this new epistemic tool and, for many years, neglected empirical studies. Throughout this development, computer-graphical visualizations played a prominent role, for they make it possible to engage in an intuitive manner with dynamic models and they also make it possible to produce scenario-based (re)constructions of natural swarming processes in digital models. In a productive way, biological swarms have been written and described with swarm-inspired agent-based computer models, and it is in this sense that swarms have been written in their own medium. In conjunction with enhanced tracking methods and

189 See, for instance, Bonnet et al., 'Closed-Loop Interactions between a Shoal of Zebrafish and a Group of Robotic Fish'; and Polverino et al., 'Zebrafish Response to Robotic Fish.'
190 Ioannou et al., 'Predatory Fish Select for Coordinated Collective Motion in Virtual Prey.'

automated matching processes, the further development of image-analysis algorithms and image-generating processes represented a demonstrable improvement over the quantitative methods of observation described in the previous chapter and led – *second* – to an additional recursive loop. With these refined methods, new and more detailed empirical data were used to modulate and fine-tune existing computer-simulation systems. *Third*, 'realistic' agents such as the Robofish, data from computer models, and the computer-controlled behavior of artificial agents were all integrated with the wetware of natural schooling fish. The study of swarms as biological *objects of knowledge* converged with the study and development of the media techniques used in agent-based computer simulations and their order of knowledge. The line between agents and fish blurred all the more when hardware, software, and wetware were brought together in concrete interactions in order to gain more accurate knowledge about the organization of fish schools.

VI. Zootechnologies

Abstract

This chapter develops the concept of *zootechnologies* and of swarming as a cultural technique with regard to four decisive application areas. First, it discusses the development of drone swarms under the hypothesis that these create a multifold 'spatial intelligence.' Second, it highlights the importance of a variety of agent-based modeling toolkits for the dissemination of 'swarm-intelligent' applications throughout different scientific disciplines. Third, it investigates the impact of 'swarm intelligence' on the field of architectural design and urbanism and discusses attempts to conceptually exploit swarming for architectural theory. Finally, it turns towards the research field of crowd control where ABM 'pre-mediates' human crowd dynamics and turns traditional concepts of 'the mass' upside down.

Keywords: drone swarms, swarm architecture, agent-based modelling, swarm intelligence, crowd simulation, parametric design

> Science has done all the easy tasks – the clean simple signals. Now all it can face is the noise; it must stare the messiness of life in the eye.[1]

In media-historical terms – as I discussed in my introduction – swarms have been fused into biological, computer-technical hybrids that can best be understood with the concept of *zootechnologies*. For, contrary to *biotechnologies* or *biomedia*, they are conceptualized less on the basis of *bíos* (the notion of 'animated' life) than on the basis of *zoē*, the unanimated life in the swarm – a sort of life that can be technically implemented. The knowledge gained from this hybridization has helped to establish a highly technical lifeworld that is increasingly confronted with models of complex contexts and systems. Whenever one is faced with disrupted or constantly changing

1 Kelly, *Out of Control*, 24.

Vehlken, S., *Zootechnologies. A Media History of Swarm Research.*
Taylor & Francis Group, 2019
DOI 10.5117/9789462986206_CH06

conditions, or whenever solutions need to be found for unclearly defined sets of problems, methods can now be employed that are commonly known under the umbrella term 'swarm intelligence.' Eric Bonabeau, Marco Dorigo, and Guy Theraulaz thus make the following remarks toward the beginning of their standard work *Swarm Intelligence: From Natural to Artificial Systems*:

> Researchers have good reasons to find swarm intelligence appealing: at a time when the world is becoming so complex that no single human being can understand it, when information (and not the lack of it) is threatening our lives, when software systems become so intractable that they can no longer be controlled, swarm intelligence offers an alternative way of designing 'intelligent' systems, in which autonomy, emergence, and distributed functioning replace control, preprogramming, and centralization.[2]

Because these almost universal claims are somewhat dubious, in what follows I will trace back the genealogy of swarm intelligence by focusing on just a few telling examples that delimit the circumstances in which recently developed motion-based intelligence has become operable on the basis of swarm principles. For, as I intend to argue, it was at the media-historical threshold where, by means of the epistemic fusion of fish and chips instigated by computer-simulation models and their graphic visualizations, an extensive and novel sort of knowledge about swarms' principles of control and self-organization became available that the latter became operable as *figures of knowledge* in various technical applications and other areas of use. This chapter will examine uses of the swarm as a figure of knowledge in applications that have implemented swarm principles as an 'intelligent' form of control logic. Associated so closely with such high-tech tools, the very genealogy of swarm intelligence is in some respects responsible for the fact that the concept came to be applied, around the year 2000, to describe such a wide variety of socio-political and economic phenomena. With such zootechnologies, we have not only reached the end of the history of media, technology, and knowledge told in this book – a historical arc along which swarms were regarded first as prototypical phenomena lying outside of knowledge, second as problematic objects of knowledge, and finally as operative figures of knowledge. It was also these zootechnologies that, as of 2000, caused swarms to enter the popular discourse not only by combining swarming and 'intelligent behavior' into a single concept but also by applying

2 Bonabeau et al., *Swarm Intelligence*, xi.

this very combination in technical contexts. The operativity of swarms as figures of knowledge thus exposed the popular-cultural discourse to something like the *Zeitgeist* of our media culture of intransparency.

1. Drone Swarms, or Upside-Down Evolution[3]

Supposedly, Stanisław Lem did not want to be known first and foremost as a science fiction writer. In fact, designations such as 'visionary' or 'futurist' seem more appropriate: The edginess of Lem's writing derives from his comprehensive education in literature, technology, and the natural sciences. And this was always driven by his genuine interest in conducting minute analyses of social phenomena, whether about the future, the present, or the past (though these categories can be difficult to separate). As early as 1964, Lem devoted his novel *The Invincible* to reconceptualizing technological progress.[4] In this story, the crew of the star cruiser *Invincible* is sent on a search mission to an unknown planet. After their arrival, the space explorers encounter a strange form of artificial intelligence: a giant swarm of simple but cohesively moving micro-machines. During their research, the crew discovers that these "pseudo-insects" are the last surviving 'species' of an evolutionary struggle for artificial life between competing war machines. Destroyed remnants and ruins of sophisticated weaponry make it clear to the expedition that the relatively non-complex swarming pseudo-insects had the upper hand over their monolithic technological counterparts.

Lem would revisit this idea almost twenty years later. In his humorous essay 'The Upside-Down Evolution,' he describes – as a narrator from the future, and thus in fictional hindsight – the abandonment of the complex but also error-prone and often easily targetable weapons technology of the twentieth century in favor of much simpler and smaller cooperating elements: "The experts of the day called the new military science an 'upside-down evolution,' because in nature what came first were the simple, microscopic systems, which then changed over the eons into larger and larger life forms. In the military evolution of the post-nuclear period, the exact opposite took place: microminiaturization."[5] Moreover, Lem's technological "involution"

3 This sub-chapter is the slightly revised version of an article which first was published as Sebastian Vehlken, 'Pervasive Intelligence. The Tempo-Spatiality of Drone Swarms,' in: *Digital Culture and Society* 4/1 (2018): Rethinking AI: Neural Networks, Biometrics and the New Artificial Intelligence, eds. Mathias Fuchs and Ramón Reichert, 99–124.

4 Lem, *The Invincible*. The original Polish edition was first published in 1964.

5 Lem, 'The Upside-Down Evolution,' 60.

towards communicating swarms of "synsects" (his term for synthetic insects) entailed a renunciation of traditional approaches to artificial intelligence. Given that, according to Lem's narrator, intelligence is of little importance for 97.8 percent of all human activity (blue and white collar alike), it was essentially unreasonable to put so much effort into the (futile) endeavor of simulating human-like intelligence: "What *was* necessary? A command of the situation, skill, care, and enterprise. All these qualities are found in insects."[6] Thus, simply by analyzing biological evolution closely, "professors of computer science" would have come to understand that simulating *artificial instincts* would be far more feasible and fruitful than simulating *artificial intelligence*. For Lem, as a consequence, the twenty-first century became an era of "artificial nonintelligence" featuring "micro-armies."[7] The latter would replace human soldiers and humanoid automata with swarms of tiny units "which possessed superior combat effectiveness only as a whole (just as a colony of bees was an independent, surviving unit while a single bee was nothing)."[8]

In light of current military developments, Lem's story (like so much of his work) has turned out to be remarkably prescient. Unmanned weapons systems are now being widely deployed,[9] and Germany's army has even begun to use them in training exercises (on account of a general lack of manpower and an insufficient supply of high-tech gear).[10] At the same time, popular science books such as David Hambling's *Swarm Troopers* present numerous examples of the limits and shortfalls of sophisticated weaponry in light of the growing presence of highly mobile and flexible micro-machines.[11] Several journal articles have examined the potential consequences of mass-producing such systems for the sake of showcasing military might and political power.[12] Still others have been concerned with the increasing shift in technological leadership from military developments toward commercial innovation cycles and production capacities.[13] The release of a video showing an apparently successful test of more than one hundred Perdix drones by the U.S. military attracted widespread media

6 Ibid., 50.

7 Ibid.

8 Ibid., 56.

9 See, for instance, Bhuta et al., *Autonomous Weapons Systems*.

10 Leidenberger et al., 'Wie kämpfen die Landstreitkräfte künftig?'

11 See Hambling, *Swarm Troopers*.

12 See, for instance, Goh, 'Unmanned Aerial Vehicles and the Future of Airpower'; Feng and Clover, 'Drone Swarms vs. Conventional Arms'; and Page and Tripp, 'The Perfect Swarm.'

13 See Hammes, 'Technologies Converge and Power Diffuses.'

attention.[14] Recently, too, *The Economist* featured an article about the maturing development of autonomous micro-robots.[15] Although these examples provide a fitting background, I am not primarily concerned here with describing the technical history and potential military applications of such gadgetry. I am rather interested in situating a certain type of robotics – collective or swarm robotics, to be precise – on a conceptual level that can illuminate the peculiar form of artificial intelligence involved in such systems while also taking into account the possible consequences of the widespread use of swarm robotics in unmanned aerial systems (UAS).

Drawing on Lem's notion of 'artificial nonintelligence,' this section will discuss the crucial *spatio-temporal features* of swarm intelligence and swarm robotics, thereby drawing a few historical connections to pioneering work in the field of embodied artificial intelligence. It will then pivot its attention to the specific *spatio-temporal directions* that swarm-robotics technology has taken. By examining recent examples of (semi-) autonomous drone swarms, I will provide a portrait of certain research projects that have endeavored to create operational unmanned aerial systems on the basis of simulated animal behavior. Finally, I will offer an overview of several military analyses that have considered the potential strategic consequences of the (anticipated) widespread use of swarm robotics – consequences that are surprisingly similar to those forecast in Lem's penetrating essay.

It can be argued that swarm robotics creates a sort of three-fold 'spatial intelligence.' First, it consists of the dynamically changing morphologies of such collectives; second, it incorporates the ability of swarms to organize themselves in changing environments; and third, swarm robotics produces representations of these environments as distributed 4D-sensor systems. Thus, its functionality is based on the particular 'artificial nonintelligence' of swarms, which replaces some traditional aspects of artificial intelligence with a 'street-smart' capacity to connect different movements, locations, and spatial features. This results in a sort of spatial intelligence that owes everything to the interactions and communicative processes between the members of the swarm. What is more, the massive parallelism of these collectives enables different functions – such as evaluating the surroundings or transmitting signals – to be distributed to different swarm members. Thus, as I hope to show with the example of unmanned aerial systems, robotic swarms literally penetrate and control space. In contrast to traditional forms

14 See, for instance, Baraniuk, 'US Military Tests Swarm of Mini-Drones Launched from Jets'; and Lamothe, 'Watch Perdix, the Secretive Pentagon Program Dropping Tiny Drones from Jets.'

15 'Military Robots Are Getting Smaller and More Capable.'

of surveillance or even to "sousveillance,"[16] this process could be referred to as *perveillance*. A critical examination of research projects exploring the capabilities of micro-drones to 'live' autonomously in dynamic environments could shed further light on the interactions that are taking place today between natural and technological environments. Unmanned aerial systems, for instance, make use of atmospheric layers in order glide like birds and they use existing power lines in order to recharge their batteries.[17] Endowed with a set of behaviors or 'instincts' that simulate those of predatory birds or bats – such as lurking in hideouts or perching from lookouts – they can be programmed to 'pervey' a given area autonomously for long periods of time.[18]

Fast, Cheap, and Out of Control

In light of their article 'Fast, Cheap, and Out of Control: A Robot Invasion of the Solar System,'[19] Rodney A. Brooks and Anita M. Flynn must have been dedicated readers of Lem's work. Brooks and Flynn were searching for new ways to create intelligent behavior in machines without having to rely on the cognitive approach of 'good old-fashioned artificial intelligence.' They believed that robots would only be able to develop intelligent behavior by interacting with the complexities of their environments. In this regard, their key term was *embeddedness*, and their conceptual principle was *bottom-up*. Knowledge about the world should be computed on-the-fly by small robots that are capable of sensing and reacting to specific environmental conditions in order to fulfill certain tasks. This was a complete reversal from previous efforts in the field of 'intelligent' robotics, which had sought to construct complicated robots with complex artificial brains with large, pre-programmed 'concepts' about the surrounding world.[20] It was not long before Brooks' lab at MIT came to resemble a technological zoo full of small robotic prototypes. Known as Genghis, the most popular of these robots was a six-legged insectoid based on a 'subsumption architecture' without a central controller. At least internally, Genghis already followed certain swarm principles. Its legs were driven by independent motors, and 'walking'

16 The term was coined by Steve Mann in his article '"Sousveillance": Inverse Surveillance in Multimedia Imaging.'

17 See, for example, Langelaan and Roy, 'Enabling New Missions for Robotic Aircraft'; and Gupta et al., 'Energy Harvesting from Electromagnetic Energy Radiating from AC Power Lines.'

18 See Hambling, 'Drones with Legs Can Perch, Watch and Walk Like a Bird.'

19 Brooks and Flynn, 'Fast, Cheap, and Out of Control.'

20 See Brooks, 'Elephants Don't Play Chess.'

was not programmed into the robot but rather arose from its legs constantly exchanging information about their respective positions. Interestingly, it was therefore not the *robot* that was walking with its legs; rather, it was a case of the *legs* walking the robot. However clumsy these first steps might have been, Brooks and Flynn had grand visions for such machines:

> Complex systems and complex missions take years of planning and force launches to become incredibly expensive. The longer the planning and the more expensive the mission, the more catastrophic if it fails. The solution has always been to plan better, add redundancy, test thoroughly and use high-quality components. Based on our experience [...] we argue here for cheap, fast missions using large numbers of mass produced simple autonomous robots that are small by today's standards (1 to 2 kg). We argue that the time between mission conception and implementation can be radically reduced, that launch mass can be slashed, that totally autonomous robots can be more reliable than ground-controlled robots, and that large numbers of robots can change the tradeoff between reliability of individual components and overall mission success. Lastly, we suggest that within a few years it will be possible at modest cost to invade a planet with millions of tiny robots.[21]

Though written in the 1980s, this passage mentions nearly all of the factors that, today, make swarm robotics a promising approach to coping with the complexities of unpredictable environmental conditions: durability, flexibility, reliability, and scalability.[22] In short: "[U]sing swarms is the same as 'getting a bunch of small cheap dumb things to do the same job as an expensive smart thing'."[23] In an indirect way, Brooks and Flynn had thus outlined the basic features of "swarm intelligence," a term that happened to be coined during the same year that their article appeared, namely by Gerardo Beni and Jing Wang at a NATO-sponsored conference on robotics.[24] Although Beni and Wang were referring specifically to computerized modeling and simulation techniques based on cellular automata, their invention of the term 'swarm intelligence' in the context of a robotics conference prompted the development of swarm-intelligent systems in a wide variety of scientific fields.

21 Brooks and Flynn, 'Fast, Cheap, and Out of Control,' 478.

22 See also Brooks, 'Elephants Don't Play Chess.'

23 Corner and Lamont, 'Parallel Simulation of UAV Swarm Scenarios,' 355.

24 See Beni and Wang, 'Swarm Intelligence in Cellular Robotic Systems.'

Whereas much of the revived interest in artificial intelligence today is based on so-called 'deep-learning' techniques in artificial neural networks, which benefit from previously inaccessible or non-existent quantities of digital data and abundant computing power, swarm intelligence is a type of intelligence that can be used to coordinate motion in real time. Because of their 'conspatiality' in local neighborhoods, systems based on swarm intelligence and swarm robotics have been described as "well-suited for tasks that are concerned with the state of a space."[25] In the case of dealing with logistics or routing problems, several swarm-intelligence applications have in fact already proven to be more effective than competing approaches. In the routing protocol known as AntNet, for instance, packets of information hop from node to node, leaving a digital signature indicating the 'quality' of their trip as they do so. Other packets evaluate the trails created in this way and choose their own paths accordingly: "In computer simulations and tests on small-scale networks, AntNet has been shown to outperform existing routing protocols. It is better able to adapt to changed conditions (for example, increased traffic) and has a more robust resistance to node failures."[26] Despite substantial corporate interest in such routing algorithms, the huge costs of hardware replacement that they required prevented their widespread implementation. However, Paolo Gaudiano and his colleagues could not help but remark that the technology "looks promising [...] for *ad hoc* mobile networks like those used by the armed forces and civil-protection agencies. And thus, their applicability points exactly in the direction of interconnected autonomous mobile robots and Swarm UAS – systems where more centralized approaches frequently lead to exponential increases in communication bandwidth requirements and in the size of the controlling software, as well as are dependent on the availability of global information."[27]

Swarms create information by means of formation. They generate a specific secondary environment – the moving collective – that surrounds the swarming individuals and facilitates adaptive processes by means of rapid, nonlinear transmissions of information between these individuals in local neighborhoods. Like the space invaders imagined by Brooks and Flynn, such collectives can take advantage of the allegedly superior capabilities of swarming collectives to explore unknown environments or areas that are difficult to access. In contrast to larger and more sophisticated single robots, drones, or rovers, the individual members of a swarm are able to

25 Şahin, 'Swarm Robotics,' 17.

26 'Riders on a Swarm.'

27 Gaudiano et al., 'Control of UAV Swarms: What the Bugs Can Teach Us,' 1.

scan different areas swiftly, communicate decisive information to their neighbors, and thus contribute to a collaborative and parallel process of collective intelligence. According to Beni, swarm robotics benefits from the fact that "[t]he production of order by disordered action appears as a basic characteristic of swarms."[28] This is due to the ability of swarms to function even though its units "operate with no central control and no global clock."[29] Swarm systems update themselves in a *partially synchronous* manner:

> In fact, during an UC [= updating cycle], any unit may update more than once; also it may update simultaneously with any number of other units; and, in general, the order of updating, the number of repeated updates, and the number/identity of units updating simultaneously are all events that occur at random during any UC. We call the swarm type of updating *Partial Random Synchronicity* (*PRS*).[30]

This allows the individual swarm members to adapt more flexibly to external factors; owing to their restricted range of interaction, moreover, each can only stimulate a limited number of neighbors to engage in an increased or decreased amount of similar activity. The specified size of a neighborhood and the resulting spatial structure and morphology of mobile collectives therefore format their synchronization processes. As a consequence, time-lags do not automatically lead to less sustainable systems; on the contrary, it is precisely the synchronization (not simultaneity) caused by the local transmission of information that enhances or weakens a collective's reactions to external influences. Asynchronicity in swarms – 'order by disordered action', as the title of Beni's article has it – thus becomes operative by means of spatial dispositions. As a result, the dynamic equilibrium of robotic collectives depends not only on its immediate reactions to external stimuli but also on its 'conspatial' arrangement, which is based on the parameters of its sensors for measuring position, distance, and speed. It is thus not a coincidence that so much of the research devoted to swarm robotics today is concerned with unmanned aerial systems, for the latter are able to navigate easily in three dimensions and do not have to cope with the complexities faced by vehicles grounded in two-dimensional environments.

The potential maneuverability of swarming unmanned aerial systems adds an additional aspect to swarm intelligence and its concern with space.

28 Beni, 'Order by Disordered Action in Swarms,' 153.
29 Ibid., 154.
30 Ibid., 157 (emphasis original).

The spatial intelligence of swarms not only constitutes their capacity for self-organization – their ability to move as cohesive collectives while adapting to external forces; it also provides them with the means to 'read' or 'record' various data about their surrounding conditions. In this sense, swarm intelligence is above all a form of *reconnaissance*. For instance, if a member of a swarm identifies something of interest with regard to a pre-defined goal, it could attract additional members to the area in question. As the computer scientist Erol Şahin has put it: "[D]istributed sensing by large numbers of individuals can increase the total signal-to-noise ratio of the system."[31] As a group, the members of a swarm can deliver a detailed view of a given target (from different angles, for instance) and they can do so more quickly than any individual. In a way, unmanned aerial swarming systems are thus physical materializations of a mathematical search procedure from the branch of swarm intelligence known as 'particle swarm optimization' (PSO). PSO algorithms were inspired by the strategies employed by flocking birds when searching for feeding sites. By implementing the 'cornfield vector' developed by the ornithologist Frank Heppner, the PSO developers James Kennedy and Russell Eberhart modeled a search algorithm for identifying the maxima and minima of nonlinear functions and for solving multi-objective optimization problems. Their model uses randomly dispersed particles to search space, compare their relative positions within it, and eventually converge around local or – in ideal circumstances – global optima.[32] Again, the central idea of this model is based on the question of how a flock of birds can come together and search for food within a given area. The target of the search motivates the movements of the simulated swarm, which then finds its goal quickly by its individuals transmitting local information to their nearest neighbors – one of the clear evolutionary advantages that the formation of swarms creates. Inspired by examples from biology, Kennedy and Eberhart devised a problem-solving method that makes use of the inter-individual relations within a swarm and their influence over the behavior of the swarm as a whole. By means of evolving and 'self-learning' algorithms, moreover, their model became more and more efficient over time. Their broad definition of a swarm is as follows:

> A swarm is a population of interacting elements that is able to optimize some global objective through collaborative search of a space. Interactions

31 Şahin, 'Swarm Robotics,' 11.

32 See Kennedy and Eberhart, 'Particle Swarm Optimization,' 1942–1948; Heppner and Grenader, 'A Stochastic Nonlinear Model for Coordinated Bird Flocks,' 233–238; Engelbrecht, *Fundamentals of Computational Swarm Intelligence*; and Poli et al., 'Particle Swarm Optimization.'

> that are relatively local (topologically) are often emphasized. There is a general stochastic (or chaotic) tendency in a swarm for individuals to move toward a center of mass in the population on critical dimensions, resulting in convergence on an optimum.[33]

Much like the use of swarms in the field of animation, the concept could be implemented effectively here without overtaxing the storage capacity and processing power of available computers. In addition to calculating the maxima and minima of nonlinear functions, the algorithm could also be used to solve problems in which a system of functions had to be optimized.[34] In doing so, the optimization of one function was often conflated with that of another, so that it was necessary to find a solution by examining the context of different target definitions. One application of PSO would be to model a manufacturing process in which all of the involved parameters influence one another and their combination affects the ratio between the quantity and quality of the production. In order to identify the perfect setting, all of the possible combinations of parameters would have to be tested, and even in the case of just a few parameters, the total number of combinations can soon become exorbitantly large. What is more, the parameters are often not integers but rather real numbers, so that it would be out of the question to calculate all of their possibilities.[35]

PSO made it possible to address such problems by exploring the space of all possible combinations of conditions in a 'swarm-like' manner. Its optimization algorithm is a simplified Reynolds model with two initial parameters: velocity and 'craziness.' The particles are then randomly distributed throughout the search space, where they remain in contact with a defined number of neighboring particles, and each respective position is simultaneously a suggested solution for the target function. Over the course of iterative steps, both the personal best positions (whose optimum is remembered as a sort of 'memory') and the various neighborhoods' best positions (a neighborhood consists of a fixed number of nearest neighbors) are determined. These are then compared with one another, and the established distances in turn define the new flight path and speed for every individual particle. A crucial aspect in all of this is the swarming of the particles, which is made dynamic by the 'craziness' parameter. It is this parameter that disturbs and alters the direction and speed of a few randomly chosen particles, and thus it serves to

33 Kennedy and Eberhart, *Swarm Intelligence*, xxvii.

34 Kennedy and Eberhart, 'Particle Swarm Optimization,' 1942.

35 See Ziegler, 'Von Tieren lernen.'

simulate the effects of external influences, incomplete information, and the resulting uncertain behavior on the part of certain swarming individuals. This variation results in realistic movements and it prevents the swarm from concentrating too quickly around a particular point (that is, a solution) and running the risk of detecting just a single local maximum. In this process, moreover, a significant role is played by the size of the neighborhood under consideration. If it is too large, the system will tend to converge too early on a single point, yet if it is too small, then this will prolong the calculation time. Like the swirling formations and ultimate congregation of birds around a food source, a particle swarm gradually condenses around one location, which is the maximum or minimum of the desired function. In the end, the manner in which the particle swarm comes together and the interrelations between the individual particles point to a nearly optimal solution, thus creating information through formation.

The PSO model has since been adjusted in various ways. Because the focus has shifted toward goals that are entirely different from studying or visualizing realistic swarming behavior (goals such as optimizing neural network, arranging sequences of genes, or analyzing the stability of electricity grids), it has been possible to make more flexible assumptions about how information is transferred within the model. Something called 'global best positions' has been introduced, for instance, and this involves suggesting the best solution during every iteration of the model, suggestions that in turn influence the positions and behavior of the particles during the subsequent interval. The craziness parameter has in large part been eliminated, as has the ability of models to match the velocities of neighboring particles, with the result that "optimization occurs slightly faster [...] though the visual effect is changed. The flock is now a swarm, but it is well able to find the cornfield."[36] With or without craziness, however, swarms and swarming have shed their image as traditional embodiments of the irrational.[37] In the case of PSO and other recent applications, they have instead become effective models for problem-solving.

Again, what characterized swarm intelligence at first as a fruitful subfield of artificial intelligence was its multifaceted *spatial intelligence* and its ability to solve complex problems through the relatively simple act of

36 Kennedy and Eberhart, 'Particle Swarm Optimization,' 1944. For an overview of further developments in the field of particle swarm optimization, see Hu et al., 'Recent Advances in Particle Swarm.'

37 See, for example, Hinske, 'Die Aufklärung und die Schwärmer: Sinn und Funktion einer Kampfidee,' 4; and Böhme and Böhme, *Das Andere der Vernunft*, 238.

self-organization. However, in light of recent developments in robotics hardware – and particularly the hardware used in unmanned aerial systems – its applications have expanded considerably and it has even become possible to design bionic robots that mimic certain facets of 'mother nature.' Here I would like to focus on spatial intelligence and its applications for *reconnaissance*. In particular, I will argue that recent research projects devoted to swarming unmanned aerial systems have interwoven technological and ecological environments in novel ways and have thus reconceptualized the idea of 'controlling' space.

Swarm Robotics

Whereas swarm intelligence and agent-based modeling began to flourish in the mid-1990s, the development of swarm robotics lagged somewhat behind. Edited by Erol Şahin, the seminal anthology *Swarm Robotics* was not published until 2008; here, swarm robotics is defined as "the study of how a large number of relatively simple physically embodied agents can be designed such that a desired collective behavior emerges from the local interactions among agents and between the agents and the environment."[38] In his introduction to the volume, Gerardo Beni offered the following explanation for why the field did not take off as quickly as one might have expected: "[T]he original application of the term [swarm intelligence] (to robotic systems [...]) did not grow as fast. One of the reasons is that the swarm-intelligent robot is really a very advanced machine and the realization of such a system is a distant goal (but still a good research and engineering problem). Meanwhile, it is already very difficult to make small groups of robots do something useful."[39] Although the book includes reports on pioneering work such as the Swarm-Bots and I-SWARM projects,[40] most of the chapters are concerned with preliminary questions about how to build functioning robotic collectives in the first place. In short, they are interested in little more than making various sorts of computer simulation software more realistic by introducing a degree of 'hardware realism,' a task for which it is necessary to take into account such things as the specific bandwidth restrictions and positioning routines of existing robotic systems. In their thoughts about "future developments," the researchers imagined a

38 Şahin, 'Swarm Robotics,' 12.

39 Beni, 'From Swarm Intelligence to Swarm Robotics,' 7.

40 See Groß et al., 'Autonomous Self-Assembly in Swarm-Bots'; and Seyfried et al., 'The I-SWARM Project.'

whole range of potential applications, including collective minesweeping and the distributed monitoring of geographic spaces and ecosystems. It was also believed that swarming robots might someday be able to monitor and anticipate environmental disasters and perhaps even join together to block leaks of hazardous material.[41]

Aside from the challenges of engineering functional physical systems, the delayed development of swarm robotics was also due to a reconceptualization of swarm intelligence. In the year 2000, Sanza Kazadi coined the term "swarm engineering" as a way to acknowledge that "the design of predictable, controllable swarms with well-defined global goals and provable minimal conditions" was a pressing need in the field of robotics. "To the swarm engineer," he wrote, "the important points in the design of a swarm are that the swarm will do precisely what it is designed to do, and that it will do so reliably and on time."[42] The idea that the robots could be 'out of control' (as in the title of Brooks and Flynn's provocative article) was dismissed in favor of establishing firm objectives and behavioral controls in order to prevent undesired outcomes. In practical terms, it is simply far more expensive and time-consuming to cope with unmanned aerial systems crashing to the ground in real-world experiments than it is to work with boids crashing into computer-simulated obstacles.

Although he was attempting to circumvent certain unproductive aspects of collective animal behavior, such as the circular milling observable in ants, Kazadi's perspective also altered the initial conceptual relation between swarm robotics and biological swarms. In swarming unmanned aerial systems, physical proximity (which is necessary in flocks of birds and schools of fish) and stigmergy (such as the pheromone trails left by social insects) are often replaced by intra-platform datalinks or backlinks to ground control stations. As long as everything remains within the range of the communication devices, such systems are thus able to maintain cohesion even if the individuals are widely distributed over a given area. This feature makes them all the more attractive as distributed sensor systems, which cause 'swarm space' and environmental space to overlap. Furthermore, swarm-engineered unmanned aerial systems – like the Perdix drone system mentioned above – are based on a principle called 'directed autonomy,' which sounds like an oxymoron.[43] While monitoring the drones' movements on a tactical screen, human operators will enter certain general commands,

41 See, for instance, Beni, 'Order by Disordered Action in Swarms.'

42 Quoted from Brambilla et al., 'Swarm Robotics,' 2. See Kazadi, *Swarm Engineering*.

43 Page and Tripp, 'The Perfect Swarm.'

so-called 'plays' such as 'encircle area X' or 'follow object Y.' The Perdix system is designed to interpret these orders and execute them autonomously according to the local circumstances at hand.[44] In another approach, the conspatiality of swarms enables the operator of an unmanned aerial system to direct an entire collective by taking control over just one swarm member. The others will then automatically follow its lead.[45]

To be sure, such efforts to keep humans 'in the loop' have a number of technical and ethical repercussions that have to be considered.[46] Also, given that a search on *IEEE Xplore* for 'swarm robotics' will generate around 1,500 hits, it seems necessary to separate the wheat from the chaff, not only in terms of the level of autonomy involved but also in terms of the functionality of the systems in question under realistic conditions. That said, it must be acknowledged that a profound transformation has taken place over the last ten years, in that swarm robotics is no longer just a technological niche idea but has instead come to occupy a prominent position in the current discourse concerning autonomous robotics. Together with increased (military) investment in micro-unmanned aerial systems and the greater availability of lightweight and reliable building components, sensors, communication devices, and software, the newfound prominence of swarm robotics has perhaps also done much to reinforce a particular interpretation of its potential political, social, and technological implications. According to the assessment in question, swarm robotics combines the existing *politics of swarms* with their physical embodiment as mechanical collectives and thus it is all the more important to discuss the *spatial intelligence* of such systems as a significant feature of today's media ecology.[47]

The objective of swarming unmanned aerial systems to gather information about their environments underscores the *reconnaissance* aspect of 'intelligence.' Understood as distributed sensor systems, these systems are designed to collect data "across spatial, temporal, and spectral domains,"[48] with many of them developed for the purpose of *pervading* space in three dimensions and time. In this way, systems of this sort could in principle provide more detailed information about a given area than any single drone or satellite ever could.[49] Moreover, recent technological developments might be able to overcome the main shortcoming of today's unmanned aerial

44 See Feng and Clover, 'Drone Swarms vs. Conventional Arms.'
45 See Scharre, *Robotics on the Battlefield Part II.*
46 See Bhuta et al., *Autonomous Weapons Systems.*
47 See Parikka, 'Politics of Swarms.'
48 Page and Tripp, 'The Perfect Swarm,' 4.
49 See, for instance, Colomina and Molina, 'Unmanned Aerial Systems for Photogrammetry,' 79.

system, namely their limited battery power. Much research is currently devoted to constructing robots that could become *permanent* elements of a hybrid techno-ecological environment, either by mimicking animal behavior like the "synsects" in Stanisław Lem's story or by integrating technical infrastructures into this set of behavior. Finally, swarming aerial systems could be used to affect certain environments on the electromagnetic level, be it as wireless communication networks,[50] in the military field of electromagnetic warfare, or as jammers and distributed beamforming radar platforms.[51]

Returning to Brooks and Flynn's article for a final time, we see that the authors emphasized yet another phenomenon that is of the utmost relevance today: economies of scale. Small robots can be mass-produced; they can be built out of parts that are readily available,[52] or they can even be 3D-printed on site because their capabilities depend on their software and communication routines and not on the sophistication of their hardware.[53] Indeed, technologies developed for smartphones – such as miniature cameras, GPS navigation, radio communication, data-processing power, and sensors for measuring their relative position, acceleration, and environmental information (sound, pressure, humidity, etc.) – are well-suited to the capabilities of miniature aerial swarms,[54] and they share the latter's need for minimal weight, size, and power.[55] The immense investments on the part of the smartphone industry into advancing such technologies has thus led to improvements in unmanned aerial systems, and these developments have in turn resulted in a rapidly growing market for commercial drones. That said, it would be an oversimplification to speak of micro-drones as winged smartphones, with "the wings [being] the cheap part."[56] Recent research devoted to propulsion technologies – from winged layouts and multicopter technology to bio-inspired and insect-like flapping techniques and cyclocopter technology (which lacks any biological analogy) – suggests that different objectives would require vastly different types of drones.[57]

50 See Kruzelecki, 'Flying Ad-Hoc Networks'; and Jimenez-Pacheco et al., 'Implementation of a Wireless Mesh Network of Ultra-Light MAVs with Dynamic Routing.'

51 See Kocaman, *Distributed Beamforming in a Swarm UAV Network.*

52 See Scharre, *Robotics on the Battlefield.*

53 See, for example, Marks, '3D Printing Takes Off with the World's First Printed Plane'; and Balazs and Rotner, 'Open, Commercial Technologies Lead to Cost-Effective Reconnaissance Solutions.'

54 See Cevik et al., 'The Small and Silent Force Multiplier,' 602.

55 See Hambling, *Swarm Troopers*, 4.

56 Ibid.

57 See Vásárhelyi et al., 'Outdoor Flocking and Formation Flight with Autonomous Aerial Robots'; De Croon et al., *The DelFly: Design, Aerodynamics, and Artificial Intelligence of a Flapping Wing Robot*; and 'Military Robots Are Getting Smaller and More Capable.'

As of today, unmanned aerial systems have already been used in a wide variety of applications, including public safety and policing via infrastructure surveillance, environmental or wildlife surveys, and even Lady Gaga's halftime show at the 2017 Super Bowl. Combined with low entry costs and rapid evolution, the availability of off-the-shelf solutions makes it attractive to harness the advantages of swarming by customizing commercial drones with swarm-intelligence algorithms. Swarming unmanned aerial systems offer substantial advantages over individual drones in research fields where the parallel coverage of wide areas is of paramount importance, as in the creation of spatial data and maps for general use,[58] environmental or wildlife monitoring, agriculture, urban studies, and military reconnaissance. Following the principles of particle swarm optimization, areas of interest can be identified easier and faster; by collecting data in parallel, moreover, the tasks of scanning and surveillance can be accomplished more quickly: "Furthermore, fault-tolerance is inherently provided by the use of swarms, because a single drone can be removed with a limited impact on the overall formation. Swarms can also provide scalability, i.e., adding or removing drones from a swarm, in order to better adapt to changing conditions or to simply replace one or more UAVs experiencing issues or battery depletion."[59]

By producing low-altitude remote sensing (LARS) in high definition – be it optical, infrared, acoustic, or otherwise – small drones can also reduce operation costs and provide data that would be unattainable by larger drones or reconnaissance satellites. The data collection conducted by small drones is unaffected by atmospheric interferences, clouds, or other objects that might block top-down surveillance, and because drones of this sort fly at much lower speeds, they can record images in higher resolutions. Some systems are able to stitch together high-resolution images into a mosaic map, with data processing carried out on board. Only when something of interest is identified will such data be relayed back to ground control for further analysis. In this way, the bandwidth requirements can be kept quite low.[60] A specific example of such a system is Carbomap's successful attempt to measure the canopy height of rainforests with airborne LIDAR ("light detection and ranging"). This technology, which recently received popular attention for having produced an airborne scanning of a canopied Maya city, can create three-dimensional maps and can calculate or estimate relevant metrics such as the levels of carbon in a given forest. It can also be used to explore the interiors of buildings from the outside.

58 Colomina and Molina, 'Unmanned Aerial Systems for Photogrammetry,' 79.

59 Bacco et al., 'UAVs and UAV Swarms for Civilian Applications,' 116.

60 Hambling, *Swarm Troopers*, 109.

As mentioned above, however, swarming unmanned aerial systems still suffer from one crucial disadvantage: the highly limited time that they are able to operate. In the case of the micro-systems, this typically ranges from one hour to just a few minutes. Researchers have therefore been attempting to improve power supplies by turning to novel battery technologies, fuel cells, or solar power, the latter being especially suitable for small drones because of their high surface-to-weight ratio.[61] Some developers have also taken certain animal behaviors as a starting point and tried to simulate them with their systems. The examples described below certainly call to mind Rodney Brooks' early research, given that they can be depicted as technologies that mediate between and capitalize on both natural and technological environments.

Researchers, for instance, have simulated the flight patterns of seabirds (with data taken from GPS trackers planted on live albatrosses) in order to extend operation times. Drones of this type can learn to orient themselves in relation to airflows and plan their trajectories through different layers of air by continuously calculating a wind-field model based on data from computer simulations and multiple sensors.[62] The company Vishwa Robotics builds unmanned aerial system that have the ability to perch. They save power between flight times and gather intelligence from more stable viewpoints instead of hovering constantly in mid-air. By analyzing the precise behavior of landing birds, the researchers identified different landing strategies and developed simplified artificial legs that allowed for bird-like landing maneuvers both on flat surfaces and on branches.[63] The ability to perch autonomously, however, also requires the ability to identify suitable locations and to steer toward them. Visual sensors were installed to identify, select, survey, and exploit such places by creating three-dimensional models of possible landing areas. Another approach has been to let the drones spiral downward and use cameras to detect possible perches by the shadows that they cast.[64]

The concept of 'resting' drones has been taken even further. If drones can be made to perch in particular places – such as power lines – then they can not only save power but in fact recharge their batteries. A product called Bat Hook by Design Research Associates can be tossed over a power line,

61 Ibid., 133–136.

62 See Hambling, 'Drones with Legs Can Perch, Watch and Walk Like a Bird'; and Langelaan and Roy, 'Enabling New Missions for Robotic Aircraft.'

63 Gajjar, 'Parallel Kinomatics Micro-Positioning System.'

64 See, for example, Bosch et al., 'Autonomous Detection of Safe Landing Areas for an UAV from Monocular Images.'

where its sharp edge will cut through the insulation. The device also works as an AC/DC converter that can reduce high-voltage power for the purpose of charging electronics.[65] An apparatus known as Urban Beat Cop, which is a sort of mobile closed-circuit television, is equipped with a similar system, with which it could conceivably carry out missions indefinitely and become a permanent feature of urban landscapes.[66] As a result, "drones [have become] not just tactical devices for patrolling or dealing with a particular incident [...]. The Urban Beat Cop design includes software to carry out some types of pattern-of-life monitoring automatically. It could keep track of the comings and goings of specific vehicles in an area and potentially even individuals. It may not be Big Brother watching you in the future, but a small perching drone."[67]

These latter projects are thus aiming to add durability to the spatial and behavioral intelligence of unmanned aerial systems, a feature that would further integrate them into their ecological and technological environment – from airflows to electric currents. Along with the spatial-intelligent aspects of dynamic self-organization and the advanced reconnaissance capabilities of distributed sensor networks, this level of durability fosters an understanding of such systems as zootechnological hybrids that mediate between and actively intertwine biological and technological ecologies. This aspect is further bolstered by certain unmanned aerial systems that could be said to function as media ecologies in their own right. For instance, the "Swarming Micro Air Vehicle Network" (SMAVNET) project that is underway at the École Polytechnique Fédérale in Lausanne (EPFL) explores the benefits of so-called flying ad-hoc networks (FANET), which can remain functional during catastrophic events when ordinary communication networks might be out of service. Flying ad-hoc networks can locate Wi-Fi devices by detecting their Wi-Fi packets. Because commercial Wi-Fi devices such as smartphones, tablets, or laptops periodically transmit signaling packets, the drone swarm can learn and process various parameters by accessing them. By comparing their respective power levels, the drones can estimate the position of transmitters and thus assist rescue teams in locating victims.[68] It goes without saying, however, that such technology could be just as useful for military reconnaissance missions.

65 Hambling, *Swarm Troopers*, 129.

66 See Anonymous, 'The New Beat Cop: Transforming a Small Drone into an Unattended Urban Surveillance System.'

67 Hambling, *Swarm Troopers*, 130.

68 See Kruzelecki, 'Flying Ad-Hoc Networks.'

Of course, it has to be kept in mind that most of these examples are in rather early stages of development and that, as mentioned above, the technology behind swarming unmanned aerial systems is still in its infancy. That said, it is clear to see that rapid advances in robotics have had profound effects on the 'materialization' of swarm-intelligent concepts and algorithms into robot collectives. These developments, moreover, have incited a lively discourse concerning the potential socio-political consequences of the hypothetical widespread use of such 'artificial nonintelligence' – a discourse that, with its pessimistic predictions about the behavior of autonomous robots, tends to revive earlier notions about the "eeriness" of swarms.[69]

Weapons of Mass Production, or: An Abuse of Consumer Electronics

To some extent, one of the most bizarre weapons projects from the Second World War already incorporated some of the features that make swarming unmanned aerial systems such a fruitful area of research today. Initiated by a dentist and inventor named Lyte S. Adams, the American military conducted an undertaking called Project X-Ray from 1942 to 1945.[70] Its aim was to devise a "vector method of incendiary bombing" by filling bomb canisters with bats. Outfitted with explosive devices, the bats were to be released above Japanese cities, where they would – it was hoped – spread out to seek refuge beneath the roofs of houses and then incinerate the hideouts with a chemical agent that would later be known as napalm.[71] Despite these intentions, however, the weapon proved to be uncontrollable, as became evident when six armed bats escaped the laboratory at Carlsbad Auxiliary Airfield and blew up several buildings. From the perspective of military strategists, what was even worse was the fact that the effects of such a weapon could not be calculated. Ultimately, the idea of the bat bomb was retired when another device for attacking Japanese cities happened to become mission-ready ahead of time and promised greater efficiency in terms of manufacturing and destruction.[72] Yet as an incalculable, area-wide, surprising, and compact weapon, the bat bomb – in Claus Pias' opinion – structurally corresponded to the device in question: the atomic bomb.[73] That said, the bat bomb would have been terrifying because of the invisibility of its explosive agents. A

69 See, for instance, Vehlken, 'Gesichter im Sand.'

70 Pias, '"Bat Men Begin",' 306.

71 Couffer, Bat Bomb: *World War II's Other Secret Weapon*, 11.

72 Fieser, *The Scientific Method*. Quoted from Couffer, *Bat Bomb*, 213.

73 Pias, '"Bat Men Begin".' 315.

dispersed bat attack would not have been identifiable as an attack but would rather have appeared as a mere environmental disaster. This would have been quite the opposite of giant blasts and mushroom clouds.

Today, swarming aerial systems have the potential to create more refined 'vector methods' than those of its portentous historical forerunner. According to a number of military analysts, recent technological developments such as those described above are likely to have an imminent effect on the nature of military force. If these 'bunches of small cheap dumb things' could accomplish the same tasks in warfare as an expensive smart weapon, a change in military strategy and thinking would be at hand. Such possibilities upend Friedrich Kittler's idea that "the entertainment industry is, in any conceivable sense of the word, an abuse of army equipment."[74] Today it seems that the military is abusing consumer electronics.

"Quantity has a quality of its own," as Joseph Stalin once said, and the trend over the past quarter century to deploy fewer but more advanced (and expensive) weapons platforms could be reversed by swarm technology: "The next generation of weapons may see sophisticated technology systems outdone by the sheer numbers of autonomous swarms."[75] Being the manufacturing center for the commercial drone and smartphone industry could therefore be a substantial competitive advantage. It is no coincidence that private sector manufacturers in China have already been co-opted to work for the People's Liberation Army.[76] Today it is the case that militaries are attempting to take advantage of the faster innovation and development cycles of private manufacturers. And if, at some future stage of development, the mass-production of swarm-engineered aerial systems becomes feasible, smaller countries and even groups of people could then be endowed "with capabilities that used to be the preserve of major powers," thus complicating responses to crises and the ability to influence events with military force:[77]

> Swarm technology, say defense experts, is attractive [...] as it would allow [nations] to project force with a lower probability of military confrontation. Drones, unlike fighter jets or aircraft carriers, are less threatening and can be shot down or captured without triggering a military escalation. In December, China seized a US underwater drone in the South China

74 Kittler, *Gramophone, Film, Typewriter*, 96–97.

75 Feng and Clover, 'Drone Swarms vs. Conventional Arms,' n.p.

76 Ibid.

77 Hammes, 'Technologies Converge and Power Diffuses,' 8.

> Sea, which the PLA then handed back after a few days. This would have triggered a major crisis had it been a manned vehicle.[78]

This conclusion is speculative, however, as are the claims that swarming aerial systems might disprove the conventional wisdom that wars cannot be won with air power alone. Equipped with improved identification and targeting technologies, which derive from the commercial sector and thus further blur the distinction between military and civilian applications,[79] swarming drones could in fact make it possible to occupy and 'pervey' large areas for long periods of time, and they could also carry out precise strikes in far greater numbers than common drones ever could.[80] In short, this ability could herald a new age in military technology: "[T]he nuclear balance is maintained because neither side can disable the other's strategic weapons with a first strike. Swarms might change this balance and make first strikes possible – or strikes by non-nuclear powers seeking to disarm nuclear ones."[81]

Around the year 2000, 'swarming' began to be discussed as an incipient doctrine of network-centric warfare in the U.S. military. At that time, however, when authors such as Sean Edwards, John Arquilla, and David Ronfeldt referred to biological, historical, and future scenarios of swarming, they employed the term in a metaphorical sense.[82] It was applied to all sorts of cooperative, networked actions on the battlefield; the focus was primarily on tactics, and particularly on the use of special forces that could be coordinated for warfare in a network- centric manner; and thus human soldiers were still central to the discourse.[83] Swarming drone systems, on the contrary, represent a shift toward 'true swarming':

> Emerging robotic technologies will allow tomorrow's forces to fight as a swarm, with greater mass, coordination, intelligence and speed than today's networked forces. Low-cost uninhabited systems can be built in large numbers, [...] overwhelming enemy defenses by their sheer numbers. Networked, cooperative autonomous systems will be capable of true swarming – cooperative behavior among distributed elements that gives

78 Feng and Clover, 'Drone Swarms vs. Conventional Arms,' n.p.

79 Ibid.

80 Hambling, *Swarm Troopers*, 288.

81 Ibid., 302.

82 See Edwards, *Swarming on the Battlefield*; and Arquilla and Ronfeldt, *Swarming and the Future of Conflict*.

83 Kaufmann, 'Soldatische Subjekte.'

> rise to a coherent, intelligent whole. And automation will enable greater speed in warfare, with humans struggling to keep pace with the faster reaction times of machines. The result will be a paradigm shift in warfare where mass once again becomes a decisive factor on the battlefield, where having the most intelligent algorithms may be more important than having the best hardware, and where the quickening pace of battle threatens to take control increasingly out of the hands of humans.[84]

As mentioned above, this type of artificial intelligence involves autonomously coordinated movement and navigation, distributed sensing, and multi-spectral imaging. The ability of the drones to self-organize collectively and oscillate between dispersion and concentration makes them efficient weapons and reduces the danger of them being detected or shot down all at once. In the words of Paul Scharre: "Mass allows the *graceful degradation* of combat power as individual platforms are attrited, as opposed to a sharp loss in combat power if a single, more exquisite platform is lost. Offensive salvos can *saturate enemy defenses*. Most defenses can only handle so many threats at one time."[85] Films such as *The Matrix Revolutions* (2003), *The Day the Earth Stood Still* (2008) and *Star Trek: Beyond* (2016) have depicted such swarm attacks with impressive CGI sequences, sequences that were themselves designed with the help of swarm intelligence.[86] It can be imagined that the visual sophistication of such fictional sketches might have contributed their part to the portrayal of Scharre's scenarios.

Some authors, moreover, have stressed that different types of sensors could be distributed to different swarm members to create something called heterogeneous group control.[87] What this means is that the functions of failing or eliminated drones can easily be taken over by other swarm members so that the operational readiness of the system remains intact. Other researchers have explored the suitability of swarming drones as electronic devices for warfare, that is, as distributed beamforming platforms for jamming enemy radar or as electromagnetic pulse weapons.[88] Swarming unmanned aerial systems could also serve as mobile minefields in the air, on the ground, or under water. Furthermore, their small size is considered an asset: "With the advancement of radar and sensors in addition to ongoing

84 Scharre, *Robotics on the Battlefield*, 10.
85 Ibid., 14 (emphasis original).
86 See, for instance, Vehlken, 'Gesichter im Sand.'
87 See 'Military Robots Are Getting Smaller and More Capable.'
88 See Kocaman, *Distributed Beamforming in a Swarm UAV Network*; Cevik et al., 'The Small and Silent Force Multiplier'; and Hammes, 'Technologies Converge and Power Diffuses,' 8.

developments of counter-stealth technology, only systems at the micro, near-silent, and ultra-low energy levels will have any chance of operating undetected."[89] With novel types of nano-explosives, moreover, the small payload capacity of swarming drones could nevertheless have highly destructive effects,[90] somewhat like the bats of Project X-Ray but now with one or multiple clearly defined vectors.

However, if (semi-) autonomous swarming systems were to be employed to project force in such ways, this would of course intensify certain ethical objections (to such things as automated decisions to kill). Various writers have therefore asserted that the use of such systems should be regulated by international law, that they ought to be classified as weapons of mass destruction, or that they should at least be banned like landmines and cluster bombs.[91] And this, finally, brings us back to the perspicacity of Stanisław Lem. In the absence of regulations and critical discussions, it is possible that the spatio-temporal intelligence of swarming unmanned aerial systems might lead to frightening scenarios that resemble those imagined in his essay:

> The greatest problem in the unhuman stage of military history was that of distinguishing friend from foe. This task had been accomplished, in the twentieth century, by means of electronic systems working on a password principle. Challenged by radio, a plane or an unmanned missile either radioed the right answer or else was attacked as an enemy craft. This ancient method now proved useless. The new weapon-makers again borrowed from the biosphere [...]. The nonliving weapon might imitate (extremely well) floating dust specks or pollen, or gnats, or drops of water. But under that mask lay a corrosive or lethal agent. [...] Thus peace was war, and war peace. Although the catastrophic consequences of this trend for the future were clear – a mutual victory indistinguishable from universal destruction – the world continued to move in that fatal direction. It was not a totalitarian conspiracy, as Orwell once imagined, that made peace war, but the technological advances that effaced the boundary between the natural and the artificial in every area of human life.[92]

In the end, the potential significance of technologies that seek to exploit the 'artificial nonintelligence' of swarm intelligence and swarm robotics

89 Goh, 'Unmanned Aerial Vehicles and the Future of Airpower,' 46.
90 For a compelling overview, see Hambling, *Swarm Troopers*, 209–241.
91 See Chamayou, *Drone Theory*; and Hammes, *Swarm Troopers*.
92 Lem, 'The Upside-Down Evolution,' 59, 66.

depends on how the concept of 'pervasion' is interpreted. As suggested by a number of promising applications – for surveying territory, solving routing problems, or assisting in rescue missions – this can be understood as something that can create more efficient and sustainable methods for controlling space (in a benignly managerial sense of the term). As the military debate implies, however, such pervasion might also have an oppressive biopolitical downside. It is not inconceivable that the biologically inspired spatio-temporal intelligence of swarming unmanned aerial systems – of all things – might lead society to regress toward something like a Hobbesian 'state of nature.' Only this time it would be a media-ecologically enhanced version of such a condition. Although such scenarios remain speculative and the technologies in question are still under development, it would not be outlandish to predict that the particular form of artificial intelligence known as swarm intelligence might soon leave the confines of computer simulations, where it is used to control the behavior of artificial agents in artificial space, and come to transform the way that we understand and control space in the real world.

2. Swarming Out

In light of the spread of swarm intelligence as a buzzword in robotics and the utility of particle swarm optimization in a variety of contexts, it can be said that the concept of the swarm has expanded beyond its association with technical system and that the impulses for this expansion have come from engineering and the natural sciences. Of course, this broadened understanding of the term has brought with it certain distortions regarding the fundamental ways in which swarms function, and in many fields the word has come to be used in an increasingly metaphorical manner. In their book *Swarm Intelligence*, Kennedy and Eberhart devote more than a hundred pages to surveying scholarly literature from a number of disciplines – social psychology, cognitive science, artificial life, robotics, and evolutionary programming – with the goal of defining *intelligence* as a *social principle*: "We will investigate that elusive quality known as intelligence, which is considered first of all a trait of humans and second as something that might be created in a computer, and our conclusion will be that whatever this 'intelligence' is, it arises from interactions among individuals."[93] More and more, swarms are being used as a synonym for the 'social.' Human beings and their interactions

93 Kennedy and Eberhart, *Swarm Intelligence*, xiii.

have returned to the foreground, and psychological, neurophysiological, and bio-technological discourses are beginning to mix together. At the same time, certain cognitive models and 'good old-fashioned artificial intelligence' have been subjected to criticism. What remains of the *swarm* is nothing but a bare framework, cobbled together with a sort of childish metaphysics: "[W]e use the word *swarm* to describe a certain family of social processes. [...] This is a good visual image of what we talk about. [...] As you will see, [...] [a] swarm is a three-dimensional version of something that can take place in a space of many dimensions – a space of ideas, beliefs, attitudes, behaviors, and the things that minds are concerned with."[94] In broadening their definition of swarm intelligence, the authors have followed the ideas of Mark Millonas, who thought of it in terms of five basic principles:

> The *proximity* principle: The population should be able to carry out simple space and time calculations. The *quality* principle: The population should be able to respond to quality factors in the environment. The principle of *diverse response*: The population should not commit its activity along excessively narrow channels. The principle of *stability*: The population should not change its mode of behavior every time the environment changes. The principle of *adaptability*: The population must be able to change behavior mode when it's worth the computational price.[95]

Despite the computer-scientific context of these remarks, it is possible to think of a number of different structures of information and interaction that would fulfill these principles. Over the course of the computer-scientific discourse about swarm intelligence, which began in the middle of the 1990s, swarms lost some of the distinctness that they were simultaneously gaining within the field of swarm research by means of the chiasmus of agent-based computer simulations and the findings of biological research. At the moment when swarms and 'intelligence' became conceptually linked, swarms were no longer just described in terms of collective, motion-based intelligence in animals and machines. On the part of computer scientists and engineers, the discursive field began to accommodate references to a variety of different systems in which self-organization, networking, and flexible dynamics play (or appear to play) a role. They became a hot topic in popular scientific publications and journalistic reports, which associated

94 Ibid., xvi.

95 Ibid., xx (emphasis original). Kennedy and Eberhart refer here to Millonas, 'Swarms, Phase Transitions, and Collective Intelligence.'

the traditional unease about swarms with techno-euphoric regulatory fantasies and other social and political issues. During this process, swarm intelligence became a diffuse concept for anything whatsoever that might exhibit collective dynamics. Swarms came to stand for processes of self-organization in general.

This discursive expansion did not, however, undo the epistemically productive intertwinement of 'bio' and 'tech.' In fact, it might have made it tighter than ever. In 1996, a research group at the Santa Fe Institute led by Nelson Minar developed something called the Swarm Simulation System. As a development tool, this agent-based software environment was designed to encourage collaboration between scientists and computer programmers. Whereas, before, scientists themselves had typically programmed agent-based models to suit their individual interests, the Swarm Simulation System was meant to serve as a convenient standard environment for constructing all sorts of (exchangeable and comparable) computer simulations:

> Unfortunately, computer modeling frequently turns good scientists into bad programmers. Most scientists are not trained as software engineers. As a consequence, many home-grown computational experimental tools are (from a software engineer's perspective) poorly designed. The results gained from the use of such tools can be difficult to compare with other research data and difficult for others to reproduce because of the quirks and unknown design decisions in the specific software apparatus. Furthermore, writing software is typically not a good use of a highly specialized scientist's time. In many cases, the same functional capacities are being rebuilt time and time again by different research groups, a tremendous duplication of effort. A subtler problem with custom-built computer models is that the final software tends to be very specific, a dense tangle of code that is understandable only to the people who wrote it. [...] In order for computer modeling to mature there is a need for a standardized set of well-engineered software tools usable on a wide variety of systems. The Swarm project aims to produce such tools through a collaboration between scientists and software engineers. Swarm is an efficient, reliable, reusable software apparatus for experimentation [...], a well-equipped software laboratory.[96]

The Swarm Simulation System was written in the object-oriented programming language known as Objective-C. Its main innovation is that it

96 Minar et al., 'The Swarm Simulation System.'

provides specific program libraries that are compatible with one another. Among other things, these contain various system layouts, agent classes, behavioral routines, and 'tools' such as genetic algorithms and random number generators that, depending on the application at hand, can be combined to form the basis of a specific simulation environment in which the corresponding agents can be activated: "For example, an agent that is a neural network could start by taking a general purpose neural network class from the *neuro* library, adding extra methods needed for the specific type of network, and then creating an instance of it to be the actual neural network."[97] Individual agents are then connected by means of 'activity' classes; the simulation environment is initiated; so-called 'observer agents' are used to monitor the development of the simulation; and this information is output as a specific form of data. The system is set up in such a way that, during every step of the process, 'probes' can be taken of each of the program's components: "Probes allow any object's state to be read or set and any method to be called in a generic fashion, without requiring extra user code. Probes are used to make data analysis tools work in a general way and are also the basis of graphical tools to inspect objects in a running system."[98] Bringing together all of the users of the system into a single community of software developers not only entailed the constant improvement and expansion of the program libraries but also led to an interdisciplinary exchange of knowledge about the various systems that can be simulated with agent-based models.[99] Like the first multi-robot systems, the early professional stages of agent-based modeling were guided by the concept of the swarm. New program libraries like the aforementioned Ascape, RePAST, and MASON were subsequently created. Though written in Java, the latter programs were designed according to similar principles. However, over time, the community of programmers using the Swarm Simulation System has become smaller and smaller, to the point that the further development of the system has more or less come to an end.

Agent-based modeling has remained popular, though, for the simple reason that it can be used to create complete and easily manageable simulation environments. Following the example of the programming language Logo, which abused many students of computer science during the 1980s and 90s, Michael Resnick and others at MIT developed the agent-based system known as Starlogo with the explicit goal of making it easier for young programmers

97 Ibid., 7 (emphasis original).

98 Ibid., 6.

99 See ibid., 11.

to learn about the dynamics of distributed systems.[100] Regarding simulations designed for scientific purposes, the program called Netlogo, which was developed in 1999 by Urs Wilensky at Northwestern University, probably has the most comprehensive publicly accessible program library, and it combines this with a user-friendly interface.[101] For its part, the commercially marketed simulation known as Anylogic makes it possible to combine agent-based modeling with other simulation methods.[102] In each of these environments, simple swarming and flocking simulations are part of the standard repertoire. Thus, it can be said that the swarms have not only inspired animation designs, served as an example for early agent-based models, and provided the name of the first professional program library. In their many computer-supported variants, they have also continued to be used as a part of agent-based modeling environments. Whereas, for centuries, swarms were found fascinating on account of their intransparency, uncanniness, and inaccessibility, computers have now made it possible to control them with model specifications and to manipulate their dynamic behavior. In addition to the scientific application of agent-based models as a new epistemic tool for biological research, these applications have now swarmed out and are being used to solve problems in a wide variety of settings.

3. Swarm Architecture[103]

Shaken or Stirred: Do I Look Like I Give a Damn?

> "We can think about form simply as organization."[104]

Mies van der Rohe, a notoriously heavy drinker who allegedly asserted that architecture is no cocktail,[105] certainly would have been shocked by the theoretical and aesthetic mix that came along with the advent of digital technologies

100 See Starlogo's website: https://education.mit.edu/portfolio_page/starlogo-tng/ (accessed 26 July 2018).

101 NetLogo's homepage at Northwestern can be found at https://ccl.northwestern.edu/netlogo/ (accessed 26 July 2018).

102 See https://www.anylogic.com/ (accessed 26 July 2018).

103 This sub-chapter is the slightly revised version of an article which was first published as Sebastian Vehlken, 'Swarming. A Novel Cultural Technique for Generative Architecture,' in: *Footprint* 15 (2014) (= Special Issue Data-Driven Design, ed. Henriette Bier, Terry Knight), 9–17.

104 Quoted from 'Interview with Roland Snooks.'

105 See Hine, 'One Architect Who Left His Mark on Cities.'

in architectural design and construction. The early 1990s saw the rise of novel approaches such as 'digital tectonics,' which appeared alongside the invention of 'spline modeler' software tools. Architects began to manipulate continuous curved lines directly on their computer screens. They began to mass-produce blob-like forms and challenge modernist concepts of ordering space.[106] Informed by the poststructuralist ideas of Gilles Deleuze, Bernard Cache, and Manuel DeLanda, the digital turn in architecture fostered a fascination with time-based, interconnected, and evolutionary processes.[107] Now, the crucial design choice involved setting adequate limits for variations, and thus the architect's role changed from designing static structures to arranging dynamic processes with various potential outcomes. As an effect, so-called 'parametricism' was hailed as the "new global style for architecture and urban design."[108] This early and influential cocktail of poststructuralist philosophy and digital architecture, however, often underplayed the specific 'materialities' of computer technology, design software, and animation tools, which were simply used to manage complex sets of data. The much-celebrated conceptual shift toward emergent characteristics, self-organizing systems, and the generative aspects of nonlinear feedback processes was more metaphorical than anything else.

This became all the more obvious when architects such as Kas Oosterhuis decided to take these approaches even further during the last decade. He not only emphasized the ongoing gamification of architectural design but, in a rather counterintuitive way, he also referred to *swarming* as a novel way of conceptualizing architectural design.[109] According to Oosterhuis, 'swarm architecture' would replace previous forms of design with an all-encompassing notion of architecture as a flow of information. His approach revolved around structuring various motion vectors within a distributed system of interacting agents (people, materials, environmental factors, etc.). And with its appeal to the bottom-up principles and emergent global behavior of agent-based modeling and simulation, swarm architecture also transcended the generative principles of spline modeling and parametric design. As the Australian architect Roland Snooks remarked:

> I consider parametric and emergent as polar opposites. Within parametric hierarchical tools all possibility is given within the starting condition,

106 See Carpo, The Digital Turn in Architecture, 9.

107 See Perella, 'Bernard Cache/Objectile'; and Manuel DeLanda's lecture series on the "biology of cities," which can be viewed on YouTube.

108 Schumacher, 'Parametricism.'

109 See Oosterhuis and Feireiss, eds., *Game, Set, and Match II.*

> while emergent conditions arise from nonlinear systems such as multi-agent models. […] [W]hat we are interested in is looking at design from the smallest element and the way that generates order at the macro level.[110]

Or, in Oosterhuis's own words: "An individual architect will no longer be tempted to have the illusion of complete control over the process. […]. Now in the beginning of the twenty-first century architecture is going wild […]."[111] These architectural concepts are part of the recent boom of swarming phenomena that has been taking place in many cultural and socio-historical contexts. This ongoing discourse has given rise to the trend of mixing together, in a sort of willy-nilly fashion, architectural theory with concepts such as emergence, rhizomatic networks, socio-political multitudes, and the social swarming phenomena of human beings. The architect Neil Leach, for instance, has outlined the potential of "swarm urbanism," but in doing so he neglects important differences between the concepts that he employs.[112]

So as not to create another imprecise philosophical cocktail infused with a metaphorical understanding of swarming and other collective dynamics, I would like here to examine swarm architecture and urbanism from another angle. In particular, I intend to adopt a media-technological perspective from which to analyze the philosophy of simulation and its significance to contemporary architectural theory.[113] My hypothesis is that swarm intelligence and agent-based modeling have become fundamental cultural techniques for understanding and governing dynamic processes, and that these techniques have tremendous potential for the field of (generative) architectural design.

Architectural design can benefit from the algorithmic logic of swarm intelligence and agent-based modeling in the following ways: *First*, these types of software expand the possibilities of managing and optimizing the complex interplay of various input variables during building processes. They can be used to integrate the individual movements of particles (simulated humans, traffic flows, winds, etc.) with the mid-level scale of single buildings and with global level of urban landscapes as well. *Second*, if given the right parameters, agent collectives will self-organize into a number of interesting or desirable forms, thereby transforming our understanding of planning and

110 See 'Interview with Roland Snooks.'

111 Oosterhuis and Feireiss, *Game, Set, and Match II*, 76.

112 Leach, 'Swarm Urbanism.'

113 See DeLanda, *Philosophy and Simulation*.

construction processes. From this new perspective, architecture can become primarily a matter of motion. Moreover, this creation of forms develops in ways that would not be possible without the media-technological means of agent-based computer simulations. *Third*, such models introduce a new sort of futurology to architecture. Experiments conducted with agent-based models make it possible to test a number of different scenarios against each other and therefore open up a view into a variety of potentially desirable futures. *Fourth*, such models allow a zootechnological and post-humanist element to enter into the design process by combining traditional (human) design practices with novel media technologies. *Fifth* and finally, the ability to add more and more elements to agent-based models facilitates the synthesis of multiple ideas and makes it easy to integrate the opinions and feedback from customers or future users during ongoing design processes.

Cultural Techniques and Architecture

Given the vast number of applications that have already been discussed in this book, it is possible to understand zootechnological swarming as a more general technique for solving previously unknowable and indistinct sets of problems. Furthermore, if we acknowledge that swarm intelligence and agent-based modeling represent a shift from an analytical to a synthetic approach, then they could in fact be regarded as novel cultural techniques with which to rearrange the world we live in. Yet why should architectural theory care about this sort of cultural analysis? The answer to this question becomes clear in the work of the sociologist Dirk Baecker, who, following Niklas Luhmann's systems theory, believes that the main role of architecture is to distinguish between the inside and the outside.[114] If this is considered the basic (cultural) function of architecture, then it is possible to examine the different media and cultural practices that process this distinction – a distinction that involves certain material and technical aspects as well. For instance, a fence could be perceived as an architectural invention that distinguishes between the inside and the outside, and it could also be seen as one of the initial techniques that transformed early nomadic culture into a culture of settlers. As Bernhard Siegert has argued, moreover, a simple door can give rise to an entire system of cultural operations involving symbolic, epistemic, and social processes. According to Siegert, a door not only connects two rooms but also defines the relation between an inside and an outside. In Georg Simmel's opinion, a closed door not only separates

114 Baecker, 'Die Dekonstruktion der Schachtel,' 83.

two rooms; at the same time, it also functions as *a sign* of that separation. As a consequence, it discriminates not only between physical spaces but also designates such things as arcane or private spheres. Finally, doors can be operated in various ways to induce different cultural practices: either in an anthropomorphic sense, as in the act of quietly closing an office door, or in a mechanical sense, as in the case of automated doors. Doors can therefore be regarded as a (material) architectural medium that becomes a medium of cultural codes and modes of operation.[115] One reason, then, for architectural theorists to take into account cultural techniques is the latter's ability to connect all of these material, social, symbolic, and practical aspects.

A second reason concerns the relationship between time and space. In addition to his discussion of doors, Siegert has also analyzed grids as a fundamental cultural technique with close connections to architecture. As he points out, grid patterns serve as a technique for structuring and controlling space, as is clear from the development of central perspective, from cartography, and from architectural construction. Yet they also aid in the invention and generation of future space, for instance by providing exact layouts for the organization of Roman military camps and reliable address systems in colonial city planning. As another material form for distinguishing internal and external relations, grid patterns thus function as cultural techniques that can be used to *represent* and to *generate* (architectural) realities at the same time.[116]

Like architecture as Baecker, Siegert and others have understood it, the data-driven generative techniques of swarm intelligence and agent-based modeling can be perceived as a novel and synthetic way to mediate between interior and exterior spaces. They build upon a potentially unlimited number of motion-based processes in which boundaries between interiors and exteriors only begin to emerge during simulation runs. Their synthetic character is based on an algorithmic structure that can define neighborhoods for all sorts of objects. In such environments, space no longer has to be organized or constituted by a defined geometric grid; rather, it self-generates out of the multiple local interactions between point clouds or particle swarms. Individual objects, architectural bodies of any size, their interiors and exteriors, and the urban landscapes that they populate can be modeled on the same algorithmic principle of autonomous neighborhoods interacting with one another according to simple rules. And the 'wild' architecture that emerges (in Oosterhuis's terms) can be made perceptible and manipulatable

115 See Siegert, *Cultural Techniques*, 192–206; and Simmel, 'Bridge and Door.'

116 See Siegert, *Cultural Techniques*, 97–120.

with the help of advanced CGI. An effect of this is that swarm intelligence and agent-based models can generate a number of possible future states of buildings, traffic flows, or urban spaces under changing environmental conditions. What is more, they make it possible to compare these potential futures.

Swarm intelligence and agent-based modeling are novel cultural techniques because they approach complex organizational problems by means of artificial *populations* of agents and their behavior in time. It is not geometric principles but rather the movements and vectors of populations that define this novel approach to architecture. It can be said that swarming has introduced animals into the discourse of cultural techniques (and into the discourse of architectural design as well) and that this has been a fruitful zootechnological relation. Produced between the fields of biology and computer science, an understanding of self-organizing collectives has helped us, in a way that anthropology cannot, to deal with certain problems and regulatory issues that are normally regarded as opaque. Models based on biological swarming have been used to co-create processes within our knowledge culture that would not be possible without their media-technological means. Yet how – and to what specific ends – has such software been put to use in contemporary architectural design?

From Insect Media to Bodies with a Vector

In their book *Swarm Intelligence*, Eric Bonabeau, Marco Dorigo, and Guy Theraulaz devote an entire chapter to the simulation of nest-building wasps. With three-dimensional cellular automata and a carefully constructed set of rules, they simulated the development of a nest that one could find in the wild.[117] Inspired by this work, computer scientists sought to shift the use of computer-simulation technology away from confirming scientific hypotheses toward the production of generative and semi-autonomous development, that is, toward 'swarm-driven idea models.' In models of this sort, the simulation environment functions as a virtual testing ground for 'breeding' emergent architectural constructions. In order to create designs that were suitable for solving a given architectural problem, the programmers integrated an evolutionary algorithm into the computer simulation that rated the building activity of a population of randomly chosen swarms. This would then lead to a new population based on the rate-dependent selection of the previous generation of swarms, while random changes

117 Bonabeau et al., *Swarm Intelligence*, 205–252.

and recombinations of successful swarms enabled the development of unforeseen constructions. Over the course of this iterative process, the simulation system yielded interesting architectures according to a set of predefined evaluation criteria.[118] Swarm intelligence thus made it possible to integrate simulated architecture into site-specific contexts and to model certain ecological and economic aspects of the building in question.[119] These generative approaches to creating architectural models, however, are highly dependent on the boundary conditions of the computer simulation, on the algorithm that defines the development and optimization of the generated forms, and not least on the expertise of the 'meta-modeler' (that is, the architect) .

Others have focused on creating a dynamic and mutable swarm space, which is something like an intermediate layer between local information processing and the collective adaptation to the constantly changing exterior forces of a given environment. This technique relies on the nonlinear interactions of multiple individuals to generate dynamic and previously unknowable global forms. Enhanced with sophisticated CGI techniques, agent-based software was soon embraced by a number of architectural design teams. They transformed the act of design into the mere development of suitable rules for governing the construction of a building, and thus architects became meta-designers of self-organizing systems.[120] On the one hand, control was handed over to the bottom-up self-organization of nonlinear systems of agents; on the other, it was also wielded by architects and experts, who would evaluate the generated forms according to certain criteria: "With the centrality of population thinking, the emphasis shifted from both individuals and generalized types to the primary of variation and deviation. [...] [D]ifference and process become comprehensible and hence controllable."[121]

Roland Snooks, one of the collaborators on an architectural project called 'Kokkugia,' has explained how the methods of agent-based modeling deal with explicit architectural problems and how this differs from many of the earlier approaches to digital architecture. Kokkugia, in his words, "has been focused on agent-based methodologies [...]. This started as an interest in generative

118 Von Mammen and Jacob, 'Swarm-Driven Idea Models.' See also Zeng et al., 'SwarmArchitect: A Swarm Framework for Collaborative Construction'; Carranza and Coates, 'Swarm Modelling'; and Nembrini et al., 'Mascarillons: Flying Swarm Intelligence for Architectural Research.'

119 Von Mammen and Jacob, 'Swarm-Driven Idea Models,' 122–124.

120 See, for example, Buus, *Constructing Human-Like Architecture with Swarm Intelligence*; and Nembrini et al., 'Mascarillons.'

121 Parikka, *Insect Media*, 167.

design, not necessarily as a specific interest in computational, algorithmic, or scripted work, but as an interest in understanding the emergent nature of public spaces [...] in Melbourne and how we could develop emergent methodologies. That led us to develop swarm systems and multi-agent models."[122] Yet this raises the question of how to define the architectural problem at hand. Owing to the nonlinear relationality of all objects in a public space,[123] the meta-designers sought to boil down these relations into simple rules. In this way, the micro-relations between individual agents could be connected to a mid-level scale concerning the form of individual buildings as well as to the macro-scale of generative urban planning. With agent-based software, as Oosterhuis has stated, a system of this sort can display real-time behavior, and the parameters can be changed continuously over time. The crucial point is that a comprehensive result can only be achieved by running through the processes. Using the technologies of swarm intelligence and agent-based modeling in generative architecture thus always seems to be a matter of shaping the bottom-up system behaviors in a trial-and-error process. Otherwise, reasonable results or idea models would simply be a matter of luck (or patience): "The challenge for the designer is to find those rules that are effective and which are indeed generating complexity. Some design rules produce death, others proliferate life. Some design rules create boring situations, other rules may generate excitement. You can only find the intriguing rules by testing them, by running the process."[124] Moreover, instead of working with black-boxed modules of commercial architecture software such as Maya or Rhino, Snooks advocates developing open source programs that are specific to a given design intention: "[T]he algorithm should emerge from the architectural problem rather than simply the architecture emerging from the algorithm."[125]

The collaborators on the Kokkugia project describe swarm-based urban planning as a simultaneous process of self-organizing agents that does not result in a single optimum solution or master-plan but rather in a "near-equilibrium, semi-stable state always teetering on the brink of disequilibrium. This allows the system to remain responsive to changing economic, political, and social circumstances."[126] In addition, the project's objective to understand urban dynamics by means of swarm-intelligent systems also involves using generative and nonlinear methodologies to

122 Quoted from 'Interview with Roland Snooks.'
123 See Thacker, 'Networks, Swarms, Multitudes,' n.p.
124 Oosterhuis and Feireiss, *Game, Set and Match II*, 25.
125 Quoted from 'Interview with Roland Snooks.'
126 Leach, 'Swarm Urbanism,' 61.

produce unexpected shapes and to develop novel construction techniques. The latter could lead to a reconceptualization of architectonic form on the basis of agent-based modeling.[127] With their focus on moving patterns and dynamic flows, swarm-intelligent agent-based models can redefine the relationship between local autonomous agents and the material composition of architectural structures, thus endowing the latter with new operational forms. As Neil Leach has stated, these computer simulations integrate the effects of spatial practices – that is, the agents' movements – into the material urban fabric while taking into account the constraints imposed on these practices by their (computer-simulated) physical context.

By this point, the effects of swarm intelligence and agent-based models as cultural techniques should be clear: "The task of design therefore would be to anticipate what would have evolved over time from the interaction between inhabitants and city. If we adopt the notion of 'scenario planning' that envisages the potential choreographies of use within a particular space in the city, we can see that in effect the task of design is to 'fast forward' that process of evolution, so that we envisage – in the 'future perfect' sense – the way in which the fabric of the city would have evolved in response to the impulses of human habitation."[128] Swarm intelligence and agent-based modeling can be defined as cultural techniques that facilitate the ascertainment of future states of buildings or urban spaces under varying environmental conditions, and thus they have the potential to change and enhance the procedures of urban planning in a profound way. It must be kept in mind, however, that such forms of scenario building also become part of the reality that they are trying to model. And unlike weather simulations, the systems being modeled – think of the people using an urban plaza in Melbourne, for instance – would certainly react to the scenarios in question if they were to be shown, say, at a community meeting. Such interactions between the public and computer simulations that *model* this very public might add yet another layer of unpredictability to the process.

Constructing Collectives

As of 2005, a number of researchers have begun to direct their attention to the potential use of swarm robotics to assist construction projects.[129] The

127 See 'Interview with Roland Snooks.'

128 Leach, 'Swarm Urbanism,' 62.

129 See, for example, Von Mammen et al., 'Evolving Swarms that Build 3D Structures'; Saidi et al., 'Robotics in Construction'; Werfel et al., 'Distributed Construction by Mobile Robots with

expectation is that swarming robots "not only can [...] lead to significant time and cost savings, but their ability to connect digital design data directly to the fabrication process enables the construction of non-standard structures."[130] At least theoretically, moreover, "robotic constructive assembly processes are by nature 'additive'; they are scalable and can incorporate variation in the assembly to accommodate not only economic and programmatic efficiency, but also complex information about individual elements and their position."[131] Finally, swarm robotics have several advantages over already existing platforms. *First*, unlike common robotic building systems, which are still centered on human involvement, swarm robotics could be employed in contexts where such involvement would be impractical or too dangerous. *Second*, swarm robotics is not constrained by the stationary methods of common robotic building platforms. Unlike the latter, they are not restricted by the size of the platform, which in common systems have a footprint that can be larger than the building under construction. And *third*, multi-robot assembly makes use of parallelism to provide greater error tolerance, given that many sub-tasks can be carried out by any robot in the collective.[132]

Approaches to using swarm robotics for construction projects can roughly be subdivided into a four-field matrix depending on whether they involve (1) grounded or (2) aerial robots and whether the latter use (3) rigid or (4) amorphous building materials. The typical grounded robot is small, lightweight, and maneuverable, and it is equipped with sensors that allow it to orient itself within its environment and to interact with other robots and with the building materials in question. Basic challenges for operating such systems include dealing with the power supply (the battery charging periods), avoiding collisions or blockages in a given environment, calculating the shortest paths from place to place, and developing reliable mechanisms for identifying, grabbing, and deploying building materials.[133]

Enhanced Building Blocks'; Werfel et al., 'Collective Construction of Environmentally-Adaptive Structures'; Werfel et al., 'Designing Collective Behavior in a Termite-Inspired Robot Construction Team'; Magnenat et al., 'Autonomous Construction Using Scarce Resources in Unknown Environments'; Stroupe et al., 'Behavior-Based Multi-Robot Collaboration for Autonomous Construction Tasks'; Augugliaro et al., 'Building Tensile Structures with Flying Machines'; Helm et al., 'Mobile Robotic Fabrication on Construction Sites: DimRob'; Wawerla et al., 'Collective Construction with Multiple Robots'; and Soleymani et al., 'Autonomous Construction with Compliant Building Material.'

130 Willmann et al., 'Aerial Robotic Construction,' 441.

131 Ibid, 446.

132 See Petersen, *Collective Construction by Termite-Inspired Robots.*

133 See Gerling and Von Mammen, 'Robotics for Self-Organized Construction.'

State-of-the-art systems such as marXbot,[134] the "Swarm Robotics Construction System" (SRoCS),[135] or TERMES (alluding to and following the biological example of termites) therefore use highly standardized, rigid building materials like cubes or – in case of TERMES – blocks designed specifically to suit the robots' manipulators and lifting devices.[136] TERMES, which are currently the most sophisticated form of swarm robotics in architectural construction, were inspired by the decentralized communication structure and collective behavior of termites. The team responsible for them, which developed an interaction algorithm for a multi-agent system, was motivated "by the goal of relatively simple, independent robots with limited capabilities, able to autonomously build a large class of nontrivial structures using a single type of prefabricated building material."[137] After running their algorithm with software agents, the research group implemented it in a group of physical robots to test its functionality in real life. Surprisingly enough, the TERMES commenced at once to put together the building blocks collectively, employing stigmergy to locate them. As for the blocks themselves, they needed to be able to adhere to one another or to be mechanically joined together because the use of a secondary material would have further complicated the overall process.

Other approaches, however, have involved amorphous materials. Some researchers have experimented with sandbags,[138] while others have used amorphous foam to build ramps on uneven terrains (thus exploiting one advantage of non-rigid materials).[139] Whereas the flexibility and adaptability of the amorphous materials facilitated the construction process in such environments, their viscosity and tendency to expand resulted in a great deal of imprecision.[140] Victor Gerling and Sebastian von Mammen have therefore proposed a combined process involving the use of amorphous materials to even out irregular terrain and then the subsequent use of rigid materials "for precise and swift construction." Although this would pose certain challenges when it comes to building tall structures, most systems in general are limited by the range of their lifting devices. On their own, however, TERMES are able to pile up their buildings blocks to

134 Bonani et al., 'The marXbot, a Miniature Mobile Robot.'

135 Allwright et al., 'SRoCS: Leveraging Stigmergy on a Multi-Robot Construction Platform.'

136 Werfel et al., 'Designing Collective Behavior,' 754–758.

137 Ibid., 755.

138 Napp et al., 'Materials and Mechanisms for Amorphous Robotic Construction.'

139 See, for example, Napp and Nagpal, 'Distributed Amorphous Ramp Construction in Unstructured Environments'; and Hunt et al., '3D Printing with Flying Robots.'

140 See Gerling and Von Mammen, 'Robotics for Self-Organised Construction.'

create temporary ramps, which they are then able to climb for the sake of constructing taller structures.[141]

Compared to grounded robots, aerial robots obviously have more freedom to navigate and also – especially with today's sophisticated quadrocopters – a high degree of precision. Aerial robots can operate dynamically in three dimensions, but do run the risk of crashing more easily, and thus their battery power needs to be monitored closely. Moreover, they are only suitable for transporting relatively light loads, and this fact slows down projects. This disadvantage remains despite recent attempts to increase the versatility of amorphous building material by mixing two-component polyurethane to be 'printed' by aerial robots.[142] Nevertheless, unmanned aerial vehicles are better suited than their grounded counterparts for building elevated structures.[143] For instance, the Aerial Robotics Construction Group (ARC), which is a research project based in Zurich, has created a six-meter-tall prototype referred to as a "flight-assembled architecture tower" that consists of 1,500 foam-brick modules and was assembled by a swarm of autonomous quadrocopters.[144] As in the case of TERMES, the research team in Zurich has stressed the importance of the 'nature' of suitable building materials:

> The payload of flying vehicles is very much limited, whereas materials with high strength and high density favor the use of ARC [...]. Consequently, this research focuses on the construction of elements, on lightweight material composites and on complex space frame structures [...]. Because the overall shape of these building modules is also determined from aerodynamic considerations, these must be designed according to the specific assembly techniques and building capabilities of the flying machines. The building modules, therefore, must have particular geometrical characteristics so as to meet the required levels of the flying vehicle's complex aerodynamics, and thus, its building performance. The consequence is a design that is never monotonous or repetitive, but rather specific and adaptable to different architectural and aerial characteristics. [...] This 'information' logic between dynamic contingencies – such as the requirements of aerial transportation and the physical constraints of production – must be seen as integral.[145]

141 See Petersen et al., 'TERMES: An Autonomous Robotic System'; and Petersen, *Collective Construction by Termite-Inspired Robots.*

142 See Hunt et al., '3D Printing with Flying Robots.'

143 See Gerling and Von Mammen, 'Robotics for Self-Organised Construction.'

144 Willmann et al., 'Aerial Robotic Construction,' 441–442.

145 Ibid., 446–447.

It is not entirely convincing that a structure designed in such a way would never be "monotonous or repetitive, but rather specific and adaptable." Even the project's prototype and its renderings of future megastructures suggest otherwise.[146] In the examples of both the ARC and TERMES, moreover, the autonomy and the adaptive capabilities of the robotic swarm collectives are highly dependent on their 'environmental interfaces,' which monitor both the external environment (air resistance, irregular surfaces, and so on) as well as the technical specifications of the respective robots (payload, the form and identifiability of building materials, the sequence of tasks, among other things). Given this fact, and the need to produce reliable outcomes, it is no surprise that most of today's swarm-robotic systems are designed to execute detailed pre-calculated blueprints. Their so-called adaptiveness is the result of a carefully pre-planned system of specifications for standardized building elements.

If it is acknowledged that the prototypes of such systems are still only able to perform in laboratory environments (and thus with a radically reduced amount of contingency), then statements like the following sound rather lofty: "While it remains to be seen whether ARC will emerge as a viable dynamic building technology, the Flight-Assembled Architecture prototype successfully illustrates how an ARC approach makes empty airspace tangible to the designer and addressable by robotic machinery."[147] Furthermore, the processes discussed above run counter to the initial idea behind swarm intelligence, as Michael Allwright and his team have pointed out in an article on their Swarm Robotics Construction System (again, SRoCS):

> Current implementations of decentralized multi-robot construction systems are limited to the construction of rudimentary structures such as walls and clusters, or rely on the use of a blueprint or external infrastructure for positioning and communication. In unknown environments, the use of blueprints is unattractive as it cannot adapt to the heterogeneities in the environment, such as irregular terrain. Furthermore, the reliance on external infrastructure is also unattractive, as it is unsuitable for rapid deployment in unknown environments.[148]

Unlike its competitors, the SRoCS avoids the use of a blueprint by enabling its robots to adapt their positions in response to visual clues from the

146 Ibid., 454.
147 Ibid., 442.
148 Allwright et al., 'SRoCS: Leveraging Stigmergy,' 167.

environment – they can, for instance, independently identify obstacles or irregularities – and in response to the building materials themselves, which are equipped with bar codes and different lights to indicate their status. After placing building blocks where they need to be, the robots update the colors of the LEDs on the blocks. Depending on the algorithm being used, these colors can be assigned various meanings; a particular color, for instance, can be used to indicate a seed block or a block that has already been placed into the structure, thereby contributing to a stigmergic building process.[149]

Despite the sophistication of architectural design and the potential for mass customization enabled by computational swarm intelligence, the physical implementation of collective building processes with swarm-robotic systems still remains rather clumsy. Rather than increasing architectural variation as part of complex design processes (involving environmental forces, random fluctuations, and the behavior of numerous agents), swarm robotics has instead been based on careful preparation and pre-planning, which largely eliminates contingency. Working with highly standardized elements and blueprints, it has reduced the lively adaptive scenarios of computer simulations to mere basic functions (such as preventing robots from crashing). Thus the already daunting task of constructing reliable robot collectives of larger sizes – such as Harvard's Kilobot project, which consists of a thousand individual robots but operates at extremely slow speeds[150] – is exacerbated even further when it comes to using them as productive construction platforms.

Compared to already existing (robotic) technologies being used in architecture, swarm robotics seems, at least for now, to involve rather too many restrictions and disadvantages (in terms of producing aesthetically and conceptually sophisticated results) and seems to offer too few advantages (such as being able to explore terrains that are inaccessible to humans). It is perhaps no coincidence, then, that the SRoCS paper happens to echo some of Rodney Brooks's ruminations from decades before about the potential uses of autonomous robots: "It is possible that a multi-robot construction system will be a practical solution in the future for building basic infrastructure, such as shelter, rail, and power distribution networks on extraterrestrial planets or moons, prior to the arrival of humans."[151]

149 Ibid., 163.

150 See Rubenstein et al., 'Kilobot: A Low Cost Scalable Robot System for Collective Behaviors.'

151 Allwright et al., 'SRoCS: Leveraging Stigmergy,' 158. See also Khoshnevis, 'Automated Construction by Contour Crafting.'

Superconnected Idiots Savants

While it is true that the application of swarm robotics in the field of architecture has faced and continues to face a number of obstacles, the concept of swarm architecture has always been closely associated with the novel collaborative working practices of contemporary digital architects.[152] The collective intelligence of computational agent-based systems seems to be reflected on the level of everyday architectural practices. This can be seen in the software that has been developed to facilitate computer-supported cooperative work and in the ability and proclivity of architects to collaborate from all over the globe. It is also reflected in the mutable and open-ended design and construction processes that allow for ongoing feedback and adjustments throughout the realization of a project. It can be seen, too, in the object-oriented programming logic of architectural design and in certain construction tools whose use could even be described as a stigmergic process in itself.[153] As with swarm-intelligent technologies in general, the aim, as Oosterhuis has written, is "not just being creative individuals, but building creative relationships" in which the design process can become an "on-line and on-site testing [...] in the swarm of flocking stakeholders."[154] Thus the computational cultural technique of swarming has also penetrated the working culture of contemporary architects, architects who, as "hyperconscious idiot savants" (in Oosterhuis's words), engage with a constant flux of information and act as "assistants" to their self-organizing computational tools.[155] There might also, as Oosterhuis suggests, be a fundamental democratic function to all of this: by making interfaces open to the public, ordinary citizens could also become participants in this "design game,"[156] thereby extending the cultural technique of swarming to an even wider sphere.

In contrast, however, to the techno-euphoria that characterized the swarm discourse's heyday, such socio-political implications remain an ongoing matter of debate. First of all, it is important to consider Eugene Thacker's crucial distinction between *pattern* and *purpose* in swarm intelligence. Thacker wonders whether swarm collectives might be able to define a strategic agenda on their own, as opposed to their unquestionable ability to react to existing environmental conditions.[157] To what extent, in other

152 See Oosterhuis and Feireiss, *Game, Set and Match II*, 61–63.

153 Christensen, 'The Logic of Practices of Stigmergy.'

154 Oosterhuis and Feireiss, *Game, Set and Match II*, 62.

155 Ibid., 58.

156 Ibid., 62.

157 See Thacker, 'Networks, Swarms, Multitudes,' n.p.

words, are the generative technologies of swarm intelligence and agent-based modeling dependent on top-down definitions of *ex-ante* boundary conditions and target functions, not to mention the *ex-post* evaluations by experts? Are they not always embedded in classical hierarchies of decision-making and thus function as 'tactical' problem-solving tools (instead of generating any original purpose of their own)? *Second,* one has to pay close attention to the precise ways in which swarm intelligence and agent-based models are being applied in architectural, engineering, or scientific processes and how they correspond to other organizational formats. How closely do certain idea models correspond to actual fabrication and manufacturing technologies? How, exactly, should the parameters of urban planning be evaluated? *Third,* unlike the use of swarm intelligence in architectural design, the use of swarm robotics in architectural construction is based on careful preparation and pre-planning. For the most part, this eliminates contingency. As mentioned above, the act of working with blueprints or centrally planned modules diminishes the secondary adaptive environments of computational approaches. The question remains whether the application of such robotic building technologies will continue to be restricted to extreme physical environments, which are inaccessible to traditional construction methods, or whether they will live up to their optimistic promise and develop into teeming swarms of durable and mobile 3D-printers – a vision that would truly be revolutionary for the construction industry.

These issues aside, swarm intelligence, agent-based modeling, and swarm-robotic applications can be regarded as cultural techniques that are well-suited for dealing with complex planning problems in architectural design and construction. Swarm architecture takes advantage of the problem-solving intelligence of self-organizing collectives and thus introduces novel human-zootechnological hybrids into the architectural design process. Computer graphics make it possible to compare various universal structures, both with respect to parameter adjustments within the rule sets of simulations as well as in terms of empirical data taken from real architectural sites. The underlying function of this scenario-based knowledge is the act of *seeing in time.* Computer science is capable of animating mathematical models, that is, endowing them with 'life' in real time. And instead of coalescing into architectural master-plans, they maintain the potential to generate a spectrum of opinions, viewpoints, and 'near-equilibriums.'

The extent to which the proponents of generative swarm architecture are able to claim that the bottom-up potential of their agent-based models applies just as well to their working practices might simply be a question of their environment. A participatory and democratic perspective, enabled by dynamic architectural models of buildings or urban sites, might one day be a realistic component of generative architecture. It sounds both feasible and necessary to integrate critical public opinions into the decision-making processes of urban planning, but of course this will only be possible in societies that are democratic to begin with – and rather not in those countries that have regrettably attracted a good number of the idiots (savants) of digital architecture in recent years.

4. Calculating Survival: Crowd Control

From Mass Panic to Crowd Dynamics

To offer a final example, let me return once more to the software package called MASSIVE, which is well known as a program for animating digital crowd scenes in popular films. From that context, however, it was just a small step to applying agent-based simulation models to real-life phenomena. Under the slogan "Simulating Life," the MASSIVE Software company has since been offering software solutions not only for architectural visualizations and simulations of consumer behavior; it also sells a modeling tool called MASSIVE Insight. This is a technical programming packet for simulating and designing "[l]ife safety, pedestrian planning, transportation, and infrastructure."[158] For the parameter settings of simulation software, it is all the same whether it is being used to animate orc battles or for studying the 'anarchic' dynamics of human crowds in critical situations. Studies of the latter sort are of course concerned with understanding crowd control by simulating the characteristic patters of seemingly haphazard behavior in panic situations and modulating or optimizing this activity by testing various scenarios on a trial-and-error basis. Quite literally, then, this is a matter of calculating survival, and it represents yet another application of swarms and swarm intelligence that has come about in the wake of their rise to prominence in the popular discourse.

The transformation from MASSIVE to MASSIVE Insight involved three essential changes. *First*, crowd control simulations like MASSIVE Insight not only take into account every individual agent's will to survive; they also make *survival* itself numerically calculable. The simulation and visualization of crowd dynamics within certain settings – in concert halls or football stadiums, for instance – are intended to make it easier to anticipate so-called

158 Quoted from http://www.massivesoftware.com/applications.html (accessed 1 August 2018).

crowd disasters; after all, the dynamics of crowds can be managed in different ways by different forms of panic-absorbing architecture. *Second*, this approach is based on an entirely new understanding of crowds. In traditional crowd psychology, they are described as irrational, energetically charged, and even barbaric collectives that pose a dangerous threat. Based on the dynamic activity of individual agents, agent-based simulation models have replaced this traditional psychological understanding with one that focuses on the multiple, relational, and parallel connections that might take place between the potential movements of individuals. The irrational, uncontrollable, and therefore threatening image of the crowd has transformed into that of an operationalizable and optimizable multitude. For, in particular crowd situations, people happen to behave like simplified swarming individuals by relying less on their cognitive and reflexive capabilities but instead coordinating their behavior with that of their nearest neighbors. In such situations (and with their relatively limited sensory faculties), people make use of a distributed information infrastructure that is similar to that in animal collectives. And it is therefore possible to model such behavior with similar means.

With this new understanding of crowds, the goal is no longer to control or manipulate them. Rather, the aim is to optimize their potential movements, and it is motivated by a genuine feeling of *concern* for the safety and survival of the individual members of a multitude, especially in critical situations such as instances of mass panic. The underlying model is therefore not that of an animalistic collective without any individual differentiation but rather that of biological swarms in which individuality and collectivity are in a constant state of flux. And *third*, this raises the issue of so-called 'social swarming,' which can contribute to the subversive potential of new collectives informed by mobile technical media.

Because of the nature of our sensory faculties and our ability to move in just two spatial dimensions, people form rather unimpressive swarms. Whereas, in moments of great danger, flocks of birds and schools of fish will still exhibit individual-based collective dynamics without having to touch one another, people who find themselves in such perilous situations tend to arrange themselves in such a way as to lead to crowd disasters. In his book *Crowds and Power*, Elias Canetti described the rise of panic as a transition from a community-oriented collective activity to individual-based swarming behavior. In the case of human beings, however, the latter situation results in uncoordinated and violent activity:

> Panic is the disintegration of the crowd *within* the crowd. The individual breaks away and wants to escape from it because the crowd, as a whole,

> is endangered. [...] The common unmistakable danger creates a common fear. [...] The transmutation shows itself in violent individual action: everyone shoves, hits and kicks in all directions. [...] In such a moment a man cannot insist too strongly on his separateness.[159]

Social crowds have been described as highly combustible conglomerations that might behave as "irrationality itself" at any given moment,[160] but even when such gatherings begin to individualize in response to mass panic, according to Canetti, they will display just as much if not more irrationality, characterized by shoving, hitting, and kicking. Panic situations typically erupt in (spatial) environments with scarce or dwindling resources, and they have long been studied by social psychologists. The latter have usually characterized panic as infectious, egoistical, asocial, or even irrational behavior that can affect crowds.[161] Such descriptions are based on a definition of crowd psychology as "the study of the mind (cf. group mind) and the behaviour of masses and crowds, and of the experience of individuals in such crowds."[162] In step with the writings of Le Bon, Tarde, and Scipio Sighele, who described the degeneration of individuals into crowds as well as the latter's manipulability and primitive collective spirit (a concept that would be revived by Serge Muscovici),[163] social psychology has in general been more concerned with the potential danger of crowds than in securing the safety of the individuals *in* the crowds. Or, as Clark McPhail observed: "Students of the crowd, with certain exceptions, have devoted far more time and effort in criticizing, debating, and offering alternative explanations than they have to specifying and describing the phenomena to be explained."[164]

One such exception was an approach formulated in the 1950s that shifted attention away from the collective consciousness of crowds (or collective *unconsciousness*, depending on one's point of view) toward the level of the individual:

> When people, attempting to escape from a burning building, pile up at a single exit, their behaviour appears highly irrational to someone who learns after the panic that other exits were available. To the actor in the

159 Canetti, *Crowds and Power*, 26–27 (emphasis original). For a general overview, see also Borch, *The Politics of Crowds*.

160 Vogl, 'Über soziale Fassungslosigkeit,' 179.

161 Keating, 'The Myth of Panic.'

162 Hewstone et al., eds., *Introduction to Social Psychology*, 448.

163 See Muscovici, *The Age of the Crowd*.

164 McPhail, *The Myth of the Madding Crowd*, xxiii.

> situation who does not recognise the existence of these alternatives, attempting to fight his way to the only exit available may seem a very logical choice as opposed to burning to death.[165]

In the same vein, Jonathan D. Sime remarked: "A number of disaster sociologists from the 1950s and 1960s onwards, notably Quarantelli (1957) who prefers the term nonrational to irrational flight behaviour, have argued that the notion that people panic, in the sense of irrational asocial or nonsocial behaviour, is a myth or at least greatly exaggerated."[166]

An individual-based perspective on crowd dynamics thus changes the way in which crowd disasters are represented, evaluated, and addressed. It frees them from previous crowd-psychological descriptions. Mass accidents now appear as conglomerations of potential individual behavior, as dynamic *aggregations* of particles or agents that transform themselves in particular ways. In the case of panic situations, of course, it can be difficult to analyze such individual behavior, though psychological laboratory and group experiments have been conducted on the effects of cooperative or competing types of behavior in circumstances where there are just a few ways for people to flee from a dangerous scenario. Such studies were thus meant to evaluate the 'rationality' of individual behavior in cases of panic.[167] Yet, according to Sime, these experiments were inadequate in many ways, particularly in their insufficient treatment of the problem of scalability:

> The experiments have failed to explore the social dynamics of crowd movement directly, why and where flight behaviour and/or crushing occurs and how it can be prevented. The single group in the psychological experiments has been assumed to possess the essential properties of the far larger crowd. Ways in which a crowd's composition will vary [...] in different types of settings and situations [...] are not represented in the laboratory-based psychology experiments.[168]

It is difficult to experiment on panicking crowds, and this is not even to mention the ethical problems that such an empirical approach involves.

165 Turner and Killian, *Collective Behaviour*, 10.

166 Sime, 'Crowd Psychology and Engineering.' Sime refers here to Quarantelli, 'The Behaviour of Panic Participants.'

167 See Mintz, 'Non-Adaptive Group Behavior'; Kelley et al., 'Collective Behaviour in a Simulated Panic Situation'; and Guten and Vernon, 'Likelihood of Escape, Likelihood of Danger, and Panic Behavior.'

168 Sime, 'Crowd Psychology,' 7.

Moreover, it is doubtful whether experiments with mice or ants can tell us anything at all about the behavior of panicking human beings.[169] Other approaches have proved deficient as well, such as technical models that treat the movements of human crowds as analogous to physical phenomena such as hydraulic fluid dynamics or granular particles in tube systems or containers. Not only do they smooth out the potentially deviating behavior of individual particles by reducing them to identical elements; they also corroborate the "notion that people can be equated with *non-thinking objects* [...], an emphasis on crowd control through centralized (autocratic) building control systems, rather than crowd management through distributed (democratic) building intelligence."[170] Or, in other words: "'Engineering for Crowd Safety' requires people in crowds to be treated as human beings, rather than as ball bearings."[171]

Since the middle of the 1990s, the collective dynamics of crowds and critical phenomena such as mass panic have also been examined by the fields of physics and computational studies. The goal of these investigations has been to enrich the less systematic psychological studies with computer models that make it possible to determine and predict the typical parameters of crowd dynamics. Previously criticized analogies to 'non-thinking objects' and experiments with human subjects were replaced by computer simulations, most of which operate on the basis of distributed agents.[172] In this way, the notion of collective behavior as a crowd-psychological phenomenon was epistemically connected to emerging physical and biological patterns of motion – to such things as accumulated individual velocities, collision probabilities, the ability to accelerate, and forces of compression. With advanced software models, these studies furthered the move away from behavioral concepts based on the supposed asocial and irrational nature of crowds. Ultimately, they promoted ways to regulate the dynamics of so-called 'non-adaptive behavior.'[173] Such simulations ultimately made it possible to connect two previously separate spheres: the psychological observation of individuals behaving in crowds and the act of engineering

169 See Musse et al., 'Groups and Crowd Simulation'; and Shao and Terzopoulos, 'Autonomous Pedestrians.'

170 Sime, 'Crowd Psychology,' 11 (emphasis original).

171 Ibid., 12.

172 See Musse et al., 'Groups and Crowd Simulation'; and Shao and Terzopoulos, 'Autonomous Pedestrians.' For a broader overview, see also Schreckenberg and Sharma, eds., *Pedestrian and Evacuation Dynamics*; and Helbing and Johansson, 'Pedestrian, Crowd, and Evacuation Dynamics.'

173 See Helbing et al., 'Simulation Dynamical Features of Escape Panic.'

regulatory mechanisms based on physical or biological models. They made it easier to understand the relations between certain spatial organizations and human behavior and they made it possible to subject mass panic to *quantitative* evaluations.

Here, as in the case of biological swarm simulations, two approaches are especially noteworthy. In a number of research projects, Dirk Helbing and his collaborators have simulated the dynamics of human crowds with hydrodynamic and other physical flow equations. These simulations used the forces of attraction and repulsion to model inter-individual behavior, adjustments to the simulation environment, and the prevention of collisions.[174] In these models, all agents are assigned identical or very similar attributes; they behave according to universally applied social or socio-psychological forces. Here, the psychological factors involved in the creation of mass panic are quantified and formalized into physical equations, and thus made mathematically manageable. Yet these physical models are also infused with a biologically inspired form of programming in that 'standard' or preferred behavioral conventions (such as the preference of pedestrians to walk on certain sides to avoid others) are taken into account by simulating learning processes with evolutionary algorithms.[175] The result is a 'behavioral force model' that allows the problem to be played through with a large number of particles.

The second approach, which also happens to be that of the MASSIVE software, is based on defining local rules for the behavior of agents. Demetri Terzopoulos, Daniel Thalmann, and others, for instance, have endowed agents with increasingly detailed simulated senses, which have enabled them to act realistically in relation to other agents and to the simulation environment.[176] The thought behind this approach is that, in certain situations, a large number of such 'lifelike autonomous agents'will automatically exhibit a form of collective behavior that is comparable to that in real-life situations and that, by modulating certain parameters on a trial-and-error basis, it should be possible to identify the decisive factors in this process. In the case of agent-based modeling, that is, behavioral repertoires are implanted 'into' the agents themselves instead of being generated by global forces. With these simulations too, however, it is not a matter of creating a sort of artificial psychology. 'Inner' processes are only relevant to the extent

174 See Helbing et al., 'Simulation of Pedestrian Crowds in Normal and Evacuation Situations.'

175 See ibid., 29.

176 See Terzopoulos, 'Artificial Life in Computer Graphics'; and Terzopoulos et al., 'Artificial Fishes.'

that they result in certain movements, and it is only these movements, which are visualized over the course of the simulation, that make it possible to modify the program retroactively. The frantic, panicked activity of the agents thus converges with the run-time of the program. In the case of these computer simulations, the knowledge that biologists have developed about the collective organization of animals is harnessed into a technically implemented programming basis for studying panic behavior and its regulation. In these programs, collective movement behavior based on swarm logic serves as both an *object of research* as well as a *modeling tool.* Human panic behavior is not described here as a sort of degeneration into the animalistic; rather, it is the behavior of animals – implemented with computer-technology – that has made it possible to describe the flurry of panicking crowds in the first place.

Since the year 2000, both approaches have been implemented to 'calculate survival' (in the literal sense of the words). Helbing, for example, has simulated the behavior of pedestrians in various environments and with various densities and speeds. His goal was to gain insight into how architectural interventions might be made to ease evacuations, absorb a sudden onrush of many people, and minimize the negative effects of these and similar sorts of events. Among other things, his model demonstrated the so-called "faster-is-slower paradox," according to which it will take longer to leave a given space the faster the individuals involved attempt to do so.[177] Thermodynamic laws also came into play. If, for instance, the motion dynamics of pedestrians in a tunnel increases and those moving in opposite directions respond by switching from side to side – a type of behavior that is analogous to phase transitions from a fluid to a gaseous state – the result will not be greater disorder but rather metastable stasis: the entire tunnel will be clogged by a sort of 'crystalized' increase in order.[178]

As with research on swarming in the animal kingdom, scholars here have also paid attention to the changing nature of inter-individual communication. Whereas, in narrow pedestrian zones, body language is normally sufficient to ensure that collisions are avoided, collective panic is characterized by the total breakdown of inter-individual communication in general: "The fundamental unit of a crowd is not the individual but the cluster, because the first thing we do in an emergency situation is look to each other for support and information."[179] This type of behavior can slow things down dramatically

177 Helbing et al., 'Simulation of Pedestrian Crowds,' 37.
178 Ibid., 35.
179 See Bohannon, 'Directing the Herd,' 221.

and it often causes individuals to move unthinkingly along with the crowd (so much so that people will walk right by open exits in clear view). And this is especially true in cases where additional environmental factors enter the picture. Helbing has thus played through scenarios in which smoke or fog reduces visibility and thus reduces the ability of agents to orient themselves.[180] Since 2005, moreover, increasingly detailed methods of 'crowd capturing' – that is, the automated analysis of digital videos of crowd phenomena – have made it possible to compare different computer-simulation models and to modify them in a process of mutual optimization.[181]

Crowd Sensing and Foggy Logic

Even more sophisticated systems are underway. At the German Research Center for Artificial Intelligence in Kaiserslautern, for instance, scientists have created models of pedestrian behavior by inferring and visualizing crowd conditions on the basis of pedestrians' GPS locations. Known as 'crowd sensing,' this technique was tested in 2011 and then applied at the 2012 London Olympics. The system is able to infer and visualize crowd density, crowd turbulence, crowd velocity, and crowd pressure in real time, and it works by collecting location updates from festival visitors. The researchers distributed a mobile-phone app that supplied users with event-related information but also periodically logged the device's location, orientation, and speed of movement by GPS. It then sent the data back to the running model. The system allegedly helped to assess crowd conditions and to spot critical situations faster than traditional video-based methods.[182] The emerging field of mobile crowd sensing,[183] or "multimodal crowd sensing,"[184] employs the multiple sensory capacities of today's smartphones for a variety of sensing applications at the level of individuals, groups, and entire communities. Such programs collectively share data and extract information to measure and map phenomena of personal and common interest. A survey article by a team from the IBM Research Center distinguishes three different types of data that such systems can collect: *first*, environmental data about things such as air pollution; *second*, infrastructural data such as traffic

180 See Helbing et al., 'Simulation of Pedestrian Crowds.'

181 See Johannson et al., 'From Crowd Dynamics to Crowd Safety'; and Helbing et al., 'The Dynamics of Crowd Disasters.'

182 See Wirz et al., 'Inferring Crowd Conditions'; and Pluta, 'Crowd Management: Smartphone soll Massenpanik verhindern.'

183 Ganti et al., 'Mobile Crowdsensing: Current State and Future Challenges.'

184 Roitman et al., 'Harnessing the Crowds for Smart City Sensing.'

congestion, road conditions, or honking levels on city streets; and *third*, data concerning social processes, including crowd management, healthcare, or sports activities.[185] All of these apps tend to be used in contexts with limited resources and high population densities; therefore, they not only contribute to a more sustainable use of resources but also hint at the logic of optimization that lies behind their interface.

The extent to which individuals themselves participate in crowd sensing depends on two different approaches: participatory sensing and opportunistic sensing.[186] The former requires the active involvement of individuals to contribute sensory data related to large-scale phenomena (by taking photographs, providing context information, and so on). This brings 'human sensory capacities' into play, which can help applications to collect higher-quality or semantically complex data that would otherwise require sophisticated hardware and software for such things as pattern recognition. Information collected in this way – the technique is often referred to as 'sousveillance' – can also be gathered indirectly from social networks like Twitter. However, as Kamel Boulos and his coauthors have remarked, the "variable amounts of 'noise', misinformation, and bias [...] usually require some advanced forms of filtering and verification by both machine-based algorithms and human experts before becoming reliable enough for use in decision-making."[187] Because of privacy concerns, moreover, some incentive is usually needed to entice people to participate.[188] Opportunistic sensing, on the other hand, is more autonomous, and the user's involvement is reduced to a minimum to add technical sensory capacities to the network (by providing continuous location sampling, for instance). Yet this also involves persuading a critical number of users to contribute their data, either by guaranteeing their anonymity or by other means.

As a group from Dartmouth College has pointed out, moreover, there are other problems to cope with, such as that of a device's context. For example, a city noise-mapping app may be designed to take sound samples, but it is only able to do so if the phone happens to be out of the user's bag or pocket. This problem can perhaps be circumvented by making use of some of the phone's other sensors; an accelerometer or a light sensor could determine whether the phone is out in the open.[189] According to researchers at IBM,

185 See Ganti et al, 'Mobile Crowdsensing.'

186 See Burke et al., 'Participatory Sensing'; and Lane et al., 'A Survey of Mobile Phone Sensing.'

187 Boulos et al., 'Crowdsourcing, Citizen Sensing, and Sensor Web Technologies,' 2.

188 Lane et al., 'A Survey of Mobile Phone Sensing,' 144.

189 Ibid., 146–147.

another issue with multimodal sensing is the wide variety of mobile devices and forms of data that have to be integrated in order to generate useful information. The type of sensory data that the devices produce and the quality of such data depend on a device's mobility, on variations in energy levels and communication channels, and on its owner's preferences.[190] In order to integrate such a variety of raw sensor data, they have to be processed with local algorithms on the mobile device. At the same time, this processing involves data compression, which helps to keep the network from flooding.[191]

As a result, all of the data can be collected at data centers and visualized on interactive (heat) maps.[192] These information environments can deal with a variety of different data and dynamics, and their visual displays can integrate modeling, simulation, and sensing applications to support the actions, assessments, and decision-making of experts.[193] They can also focus on just one type of sensory data, as in the case of the CO_2 measurements taken in Copenhagen. This involved a small mobile network of just ten bicycle messengers equipped with air-quality sensors, and the project was able to produce far more detailed data about the dynamics of CO_2 pollution and its dependency on traffic congestion, humidity, temperature, and wind direction than a traditional system with many fixed physical sensors ever could.[194] Similar examples include social-health networks for forecasting sickness (see, for instance, Sickweather.com) and the crowdsourced radiation map developed after the Fukushima disaster.[195]

Information of this sort can be fed back to individual users who commute within a certain area. Moreover, many crowd-sensing applications function on a distributed basis, even when it comes to evaluation. In the case of mid-size groups such as local neighborhoods, applications can automatically feed monitored data back to individuals, who can then make decisions on their own. The recycling rate of a university campus, for example, can be optimized by collectively sharing information about the locations of garbage bins. That said, these examples also demonstrate the rather disconcerting ways in which "crowd-enabled systems are revolutionizing the way we tackle problems and allowing us to monitor and act upon almost anything,

190 Ganti et al., 'Mobile Crowdsensing,' 36–37.

191 Ibid., 35–36.

192 Boulos et al., 'Crowdsoursing, Citizen Sensing, and Sensor Web Technologies for Public and Environmental Health Surveillance and Crisis Management,' 22.

193 Goodchild, 'Citizens as Sensors.'

194 Boulos et al., 'Crowdsourcing, Citizen Sensing, and Sensor Web Technologies,' 15–17.

195 Saenz, 'Japan's Nuclear Woes Give Rise to Crowd-Sourced Radiation Maps in Asia and US.'

anywhere, in real time."[196] The ubiquitous effort to optimize the organization of "the dust of events, actions, behaviors, opinions" (in Foucault's terms),[197] which is inherent to all of these applications, tends to corroborate the more alarming inclinations of pedantic petit-bourgeois thinking. On the one hand, voluntary modes of participation in mobile crowd-sensing applications produce more precise data about environmental conditions than fixed network systems and thus contribute to the improvement of public spaces. On the other hand, they instantiate a form of governmentality that imposes a techno-savvy "hermeneutics of the subject" on all sorts of everyday practices.[198] This (semi-) automated form of normalization, which takes place within a system consisting of "the instinct of the experts, the wisdom of the crowds, and the power of algorithms,"[199] runs the risk of de-individualizing the very individuals who participate in such self-reflective crowds.

The transformation from traditional conceptions of crowds to agent-based simulation models replaced a psychological understanding of crowd behavior with the physics of bodily movement vectors in environments with scarce resources. The traditional crowd and its dangerous "thermodynamics" and explosiveness dissolved into the computable individual actions of simulated agent collectives.[200] This transformation was then followed by the development of the sensing and self-reflective crowd, in which media-technological applications no longer restrict individual actions to mere abstract (collective) movements but incorporate all sorts of 'sensor data' from both humans ('qualitative' data from social network feeds, etc.) and mobile devices ('quantitative' technical sensor data), thus mapping a whole spectrum of complex real-life behaviors and interconnected environmental dynamics. These can be externally monitored or fed back to the 'autonomous individuals' in the crowd almost in real time. 'Lifelike agents' have thus dissolved into a set of data streams. The reverse side of participating in crowd sensing is that it reduces the individual to a "dividual" state.[201] This fundamental dissolution not only makes the techno-social crowd a productive force; it also imposes a totality of governmental self-optimization.

Today, *calculating disasters* involves integrating empirical data about past catastrophes, observational data about crowd events, and information from computer-based experiments with agent-based models. It is thus a combination

196 Boulos et al., 'Crowdsourcing, Citizen Sensing, and Sensor Web Technologies,' 2.

197 Foucault, *Discipline and Punish: The Birth of the Prison*, 213.

198 See Foucault, *The Hermeneutics of the Subject*.

199 Boulos et al, 'Crowdsourcing, Citizen Sensing, and Sensor Web Technologies,' 8.

200 See Schäfer and Vogl, 'Feuer und Flamme,' 191–211.

201 Deleuze, 'Postscript on the Societies of Control,' 5.

of analytic and synthetic approaches to simulating and monitoring crowds (with the help of advanced visualization techniques). In the case of crowd sensing, the crowd itself becomes a sort of operational medium – a medium that helps to regulate its own activity through a real-time feedback loop with a computer model. In the event of mass *panic*, however, it is doubtful whether such a feedback loop would be effective; in such situations, people would likely be too fraught to follow any directives from their smartphones.

It should be noted that, in recent years, the concept of panic has done more than inspire a few computer-simulation models; it has also been embraced for its subversive political potential. In an essay published in 2001, the French author collective known as Tiqqun cited panic as a means to undermine the all-encompassing and cyberneticized methods of control that characterize neoliberal society: "Defeating the process of cyberneticization, toppling the empire, will take place through opening up a breach for *panic*. [...] Panic makes the cyberneticians panic. It represents *absolute risk*, the permanent potential threat."[202] Tiqqun's interpretation of panic opposes the idea that it is somehow deficient, asocial, and archaic. The latter attributions are based on the misunderstanding that panic only occurs in closed environments. On the contrary, Tiqqun cite the philosopher Peter Sloterdijk, who has identified in panic the possibility for *rational ecstasy*, which is needed to make living civilizations possible. In addition, they refer to a passage in Canetti's *Crowds and Power* that contradicts the traditional understanding of panic situations: "If they were not in a theatre, people could flee together like a herd of animals, and increase the impetus of their flight by the simultaneity of identical movements."[203] Panic, according to Tiqqun, could thus be regarded as a state of "confused intuition" or "con-fusion." It could be seen as a technique for fleeing away from cybernetically structured society in that it turns every individual into "the living foundation of his own crisis." In using these individual flight lines, the authors go on, there lies a special type of potential, for it is thereby conceivable to intensify the noise in the system beyond its critical threshold: "The overproduction of bad feedbacks that distort what they're supposed to signal. [...] To provoke panic first of all means *extending the background interference* that imposes itself when the feedback loops are triggered, and which makes the recording of behavioral discrepancies by the ensemble of cybernetic apparatuses costly."[204]

202 Tiqqun, 'The Cybernetic Hypothesis' (2001), n.p. (emphasis original). Quoted from https://theanarchistlibrary.org/library/tiqqun-the-cybernetic-hypothesis (accessed 10 June 2018).
203 Canetti, *Crowds and Power*, 26.
204 Tiqqun, 'The Cybernetic Hypothesis,' n.p. (emphasis original).

Beyond mere speculations and wordplay, Tiqqun turn specifically to the anti-globalization demonstrations that took place in Genoa in 2001 and ask how it might be possible to reevaluate the panic of the protesters. In doing so, however, the authors placed themselves firmly in the discourse concerning the subversive possibilities of collectives armed with mobile communication technology, a phenomenon that has elsewhere been described in terms of "smart mobs" or "social swarming."[205] According to Tiqqun, the crux of the matter is that "the revolution should consist in a reappropriation of the most modern technological tools, a reappropriation that should permit contestation of the police on their own turf."[206] At the much-discussed 'Battle of Seattle' in 1999, for instance, the use of networked mobile devices had enabled demonstrators to form a collective organization that was far more agile and quick to act than the lockstep march of the police forces.[207]

It remains to be asked, however, whether the concept of panic has anything more to offer in this context. Ultimately, 'reappropriated' mobile media can still only function within established and regulated networks and their protocols. Furthermore, instantaneous technical connections for the sake of coordinating movements stem from a concept of swarming that was itself only able to gain a technical dimension through the intermingling of biological research and computer simulations. Such connections may be able to create a higher level of collective efficiency, and they may be able to generate an effective level of interference within the system, but they can lead to no more than a temporary flight from our thoroughly cyberneticized society. Rather than leading to the 'overproduction of bad feedback,' this sort of interference seems rather to make feedback *more flexible*, and thus it is no surprise that such strategies have been studied and implemented by those in power. An example of this would be the research conducted by the RAND Corporation that, in the same year of 2001, led to proposed "swarming doctrines" for military and police strategic planning.[208]

What Tiqqun perhaps underestimated is the decentralized nature of the cyberneticization that they attempted to theorize and subvert. Instead of there being any centralized cybernetic control that can be overwhelmed by some sort of actively instigated panic situation, those in power in fact respond to realities by constantly constructing scenarios, experimenting with computer simulations, and rearranging and adjusting spaces and

205 See Rheingold, *Smart Mobs*.
206 Tiqqun, 'The Cybernetic Hypothesis,' n.p.
207 See De Armond, 'Black Flag Over Seattle.'
208 See Arquilla and Ronfeldt, *Networks and Netwars*.

multi-agent systems. With such an epistemology, it is easy to learn on a case-by-case basis how to circumvent any bad feedback. In this situation, every counterstrategy in itself provides conceivable (and even necessary) input for their strategy-making. Interfering with such a system by "prolonging the fog" of subversive panic – as many crowd simulations have already shown – will not lead to greater dynamics but rather to metastable logjams with potentially deadly results:

> The main killer when people mass is not trampling, as is commonly thought, but 'crowd crush.' When two large groups merge or file into a dead end, the density makes it impossible to fall down. But the accumulated pushing creates forces that can bend steel barriers. The situation is horrible [...]. Suddenly everything goes quiet as peoples' lungs are compressed. No one realizes what's happening as people die silently.[209]

Perhaps it is not a fog that Tiqqun wishes to spread but rather just a nebulous, metaphorical discourse. In the meantime, the revolution has been taking place *on the inside* – in the CGI and multi-agent systems that have connected the 'life' of virtual agents with the survival of real human beings.

Under the rubric of *zootechnologies*, I have provided examples of applications that outline the use of swarms as figures of knowledge and as technologies, and these examples further demonstrate the extent to which the concept of swarm intelligence has expanded in popular and scientific discourses. In the fields of robotics, architecture, and crowd control, self-organizing mathematical models such as particle swarm optimization and agent-based systems have made it possible to adapt to unclearly delineated sets of problems and clarify the operations of opaque systems. In doing so, they operate at the limits of what can be calculated; they offer performative, synthetic, and approximate solutions in cases where analytic approaches would require unrealistic amounts of effort and where it would be absurd to program detailed software applications. Above all, they function as optimization processes, whether it be to coordinate robotic collectives, to optimize nonlinear systems of equations, or to explicate the behavior of human crowds in panic situations.

As figures of knowledge, swarms have therefore been entirely removed from any fixed material or substantial basis. Today, their primary applications concern abstract questions of interaction, communication, and control

209 Bohannon, 'Directing the Herd,' 221.

whose answers depend on the self-organizational potential of distributed, interconnected, and decentralized multitudes of homogeneous or highly similar agents. It must always be kept in mind, however, that this potential is only advantageous in certain circumstances – namely, where coordination problems can be addressed by the particular 'spatial intelligence' or 'motion intelligence' of swarms. This is because the advantages of swarm logic – adaptability, resistance to interference, the ability to learn, redundancy, fault tolerance, low costs on account of simply structured agents, and the potential to offer counterintuitive solutions – are balanced out by certain disadvantages that do not burden other more rigid and centralized logics of control. These disadvantages often lie in the suboptimal types of behavior that can arise from the nonlinear interactions among swarming individuals. Such behavior introduces a degree of uncontrollability and unpredictability that, in technical applications (think of particle swarm optimization), has to be tempered with hierarchical structures and control interfaces in order for systems to function efficiently. This same functional logic also means that swarm systems occasionally operate more slowly than might be desired – a disadvantage for which there are also certain ways to compensate. As objects of knowledge, swarms evoke media-technological cultivations of intransparency, and as figures of knowledge – as zootechnological hybrids of biological and computer-technological knowledge – they are at their most productive in just such media cultures. Whenever clearly delimited problems and unambiguous goals can be formulated – that is, whenever it is possible to approach a phenomenon analytically – the logic of swarms can be of little or no use. In the end, however, the discursive yet (media-) technologically induced expansion of swarm-intelligence research has opened up the concept of swarming to a broader range of applications and may even lead to the construction of a seemingly universal model of socio-political, socio-technical, and economic conditions.

Conclusion

The self-organization of swarms – their opaque and nonlinear global effects, which are created by the local interactions of numerous agents – has both zoopolitical and zootechnological implications. Would it not be nice if human social processes and political decisions could take place as simply and quickly as the decisions of swarming collectives to change direction? And what would be more appealing than no longer having to rely on old analogies with colonies of ants and hives of bees as positive or negative examples of collective organization, but instead being able to describe the dynamics of 'human swarms' with reference to technical interfaces for communicating and making connections? Since the 1990s, swarms have been reformed into technologized, rationally implementable, and effectively visualizable zootechnologies, and thus it is easy to see why they have been applied as a powerful metaphor to describe various processes of "social swarming."[1] The condition of possibility for such metaphorical transferences, which differ from previous zoopolitical or anthropomorphic comparisons, was the reevaluation of swarms as figures of knowledge. Reference is no longer made to the biological 'life form' of various swarms but rather to their media-technological control logic and their available technical applications.

Any critical description of such transferences, however, should closely examine the extent to which swarms have been equated with networks, for instance, and where they are cited for their emancipatory potential despite the fact that human beings are fundamentally different 'agents' from those in swarm simulations, flocks of birds, and schools of fish. Perhaps it can be said that swarms should not be regarded as the most advanced form of older collectives – such as crowds or social groups – but rather as structures of organization and coordination that, in light of our media-technologically charged culture of intransparency and the ever-changing nature of so many areas of life, have become effective optimization strategies *in* these very

1 See Horn, 'Schwärme – Kollektive ohne Zentrum: Einleitung.'

Vehlken, S., *Zootechnologies. A Media History of Swarm Research.*
Taylor & Francis Group, 2019
DOI 10.5117/9789462986206_CONCL

areas. Or, better: they have become effective as *self-optimization strategies* whose specific governmentality is especially worthy of investigation.

"The Leviathan has had its day, and a swarm of hornets has appeared in its place." With these words, Bernhard Siegert summarized the precarious relationship between *polis* and *nomos* and its effects on naval warfare during the twentieth century.[2] The swarms that have since been established as figures of knowledge no longer confront governmental principles of order as something politically uncanny. They seem to have transcended this characterization by installing swarm-like processes of self-organization (or, in Foucault's terms, processes of self-technology and self-regulation) as conditions of possibility in a highly networked society that has been permeated by technical and physical enhancements of communication. Foucault's concept of governmentality differs from other notions of control, domination, and regulation in that it does not treat that which is to be governed as a given or 'natural' problem for which a necessary solution has to be found but rather as a *problematization* that, in the words of Markus Stauff, "has to be placed on one level with the processes and defined goals of regulation: The processes that create knowledge about certain procedures and conditions; the technologies that allow access to certain procedures and conditions; and the object area, with its specific 'internal' laws, are mutually constituted."[3] Strategies for encouraging constant production and for reactualizing knowledge therefore have to take the place of normative precepts. By means of adequate but indirect guidance, moreover, desired manners of behavior can be achieved by structuring the potential for self-regulation within an object area: "This mode of government is thus characterized by the ongoing problematization of object areas, strategies, and goals; it is not the installation of a stable regulatory process but rather constant modification, adjustment, and questioning that defines governmental politics of this sort, which manifests itself in an adequate form of guidance,"[4] and this adequate form is always a strategy for rationalizing government and regulatory technologies.

Yet Foucault described the prevailing governmental principle of the twentieth century as one structured according to neoliberal standards – a principle that orients not just the economy but rather all object areas toward the "model of entrepreneurial activity."[5] In order to function, such activity

2 Siegert, 'Der Nomos des Meeres,' 54.

3 Stauff, 'Zur Gouvernementalität der Medien,' 91.

4 Ibid., 92.

5 See Foucault, *The Birth of Biopolitics*; and Lemke et al., 'Gouvernementalität, Neoliberalismus und Selbsttechnologien,' 16–17.

must evade *political* rationality. Foucault described this phenomenon in light of the principle of invisibility in Adam Smith's metaphor of the 'invisible hand,' which Smith regarded as a constitutive feature of a functional economy:

> For there to be certainty of collective benefit [...] not only is it possible, but it is absolutely necessary that each actor be blind with regard to this totality. Everyone must be uncertain with regard to the collective outcome if this positive collective outcome is really to be expected. Being in the dark and the blindness of all the economic agents are absolutely necessary. The collective good must not be an objective [...] because it cannot be calculated, at least, not within an economic strategy. Here we are at the heart of a principle of invisibility.[6]

This irrationality with regard to the totality – this local orientation as an economic principle that gives rise to global order (that is, the *invisible hand* of eighteenth-century liberalism) – is akin to the global movements, structures, and orders of swarm collectives.

Like the principle behind the liberal market, swarms depend on the disqualification of the political sovereign or, in more radical terms, they depend on the "disqualification of a political reason indexed to the state and its sovereignty."[7] The rationality of the total system derives from the irrational, blind, and neighborhood-oriented behavior of its individual agents, which, *on behalf of the state*, require no more than a legal framework to grant them the greatest possible freedom (including freedom from state intervention). Furthermore, in this neoliberal twist on the concept of the market, in which the principle of self-limiting government has been abandoned for a sort of "permanent economic tribunal,"[8] it is possible for the irrational rationality of the swarm economy to coexist alongside traditional political rationalities. In the case of 'importing' biological principles into computer science or robotics, these optimization strategies are relatively obvious. Here objectives arise from such things as the orientation of swarm individuals to the *fitness functions* of a simulated environment. Yet the techno-biological genealogy of the swarm is a precondition for our ability to see more in the 'disqualification of the political sovereign' than a delineation of the outer limits of order by means of the deterritorializing functions that Deleuze and

6 Foucault, *The Birth of Biopolitics*, 279.
7 Ibid., 284.
8 Lemke et al., 'Gouvernementalität, Neoliberalismus und Selbsttechnologien,' 17.

Guattari attributed to packs, mobs, and swarms. And it also differs from the many analogies to insect colonies that were made by early ethologists. Not until swarms were conceptualized as a specific form of interconnective technology – as a four-dimensional "living network"[9] – did they become the condition of possibility for fantasizing about a new sort of socio-political and economic activity. Swarms did not appear as a novel political concept (in the sense of a governmental strategy) until the late 1990s, at a moment in the history of media by which they were no longer 'natural' but had rather become hybrids composed of biological and computer-technical elements of knowledge, programming methods, visualizations, and applications.

In such things it is still clear to see an affinity to neoliberal principles of economic optimization. Thus the *zootechnologies* of swarm collectives implement a 'self-administration of life' that derives from *zoē*, from the animal life in the swarm. And this is always done in conjunction with computer simulation and its 'behavioral science of systems' (as Bernd Mahr called it). Alongside the rational-political paradigm of *biopolitically* optimizing an object area such as a *population*, there thus emerged a new paradigm involving the *zoopolitical* self-optimization of disparate yet interconnected swarm collectives whose 'agents' participate in collective processes on a case-by-case basis and, to do so, use mobile technical interfaces for exchanging information.

The discursive expansion that, since the middle of the 1990s, went beyond the application of swarms as figures of knowledge and led to the technically induced combination of the terms *swarm* and *intelligence* was accompanied by additional transferences and discourse dynamics that turned swarms into a metaphor for just about anything. As a new umbrella term – and often with little regard for the interactive infrastructure of actual swarms – this metaphor has been used to lend a dynamic quality to the somewhat older but no less expansive discourse concerned with networks: swarms as *Network 2.0.* A recent example of this is a project initiated by Volkswagen and the energy provider Lichtblick that has been given the title 'SchwarmStrom' (that is, 'Swarm Electricity').[10] Kevin Kelly

9 Thacker, 'Networks, Swarms, Multitudes,' n.p.

10 See Kampfwirth, 'SchwarmStrom: Die Energie der Zukunft.' Cultural theorists who concentrate on networks will certainly contradict this point and say that the swarm discourse has simply been subsumed by the network discourse. As I have already noted, however, swarm research has developed independently of the network discourse both genealogically and in media-historical terms. These two discourses have only overlapped (and problematically so) as a result of the transformation of swarms into zootechnologies and the consequent prominence of swarms in the popular discourse. For cultural-theoretical approaches to networks, see, among

was one of the first authors to devote an entire chapter to swarms in a publication concerned with the potential applications of biological models in the fields of chaos research and complexity studies.[11] In 2001, Steven Johnson examined the common features in the lives of ants, brains, cities, and software.[12] In a book published the following year, Howard Rheingold outlined his notion of "smart mobs," which he used to describe phenomena such as "flash mobs," the "critical mass" movement (which involves groups of cyclists coming together to disrupt automobile traffic), acts of protest against globalization (such as the abovementioned 'Battle of Seattle'), mass protests initiated by text messages (one of which led to the ousting of Joseph Estrada, the President of the Philippines, in 2001), and the use of mobile phones by teenagers in Tokyo to plan parties.[13] In the wake of Rheingold's work, authors such as James Surowiecki and Philip Ball published books about the "wisdom of crowds" that, while making no reference to biological research, investigate the effectiveness of *collective intelligence* in diverse socio-political and economic spheres.[14] As mentioned in my introduction, moreover, choreographers, designers, subversive political groups, grass-roots networkers, military tacticians, and trend researchers began to employ the concept of the swarm in an increasingly undifferentiated manner.[15] Statements such as the following raise questions not only about the blindness of 'swarming' humans toward the socio-political and economic relations within swarm collectives and their underlying power structures but also about a sort of local blindness that occasionally reveals certain cases of alleged swarm intelligence to be examples of 'swarm stupidity' instead. One newspaper article, for instance, celebrated a swarm's potential for instantaneous action with these words: "They don't have to spend all day protesting. They just get a message telling them when it's starting, and then take the elevator down the street. They can be seen, scream a little and then go back to work."[16] In the same spirit, certain discourses about swarm-like management methods and workforce collaboration, which may at first

other works, Barkhoff et al., eds., *Netzwerke: Eine Kulturtechnik der Moderne*; Gießmann, *Netze und Netzwerke*; and idem, 'Netzwerkprotokolle und Schwarm-Intelligenz.'

11 See Kelly, *Out of Control*.

12 Johnson, *Emergence: The Connected Lives of Ants, Brains, Cities, and Software*.

13 See Rheingold, *Smart Mobs*.

14 See, for example, Surowiecki, *The Wisdom of Crowds*; and Ball, *Critical Mass*.

15 See, among other works, Miller et al., *Swarm*; Brandstetter et al., *Swarm(E)Motion*; Rheingold: *Smart Mobs*; and Medosch, 'Meshing the Future'; and Arquilla and Ronfeldt, *Swarming and the Future of Conflict*.

16 Garreau, 'Cell Biology.'

glance seem to be democratically motivated, are in fact embedded in rather traditional and hierarchical leadership structures. What lurks behind the term "swarm architecture," moreover, is not an ephemeral construction concept but rather just a strategy for developing creative ideas in technically networked and flexibly interactive working groups.[17] In the field of management, 'swarms' are typically invoked as an application that can be made operative and effective within certain structures of control and leadership.[18]

This proliferation of the use of the swarm concept to refer to human 'collectives,' which is now found in so many areas of life, converges with new social network platforms and their infrastructure. The common features of every 'social swarm' – that is, the spatial fluidity and temporal spontaneity of human gatherings in real or computer-generated environments, which distinguishes these social forms from traditional collective models such as social class, political parties, labor unions, and interest groups – are also reflected in the metaphors used to refer to the gadgets and interfaces that enable this dynamic form of communication. In certain file-sharing protocols, for instance, users are referred to as members of a swarm; people used to take advantage of the political effects of the 'blogosphere,' which is now out of fashion;[19] communication on Twitter is now routine, as is grass-roots activism on Facebook; and critical studies have highlighted the negative factors of filter bubbles and homophily in social networks and search engines[20] since certain Wi-Fi protocols and open networks have introduced swarm-like principles on the infrastructural level.[21] In this environment, swarms and swarm logic have begun to converge with dynamic, decentralized, and distributed network concepts to form a discourse that more or less ignores its material and media-technological foundations and thus tends to lack any historical perspective. Perhaps because of the disparate object areas in which this discourse has fanned out, however, it might not be feasible to undertake such an archaeology of the present. Moreover, any such effort would always run the risk of quickly becoming obsolete and appearing as outdated as the texts written by those techno-apologists who, around the year 2000, began to fuel the swarm discourse in popular-scientific books and magazines. Student protesters or Occupy activists have also attempted to organize themselves in a 'swarm-like' manner, but this has had little

17 See Oosterhuis, 'Swarm Architecture'; and idem, *Hyperbodies*.

18 See Neef and Burmeister, 'Swarm Organization.'

19 See, for instance, Lovink, *Zero Comments: Blogging and Critical Internet Culture*.

20 See, for instance, Pariser, *The Filter Bubble*, and Chun, *Queering Homophily*.

21 On networks of this sort, see Minar et al., 'Hive: Distributed Agents for Networking Things'; Medosch, *Freie Netze*, 57–83; and idem, 'Meshing the Future.'

effect (university leaders and politicians have simply been able to say that, under such conditions, it is unclear with whom they should be negotiating). And it is true that revolutionary protests have been organized in autocratic states with the help of network-based platforms, but as soon as government authorities shut down these 'information superhighways,' such movements must rapidly transform into more traditional forms of protests attempting to take over 'the street' without any technological savviness whatsoever.

In terms of media history and the history of knowledge, examples of this sort obscure the distinct developments of swarm research in the fields of biology and computer science – developments that I have traced throughout this book. Even though swarms have been released into the open sea of discourse, I have attempted to concentrate on the media-technological conditions that have made it possible to formulate and describe today's media culture of intransparency and on how these conditions are interwoven with computer simulation's order of knowledge. I have therefore not been interested in relating the story of the latest techno-social smart mob or in discussing the next 'social revolution' that might be on the horizon. Regarding today's media culture, the far more profound (and interesting) story is that of the disrupted genealogies, the esoteric venues, and the tentative models and theories that have constituted swarm research throughout its history – the story of how laboratory researchers designed their instruments; how underwater researchers developed their own diving techniques and methods of observation; how interdisciplinary approaches (cobbled together from mathematics, information theory, physics, and graphic design) have taken swarm research in new directions; and how, through a recursive epistemological process, swarms have ultimately become self-described media. Of greater interest, in short, are the conditions that have made it possible to discuss swarms as operative zootechnologies and the media histories of swarm research that guide this discussion and have made it possible for swarms to transform into figures of knowledge.

Swarms are exemplary and material indications of the ubiquitous media culture of unclarity and intransparency that has assumed the 'messiness of life' and that underlies the widespread use of abstract swarm metaphors in popular culture and in socio-economic spheres. In the reevaluation of swarms from entities outside of knowledge to figures of knowledge, there also occurred a reevaluation of knowledge itself and a transformation of epistemic strategies – a change, that is, in the condition for accessing what can be known at all. And this reevaluation is tied to a number of media-technological scenarios that enabled swarms to enter into the realm of knowledge. It is also oriented toward a number of simulation-based, trial-and-error methods in which biological principles were implemented and in which biological research gained a new level of effectiveness – not least as a component of the more comprehensive epistemology of computer simulation. The metaphorical transference of swarms to human beings was thus preceded by various interrelations between humans, swarming animals, machines, and programs. At its heart lies a conceptualization of swarms as *zootechnologies* – and hence my investigation into the media-historical transformation of swarms into technologized collectives.

Through this application of self-organizing processes to processes of self-organization, swarms transformed into an object of a type of media history that is interested in operative media, and it was my aim in this book to track the developments of these interrelations in the history of media, technology, and knowledge. As an effect, it is possible to speak of the *media-becoming* of swarms. This media-becoming is embedded in an epoch of computer simulation – in an epoch of intransparent media cultures in which various scientific fields have been transformed into behavioral sciences and within whose epistemological scope the traditional separations between induction and deduction or the established distinctions between epistemic and technical things have become obsolete.

The broader arguments of this book can thus be summarized as follows. First, a media history of swarms as zootechnologies between biology and computer simulation is of *media-theoretical* interest because, within the context of a more widely discussed media theory of disruption, it is concerned with an object of knowledge that, though concrete, materializes an irreducible disruptive moment. Though transmission events themselves, swarms disrupt events of transmission, such as the channels of the very media technologies that have been used to objectify them scientifically. To this can be added an *epistemological* interest. This is reflected in a strategy that was only able to overcome the epistemic obstacles that produce intransparent objects of knowledge such as swarms by retreating from naturalness and relying instead on control-technical, information-theoretical, cybernetic, and systemic concepts and technologies. Only by *synthesizing* swarms as dynamic collectives by means of the technical methods of computer simulation were biological researchers and their traditional analytic environments able to proceed with their object of knowledge on a new level. Thus, the third focus of this book has been to describe an *historiographical* development within which biologically inspired computer-simulation models have been recursively used to conduct research on biological swarms. In this intermingling of biology and computer technology, swarms have thus co-written themselves along with the knowledge that had already existed about swarms. Furthermore, because this writing process has essentially been dependent on graphical visualization techniques, which (and this is an epistemologically interesting constellation in its own right) make it possible to compare scenarios and produce *practical* similarity relationships between simulation models and dynamic four-dimensional objects of knowledge, the media history of swarm research also provides insight into the present era of computer simulation, which is likewise concerned with intransparent sets of problems. In investigating such problems, the production of *dynamic data images* has by now become more important than the older concepts and techniques of laboratory studies. One of the developments stimulated by the visualizations of agent-based models has been the dissolution of epistemic and technical things into the *epistemic aggregations* of parallel computer processing and differentially evaluated simulation scenarios.

The final stages of my analysis concern the media-becoming of swarms and their *transformation* into figures of knowledge, and it is at this point where it is first possible to detect the conditions of possibility for the discursive dynamics that have made swarms a nearly omnipresent topic in recent years. My last chapter outlines how computer-scientifically informed biology and biologically informed and inspired computer science have productively blended together in such a way as to establish an entirely different discourse about swarms that, in conjunction with the concept of 'intelligence,' has drawn them out of the internal discourses of biology, computer science, or robotics. Only at the end of this book, then, did the discourse dynamics come to light that happened to be its point of departure. Only through the formation and transformation of swarms into objects and figures of knowledge could they become an attractive theme for the 2005 trend conference in Hamburg, whose brochure featured a school of sardines being punctured by a feeding shark and inspired me to trace the interwoven and meandering histories of swarm research. Part biology and part computer simulation, swarm research is an essential component of today's media cultures not in the sense that they simply provide fitting metaphors. Rather, they operate as technological processes of self-organization directly at the media-technological foundations of social collective dynamics and, at the same time, as an adequate tool for investigating them – so much is clear from their diverse applications as zootechnologies in logistics, mathematical optimization, panic and traffic studies, social simulations, production planning, and robotic systems. Beyond what analysts say about the mass-media impact of so-called new social networks on socio-economic processes, every revolution is already a media revolution, and every mass movement can be dissolved into the crowd dynamics of numerous autonomous individuals.

Swarms produce disruptions and make them productive. They are figures and defigurations that intercede not only at the beginning of media theory but also at the beginning of unclear knowledge – at the beginning of epistemic strategies whose investigation here should end (or not) with an observation by Michel Serres: "Everything happens as if the following proposition were true: it works because it does not work. [...] Fluctuation, disorder, opacity, and noise are not and are no longer affronts to the rational [...]. The difference is part of the thing itself, and perhaps it even produces the thing. Maybe the radical origin of things is really that difference, even though classical rationalism damned it to hell."[22] It is this sort of noise in which biological swarm research and agent-based computer simulation have intermixed in an entirely new and unusual way. Even though the movements of this text have now come to a halt, this is not the case for swarms as four-dimensional collectives (as I have understood them). Admittedly, since Jean Painlevé's time we have known: "When movement ceases, the show is over."[23] Yet for the object of this study – for swarms as objects and figures of knowledge – the following is likewise true: "The swarm may hover, but it does not rest."[24] Swarms record themselves as writing processes; as figures of knowledge, they collaborate in their constitution as objects of knowledge. What is more, however, they also constitute and describe a media culture of unclarity and intransparency – and in this context they *mobilize* multiple application tools and problem-solving methods. All metaphors aside, their place between biological research and computer science in the history of media and technology has made it clear that swarms *move* the world. In the beginning was the noise, and so here as well.

22 Serres, *The Parasite*, 13.
23 Rugoff, 'Fluid Mechanics,' 56.
24 Stroud, 'Approaching Swarms,' 11.

Works Cited

Adamatzky, Andrew and Genaro J. Martínez, eds. *Designing Beauty: The Art of Cellular Automata*. Cham: Springer International, 2016.

Adelmann, Ralf et al. *Datenbilder: Zur digitalen Bildpraxis in den Naturwissenschaften*. Bielefeld: Transcript, 2009.

Adelung, Johann Christoph. *Grammatisch-kritisches Wörterbuch der hochdeutschen Mundart, mit beständiger Vergleichung der übrigen Mundarten, besonders aber der Oberdeutschen*. Vol. 3. Vienna: Bauer, 1811.

Allee, Warder C. 'Animal Aggregations.' In *Principles of Animal Ecology*, edited by Alfred E. Emerson et al., 393–419. Philadelphia: Saunders, 1950.

———. 'Animal Aggregations: A Request for Information.' *The Condor* 25 (1923): 129–131.

———. 'Cooperation Among Animals.' *American Journal of Sociology* 37 (1931): 386–398.

Allen, Austin. 'Interview with Iain Couzin.' *BigThink.com* (15 December 2009): http://bigthink.com/videos/big-think-interview-with-iain-couzin.

Allen, E. J. and H. W. Harvey. 'The Laboratory of the Marine Biological Association at Plymouth.' *Journal of the Marine Biology Association of the United Kingdom* 15 (1928): 734–751.

Allen, Stan. 'From Objects to Fields.' In *Architecture After Geometry*, edited by Peter Davidson and Donald L. Bates, 24–31. London: Architectural Design, 1997.

Allwright, Michael et al. 'SRoCS: Leveraging Stigmergy on a Multi-Robot Construction Platform for Unknown Environments.' In *Swarm Intelligence*, edited by Marco Dorigo, 158–169. Berlin: Springer, 2014.

Alt, Casey. 'Objects of Our Affection: How Object Orientation Made Computers a Medium.' In *Media Archaeology: Approaches, Applications, and Implications*, edited by Erkki Huhtamo and Jussi Parikka, 278–301. Berkeley: University of California Press, 2011.

Anonymous. 'Men of the Day, No. 585: Mr Frederic Courtney Selous.' *Vanity Fair Album* 24 (1894): 263.

Anonymous. 'The New Beat Cop: Transforming a Small Drone Into an Unattended Urban Surveillance System.' *Launch Stories: Air Force Funded Technology, Small Business Success* (2018): http://launchstories.org/stories/TheNewBeatCop.

Anonymous. '*Thought-Transference (Or What?) in Birds*, by Edmund Selous: Short Review.' *Nature* 129 (1932): 263.

Aoki, Ichiro. 'A Simulation Study on the Schooling Mechanism in Fish.' *Bulletin of the Japanese Society of Scientific Fisheries* 48 (1982): 1081–1088.

Arquilla, John and David Ronfeldt. *Networks and Netwars: The Future of Terror, Crime, and Militancy.* Santa Monica: RAND, 2001.

———. *Swarming and the Future of Conflict.* Santa Monica: RAND, 2000.

Asendorf, Christoph. Ströme und Strahlen: Das langsame Verschwinden der Materie um 1900. Gießen: Anabas, 1989.

Ashby, W. Ross. *Design for a Brain: The Origin of Adaptive Behaviour,* 2nd ed. London: Chapman & Hall, 1960.

Augugliaro, Federico et al. 'Building Tensile Structures with Flying Machines.' In *Proceedings of the IEEE/RSJ International Conference on Intelligent Robots and Systems,* 3487–3492. Piscataway, NJ: IEEE, 2012.

Axelrod, Robert. *The Complexity of Cooperation: Agent-Based Models of Competition and Collaboration.* Princeton: Princeton University Press, 1997.

Bacco, Manlio et al. 'UAVs and UAV Swarms for Civilian Applications: Communications and Image Processing in the SCIADRO Project.' In *Wireless and Satellite Systems: 9th International Conference, WiSATS 2017,* edited by Prashant Pillai, 115–124. Berlin: Springer, 2017.

Baecker, Dirk. 'Die Dekonstruktion der Schachtel: Innen und Außen in der Architektur.' In *Unbeobachtbare Welt: Über Kunst und Architektur,* edited by Niklas Luhmann et al, 67–99. Bielefeld: Haux, 1990.

———. 'Vorwort.' In *Ding und Medium,* by Fritz Heider, 7–20. Berlin: Kadmos, 2005.

Balazs, Michael A. and Jonathan Rotner. 'Open, Commercial Technologies Lead to Cost-Effective Reconnaissance Solutions.' *MITRE* (December 2013): https://www.mitre.org/publications/project-stories/open-commercial-technologies-lead-to-cost-effective-reconnaissance.

Ball, Philip. *Critical Mass.* London: Cornerstone, 2014.

Ballerini, Michele et al. 'Empirical Investigation of Starling Flocks: A Benchmark Study in Collective Animal Behaviour.' *Animal Behaviour* 76 (2008): 201–215.

Ballerini, Michele et al. 'Interaction Ruling Animal Collective Behavior Depends on Topological Rather than Metric Distance: Evidence from a Field Study.' *PNAS* 105 (2008): 1232–1237.

Baraniuk, Chris. 'US Military Tests Swarm of Mini-Drones Launched from Jets.' *BBC News* (10 January 2017): https://www.bbc.com/news/technology-38569027.

Barberi, Alessandro. 'Editorial: Historische Epistemologie und Diskursanalyse.' *Österreichische Zeitschrift für Geschichtswissenschaften* 11/4 (2000): 5–10.

Barkhoff, Jürgen et al., eds. *Netzwerke: Eine Kulturtechnik der Moderne.* Cologne: Böhlau, 2004.

Bartholomew, George A. 'The Fishing Activities of Double-Crested Cormorants on San Francisco Bay.' *The Condor* 44 (1942): 13–21.

Bateson, Gregory. *Steps to an Ecology of the Mind: Collected Essays in Anthropology, Psychiatry, Evolution, and Epistemology*. Chicago: University of Chicago Press, 1972; repr. 2000.

Bateson, William. 'The Sense-Organs and Perceptions of Fishes, with Remarks on the Supply of Bait.' *Journal of the Marine Biology Association of the United Kingdom* 1 (1890): 225–258.

Batty, Michael. *Cities and Complexity: Understanding Cities with Cellular Automata, Agent-Based Models, and Fractals*. Cambridge, MA: MIT Press, 2005.

Bazigos, G. P. *A Manual on Acoustic Surveys: Sampling Methods for Acoustic Surveys*. Rome: Food and Agriculture Organization of the United Nations, 1981.

Bazin, André. 'Science Film: Accidental Beauty.' In *Science is Fiction: The Films of Jean Painlevé*, edited by Andy Masaki Bellows et al., 144–147. Cambridge, MA: MIT Press, 2000.

Beare, D. J. et al. 'Spatio-Temporal Patterns in Herring (*Clupea harengus* L.) School Abundance and Size in the Northwest North Sea: Modelling Space-Time Dependencies to Allow Examination of the Impact of Local School Abundance on School Size.' *ICES Journal of Marine Science* 59 (2002): 469–479.

Bechterev, Vladimir M. *Suggestion und ihre soziale Bedeutung*. Leipzig: Georgi, 1899.

Beni, Gerardo. 'From Swarm Intelligence to Swarm Robotics.' In *Swarm Robotics*, edited by Erol Şahin and William M. Spears, 3–9. New York: Springer, 2008.

———. 'Order by Disordered Action in Swarms.' In *Swarm Robotics*, edited by Erol Şahin and William M. Spears, 153–172. New York: Springer, 2008.

Beni, Gerardo and Jing Wang. 'Swarm Intelligence in Cellular Robotic Systems.' In *Robots and Biological Systems: Towards a New Bionics?* Proceedings of the NATO Advanced Workshop on Robots and Biological Systems Held at II Ciocco, Toscana, Italy, June 26–30, 1989, edited by Paolo Dario et al., 703–712. Heidelberg: Springer, 1993.

Berg, Brigitte. 'Contradictory Forces: Jean Painlevé 1902–1989.' In *Science is Fiction: The Films of Jean Painlevé*, edited by Andy Masaki Bellows et al., 2–57. Cambridge, MA: MIT Press, 2000.

Berill, N. J. 'The Pearls of Wisdom: An Exposition.' *Perspectives in Biology and Medicine* 28 (1984): 1–16.

Bernard, Claude. *An Introduction to the Study of Experimental Medicine*. Translated by Henry Copley Greene. New York: Henry Schuhman, 1949.

———. *Philosophie: Manuscrit inédit*. Edited by Jaques Chevalier. Paris: Hatier-Boivin, 1954.

Berz, Peter. 'Die Wabe.' In *FAKtisch: Festschrift für Friedrich Kittler zum 60. Geburtstag*, edited by Bernhard Siegert et al., 65–81. Munich: Fink, 2003.

Bhuta, Nehal et al. *Autonomous Weapons Systems: Law, Ethics, Policy*. Cambridge: Cambridge University Press, 2016.

Birch, Herbert G. 'Communications Between Animals.' In *Cybernetics: The Macy Conferences, 1946–1953*. Vol. 1, edited by Claus Pias, 446–528. Zurich: Diaphanes, 2004.

Bohannon, John. 'Directing the Herd: Crowds and the Science of Evacuation.' *Science* 310 (2005): 219–221.

Böhme, Hartmut and Gernot Böhme. *Das Andere der Vernunft: Zur Entwicklung von Rationalitätsstrukturen am Beispiel Kants*. Frankfurt am Main: Suhrkamp, 1983.

Bonabeau, Eric. 'Agent-Based Modeling: Methods and Techniques for Simulating Human Systems.' *Proceedings of the National Academy of Sciences of the United States of America* 99 (2002): 7280–7287.

Bonabeau, Eric et al. *Swarm Intelligence: From Natural to Artificial Systems*. New York: Oxford University Press, 1999.

Bonani, Michael et al. 'The marXbot, a Miniature Mobile Robot Opening New Perspectives for the Collective-Robotic Research.' In *Proceedings of the 2010 IEEE/RSJ International Conference on Intelligent Robots and Systems*, 4187–4193. Piscataway, NJ: IEEE, 2010.

Bonnet, Frank et al. 'Closed-Loop Interactions Between a Shoal of Zebrafish and a Group of Robotic Fish in a Circular Corridor.' *Swarm Intelligence* 12 (2018): 227–244.

Borch, Christian. *The Politics of Crowds: An Alternative History of Sociology*. Cambridge: Cambridge University Press, 2012.

Borshchev, Andrei and Alexei Filippov. 'From System Dynamics, and Discrete Event to Practical Agent Based Modeling: Reasons, Techniques, Tools.' In *Proceedings of the 22nd International Conference of the System Dynamics Society* edited by Michael Kennedy, N.p. Albany: System Dynamics Society, 2004.

Bosch, Sébastien et al. 'Autonomous Detection of Safe Landing Areas for an UAV from Monocular Images.' In *Proceedings of the 2006 IEEE/RSJ International Conference on Intelligent Robots and Systems*, 5522–27. Piscataway, NJ: IEEE, 2006.

Boulos, Kamel et al. 'Crowdsourcing, Citizen Sensing, and Sensor Web Technologies for Public and Environmental Health Surveillance and Crisis Management: Trends, OGC Standards and Application Examples.' *International Journal of Health Geographics* 10/67 (2011): 1–29.

Brain, Robert. *The Graphic Method: Inscription, Visualization, and Measurement in Nineteenth-Century Science and Culture*. Doctoral Diss.: UCLA, 1996.

Brambilla, M. et al. 'Swarm Robotics: A Review from the Swarm Engineering Perspective.' *Swarm Intelligence* 7 (2013): 1–41.

Brandstetter, Gabriele et al., eds. *Swarm(E)Motion: Bewegung zwischen Affekt und Masse*. Freiburg: Rombach, 2007.

Breder, Charles M. 'Equations Descriptive of Fish Schools and Other Animal Aggregations.' *Ecology* 35 (1954): 361–70.

———. 'Fish Schools as Operational Structures.' *Fishery Bulletin* 74 (1976): 471–502.

———. 'Studies on the Structure of the Fish School.' *Bulletin of the American Museum of Natural History* 98 (1951): 1–28.

Brooks, Frederick. 'No Silver Bullet: Essence and Accidents of Software Engineering.' *Computer* 20 (1987): 10–19.

Brooks, Rodney A. 'Elephants Don't Play Chess.' *Robotics and Autonomous Systems* 6 (1990): 3–15.

Brooks, Rodney A. and Anita M. Flynn. 'Fast, Cheap, and Out of Control: A Robot Invasion of the Solar System.' *Journal of the British Interplanetary Society* 42 (1989): 478–85.

Brown, Richard E. and M. Roger Fedde. 'Airflow Sensors in the Avian Wing.' *Journal of Experimental Biology* 179 (1993): 13–30.

Buck, John and Elisabeth Buck. 'Mechanism of Rhythmic Synchronous Flashing of Fireflies.' *Science* 159 (1968): 1319–27.

Bühler, Benjamin and Stefan Rieger. *Vom Übertier: Ein Bestiarium des Wissens*. Frankfurt am Main: Suhrkamp, 2006.

Bülow, Ralf. 'Halb Tarzan, halb Grzimek.' *Spiegel Online* (November 23, 2007). http://www.spiegel.de/einestages/tauch-und-tv-pionier-hans-hass-a-948851.html.

Burckhardt, Richard W. *Patterns of Behavior: Konrad Lorenz, Nico Tinbergen, and the Founding of Ecology*. Chicago: University of Chicago Press, 2005.

Burdic, William S. *Underwater Acoustics Systems Analysis*, 2nd ed. Englewood Cliffs, NJ: Prentice Hall, 2003.

Burke, Jeff. et al. 'Participatory Sensing.' *ACM SenSys* (2006): https://escholarship.org/uc/item/19h777qd.

Buus, Dennis P. *Constructing Human-Like Architecture with Swarm Intelligence*. Doctoral Diss.: Aalborg University, 2006.

Camazine, Scott et al. *Self-Organization in Biological Systems*. Princeton: Princeton University Press, 2003.

Canetti, Elias. *Crowds and Power*. Translated by Carol Stewart. New York: Continuum, 1962.

Carranza Pablo Miranda and Paul Coates, 'Swarm Modelling: The Use of Swarm Intelligence to Generate Architectural Form.' *Generative Art* (2000): http://www.generativeart.com/on/cic/2000/CARRANZA_COATES.HTM.

Carpo, Mario. *The Digital Turn in Architecture, 1992–2012*. New York: Wiley, 2013.

Casti, John L. *Would-Be-Worlds: How Simulation Is Changing the Frontiers of Science*. New York: Wiley, 1997.

Cavagna, Andrea et al. 'The STARFLAG Handbook on Collective Animal Behaviour: 1. Empirical Methods.' *Animal Behaviour* 76 (2008): 217–236.

Cavagna, Andrea et al. 'The STARFLAG Handbook on Collective Animal Behaviour: 2. Three-Dimensional Analysis.' *Animal Behaviour* 76 (2008): 237–248.

Cazenille, L. et al. 'How Mimetic Should a Robotic Fish Be to Socially Integrate Into Zebrafish Groups?' *Bioinspiration & Biomimetics* 13 (2018): 1–15.

Cevik, Polat et al. 'The Small and Silent Force Multiplier: A Swarm UAV-Electronic Attack.' *Journal of Intelligent & Robotic Systems* 70 (2013): 595–608.

Chamayou, Grégoire. *Drone Theory*. London: Penguin, 2015.

Charbonneau, Paul. *Natural Complexity: A Modeling Handbook*. Princeton: Princeton University Press, 2017.

Christensen, Lars Rune. 'The Logic of Practices of Stigmergy: Representational Artifacts in Architectural Design.' In *Proceedings of the 2008 ACM Conference on Computer Supported Cooperative Work*, 559–568. New York: ACM, 2008.

Chun, Wendy Hui Kyong. 'Queerying Homophily. Muster der Netzwerkanalyse.' In *Zeitschrift für Medienwissenschaft* 18/1 (2018): 131-148

Colomina, I. and P. Molina. 'Unmanned Aerial Systems for Photogrammetry and Remote Sensing: A Review.' *ISPRS Journal of Photogrammetry and Remote Sensing* 92 (2014): 79–97.

Corner, J. J. and G. B. Lamont. 'Parallel Simulation of UAV Swarm Scenarios.' In *Proceedings of the 2004 Winter Simulation Conference*, edited by Ricki G. Ingalls et al., 355–363. New York: ACM, 2004.

Corning, Peter A. 'The Re-Emergence of 'Emergence': A Venerable Concept in Search of a Theory.' *Complexity* 7/6 (2002): 18–30.

Couffer, Jack. *Bat Bomb: World War II's Other Secret Weapon*. Austin: University of Texas Press, 1992.

Cousteau, Jacques-Yves. *The Silent World*. New York: Harper, 1953.

Couzin, Iain D. et al. 'Collective Memory and Spatial Sorting in Animal Groups.' *Journal of Theoretical Biology* 218 (2002): 1–11.

Couzin, Iain D. and Jens Krause. 'Self-Organization and Collective Behavior in Vertebrates.' *Advances in the Study of Behavior* 32 (2003): 1–75.

Cullen, J. Michael et al. 'Methods for Measuring the Three-Dimensional Structure of Fish Schools.' *Animal Behavior* 13 (1965): 534–543.

Czirok, András and Tamás Vicsek. 'Collective Behavior of Interacting Self-Propelled Particles.' *Physica A* 281 (2000): 17–29.

Damisch, Hubert. 'Die Geschichte und die Geometrie.' *Archiv für Mediengeschichte* 5 (2005): 11–25.

Daston, Lorraine and Peter Galison. 'The Image of Objectivity.' *Representations* 40 (1992): 81–128.

Dearden, Paul and Michael Akam. 'Segmentation *in Silico*.' *Nature* 406 (2000): 304–305.

De Armond, Paul. 'Black Flag Over Seattle.' *The Albion Monitor* (29 February 2000): http://www.albionmonitor.com/seattlewto/index.html.

De Certeau, Michel. *Heterologies: Discourses of the Other*. Translated by Brian Massumi. Minneapolis: University of Minnesota Press, 1986.

De Chadarevian, Soray and Nick Hopwood, eds. *Models: The Third Dimension of Science*. Stanford: Stanford University Press, 2004.

De Chiara, Rosario et al. 'An Architecture for Distributed Behavioral Models with GPUs.' In *Papers Presented at the Fourth Eurographics Italian Chapter Conference*, edited by Giovanni Gallo, 1–7. Aire-la-Ville: Eurographics Association, 2006.

De Croon, G. C. H. E. et al. *The DelFly: Design, Aerodynamics, and Artificial Intelligence of a Flapping Wing Robot*. Berlin: Springer, 2016.

Deep Blue. Directed by Alistair Fothergill. Los Angeles: Miramax Films, 2003.

DeLanda, Manuel. *Philosophy and Simulation: The Emergence of Synthetic Reason*. London: Continuum, 2011.

Deleuze, Gilles. *Foucault*. Translated by Seán Hand. New York: Continuum, 2006.

———. 'Postscript on the Societies of Control.' *October* 59 (1992): 3–7.

Deleuze, Gilles and Félix Guattari. *A Thousand Plateaus: Capitalism and Schizophrenia*. Translated by Brian Massumi. Minneapolis: University of Minnesota Press, 1987.

Dell, Anthony I. et al. 'Automated Image-Based Tracking and Its Application in Ecology.' *Trends in Ecology & Evolution* 29 (2014): 417–428.

DeMello, Margo. *Animals and Society: An Introduction to Human-Animal Studies*. New York: Columbia University Press, 2012.

Denny, Mark. *Blip, Ping & Buzz: Making Sense of Radar and Sonar*. Baltimore: Johns Hopkins University Press, 2007.

De Tarde, Gabriel. *Les crimes des foules*. Paris: Messon, 1893.

———. *The Laws of Imitation*. Translated by Elsie Clews Parsons. New York: H. Holt, 1903.

———. *L'Opinion et la foule*. Paris: F. Alcan, 1901.

———. *Penal Philosophy*. Translated by Rapelje Howell. Montclair, NJ: Patterson-Smith, 1968.

Dueck, Günther. *Schwarmdummheit: So blöd sind wir nur gemeinsam*. Frankfurt am Main: Campus, 2015.

Dugatkin, Lee Alan. *Cooperation Among Animals: An Evolutionary Perspective*. Oxford: Oxford University Press, 1997.

Edwards, Sean J. A. *Swarming on the Battlefield: Past, Present, and Future*. Santa Monica: RAND, 2001.

Emmeche, Claus. *The Garden in the Machine: The Emerging Science of Artificial Life*. Translated by Steven Sampson. Princeton: Princeton University Press, 1994.

Engelbrecht, Andries P. *Fundamentals of Computational Swarm Intelligence*. Chichester: Wiley, 2005.

Engell, Lorenz. 'Das Amedium – Grundbegriffe des Fernsehens in Auflösung: Ereignis und Erwartung.' *Montage/AV: Zeitschrift für Theorie und Geschichte audiovisueller Kommunikation* 5 (1996): 129–153.

———. 'Zur Einführung.' In *Kursbuch Medienkultur: Die maßgeblichen Theorien von Brecht bis Baudrillard*, edited by Joseph Vogl et al., 301–307. Stuttgart: DVA, 1999.

Engell, Lorenz and Joseph Vogl. 'Einleitung.' In *Mediale Historiographien*, edited by Lorenz Engell and Joseph Vogl, 1–9. Weimar: Universitätsverlag, 2001.

———. 'Vorwort.' In *Kursbuch Medienkultur: Die maßgeblichen Theorien von Brecht bis Baudrillard*, edited by Joseph Vogl et al., 8–12. Stuttgart: DVA, 1999.

Epstein, Joshua M. and Robert L. Axtell. *Growing Artificial Societies: Social Science from the Bottom Up*. Washington, DC: Brookings Institution, 1996.

Ernst, Wolfgang. 'Der Appell der Medien: Wissensgeschichte und ihr Anderes.' In *Rekursionen: Von Faltungen des Wissens*, edited by Ana Ofak and Philipp von Hilgers, 177–197. Munich: Fink, 2010.

Erra, Ugo et al. 'An Efficient GPU Implementation for Large Scale Individual-Based Simulation of Collective Behavior.' In *Proceedings of the 2009 International Workshop on High Performance Computational Systems Biology*, 51–58. Los Alamitos: IEEE Computer Society, 2009.

Espinas, Alfred. *Des sociétés animales: Étude de psychologie comparée*. Paris: Librarie Germer Baillière, 1877.

Faria, Jolyon J. et al. 'A Novel Method for Investigating the Collective Behaviour of Fish: Introducing 'Robofish'.' *Behavioral Ecology and Sociobiology* 64 (2010): 1211–1218.

Feng, Emily and Charles Clover. 'Drone Swarms vs. Conventional Arms: China's Military Debate.' *Financial Times* (24 August 2017): https://www.ft.com/content/302fc14a-66ef-11e7-8526-7b38dcaef614.

Fieser, Louis F. *The Scientific Method: A Personal Account of Unusual Projects in War and Peace*. New York: Reinhold Publishers, 1964.

Fischer, Wolfgang. 'Methodik und Ergebnisse der Erforschung des Schwarmverhaltens von Fischen mit der Tauchmethode.' *Helgoländer wissenschaftliche Meeresuntersuchungen* 24 (1973): 391–400.

Fisher, Len. *The Perfect Swarm: Science and Complexity in Everyday Life*. New York: Basic Books, 2011.

Flake, Gary William. *The Computational Beauty of Nature: Computer Explorations of Fractals, Chaos, Complex Systems, and Adaptations*. Cambridge, MA: MIT Press, 2000.

Fleck, Ludwik. *Genesis and Development of a Scientific Fact*. Translated by Fred Bradley and Thaddeus J. Trenn. Chicago: University of Chicago Press, 1979.

Forbes, Nancy. *Imitation of Life: How Biology is Inspiring Computing*. Cambridge, MA: MIT Press, 2004.

Forel, Auguste. *Les fourmis de la Suisse*. Zürich: Schweizerische Gesellschaft, 1873.

Foster, Ian. 'A Two-Way Street to Science's Future.' *Nature* 440 (2006): 419.

Foucault, Michel. *The Birth of Biopolitics: Lectures at the Collège de France, 1978–1979*. Translated by Graham Burchell. New York: Palgrave Macmillan, 2008.

———. *Discipline and Punish: The Birth of the Prison*. Translated by Alan Sheridan. New York: Vintage, 1977.

———. *The Hermeneutics of the Subject: Lectures at the Collège de France*, 1981–82. Translated by Graham Burchell. New York: Palgrave Macmillan, 2005.

———. *The Order of Things: An Archaeology of the Human Sciences*. Translated by Alan Sheridan. New York: Vintage, 1970.

Franke, Hermann, ed. *Dtv-Lexikon der Physik*. 10 vols. Munich: Deutscher Taschenbuch Verlag, 1970, III.

Franklin, Allan. 'Experiment in Physics.' In *The Stanford Encyclopedia of Philosophy*, edited by Edward N. Zalta. Stanford: The Metaphysics Research Lab, 1998; rev. 2015. https://plato.stanford.edu/entries/physics-experiment/#PDA.

Freud, Sigmund. *Group Psychology and the Analysis of the Ego*. Translated by James Strachey. London: The Hogarth Press, 1949.

Gajjar, Barghav I. 'Parallel Kinomatics Micro-Positioning System.' U.S. Patent No. 20100253275A1 (2010): https://patentimages.storage.googleapis.com/d1/85/27/e70ef102ba6dde/US20100253275A1.pdf.

Galison, Peter. 'Computer Simulations and the Trading Zone.' *The Disunity of Science: Boundaries, Contexts, and Power*, edited by Peter Galison and D. J. Stump, 118–157. Stanford: Stanford University Press, 1996.

———. *Einstein's Clocks, Poincaré's Maps: Empires of Time*. New York: W. W. Norton, 2003.

———. *Image and Logic: A Material Culture of Microphysics*. Chicago: University of Chicago Press, 1997.

Galton, Francis. *Inquiries into Human Faculty and Its Development*. New York: MacMillan, 1883.

Gamper, Michael and Peter Schnyder, eds. *Kollektive Gespenster: Die Masse, der Zeitgeist und andere unfaßbare Körper*. Freiburg: Rombach, 2006.

Ganti, Raghu K. et al. 'Mobile Crowdsensing: Current State and Future Challenges.' *IEEE Communications Magazine* 49/11 (2011): 32–39.

Garnier, Simon et al. 'The Biological Principles of Swarm Intelligence.' *Swarm Intelligence* 1 (2007): 3–31.

Garreau, Joel. 'Cell Biology.' *The Washington Post* (30 July 2002): https://www.washingtonpost.com/archive/lifestyle/2002/07/31/cell-biology/14b56c58-ed91-4334-afe5-d78b98eacc77/?utm_term=.5ad513985551.

Gaudiano, Paolo et al. 'Control of UAV Swarms: What the Bugs Can Teach Us.' In *Proceedings of the 2nd AIAA 'Unmanned Unlimited' Conference and Workshop*, 1–11. San Diego: AIAA, 2003.

Gerling, Victor and Sebastian von Mammen. 'Robotics for Self-Organized Construction.' In *Proceedings of the IEEE 1st International Workshop on Foundations and Applications of Self*Systems (FAS*W)*,162–167. Piscataway, NJ: IEEE, 2016.

Gerlotto, François. 'Gregariousness and School Behaviour of Pelagic Fish: Impact of the Acoustics Evaluation and Fisheries.' In *Proceedings of Acoustics Seminar AKUSTIKAN 2*, edited by Didier Petit et al., 233–252. Jakarta: Pusat Penelitian, 1997.

Gerlotto, François et al. 'From Two Dimensions to Three: The Use of Multibeam Sonars for a New Approach in Fisheries Acoustics.' *Canadian Journal of Fishery Aquatic Sciences* 56 (1999): 6–12.

Gerlotto, François et al. 'Waves of Agitation Inside Anchovy Schools Observed with Multibeam Sonar: A Way to Transmit Information in Response to Predation.' *ICES Journal of Marine Science* 63 (2006): 1405–1417.

Gießmann, Sebastian. *Netze und Netzwerke: Archäologie einer Kulturtechnik, 1740–1840*. Bielefeld: Transcript, 2006.

———. 'Netzwerkprotokolle und Schwarm-Intelligenz: Zur Konstruktion von Komplexität und Selbstorganisation.' In *Schwärme – Kollektive ohne Zentrum: Eine Wissensgeschichte zwischen Wissen und Information*, edited by Eva Horn and Lucas Marco Gisi, 163–182. Bielefeld: Transcript, 2009.

Gilpatric, Guy. *The Compleat Goggler*. New York: Dodd, Mead, & Co., 1938.

Goh, Gerald. 'Unmanned Aerial Vehicles and the Future of Airpower.' *Pointer* 43/2 (2017): 45–57.

Goldstein, Jeffrey. 'Emergence as a Construct: History and Issues.' *Emergence* 1 (1999): 49–72.

Goodchild, Michael. 'Citizens as Sensors: The World of Volunteered Geography.' *GeoJournal* 69/4 (2007): 211–221.

Gramelsberger, Gabriele. *Computerexperimente: Zum Wandel der Wissenschaft im Zeitalter des Computers*. Bielefeld: Transcript, 2010.

———. *Semiotik und Simulation: Fortführung der Schrift ins Dynamische – Entwurf einer Symboltheorie der numerischen Simulation und ihrer Visualisierung*. Doctoral Diss.: Freie Universität Berlin, 2000.

Graves, John. 'Photographic Method for Measuring Spacing and Density Within Pelagic Fish Schools at Sea.' *Fishery Bulletin* 75 (1977): 230–239.

Greene, Charles H. and Peter H. Wiebe. 'Acoustic Visualization of Three-Dimensional Animal Aggregations in the Ocean.' In *Animal Groups in Three Dimensions: How Species Aggregate*, edited by Julia K. Parrish et al., 61–67. Cambridge: Cambridge University Press, 1997.

Griesemer, James. 'Three-Dimensional Models in Philosophical Perspective.' In *Models: The Third Dimension of Science*, edited by Soraya de Chadarevian and Nick Hopwood, 433–442. Stanford: Stanford University Press, 2004.

Grimm, Jakob and Wilhelm Grimm. *Deutsches Wörterbuch*. 33 vols. Leipzig: S. Hirzel, 1854–1971; repr. München: Dtv, 1984.

Groß, Roderich et al. 'Autonomous Self-Assembly in Swarm-Bots.' *IEEE Transactions on Robotics* 22 (2006): 1115–1130.

Grünbaum, Daniel and Akira Okubo. 'Modelling Social Animal Aggregation.' In *Frontiers in Mathematical Biology*, edited by Simon A. Levin, 296–325. New York: Springer, 1994.

Grüne-Yanoff, Till. 'Artificial Worlds and Agent-Based Simulation.' In *The SAGE Handbook of the Philosophy of Social Sciences*, edited by I. C. Jarvie and Jesús P. Zamora Bonilla, 613–631. London: SAGE, 2011.

Gunning, Tom. 'Never Seen this Picture Before.' In *Time Stands Still: Muybridge and the Instantaneous Photography Movement*, edited by Phillip Prodger, 222–257. New York: Oxford University Press, 2003.

Gupta, Vikram et al. 'Energy Harvesting from Electromagnetic Energy Radiating from AC Power Lines.' In *Proceedings of the Sixth Workshop on Hot Topics in Embedded Network Sensors*, 1–5. New York: ACM, 2010.

Guten, S. and L. A. Vernon. 'Likelihood of Escape, Likelihood of Danger, and Panic Behavior.' *Journal of Social Psychology* 87 (1972): 29–36.

Hacking, Ian. *Representing and Intervening: Introductory Topics in the Philosophy of Natural Science*. Cambridge, UK: Cambridge University Press, 1983.

Hackmann, Willem. 'ASDICs at War.' *IEE Review* 46/3 (2000): 15–19.

Haefner, James W. *Modeling Biological Systems: Principles and Applications*, 2nd ed. New York: Springer, 2005.

Haklyut, Richard. A Selection of the Principal Voyages, Traffiques and Discoveries of the English Nation. Edited by Laurence Irving. London: W. Heinemann, 1927.

Hambling, David. 'Drones with Legs Can Perch, Watch and Walk Like a Bird.' *New Scientist* (27 January 2014): https://www.newscientist.com/article/dn24951-drone-with-legs-can-perch-watch-and-walk-like-a-bird/.

———. *Swarm Troopers: How Small Drones Will Conquer the World*. Venice, FL: Archangel Ink, 2015.

Hamilton, William D. 'Geometry for the Selfish Herd.' *Journal of Theoretical Biology* 31 (1971): 295–311.

Hammes, T. X. 'Technologies Converge and Power Diffuses: The Evolution of Small, Smart, and Cheap Weapons.' *Cato Institute: Policy Analysis* 786 (2016): 1–14.

Hannah, Nick and Alex Mustard. *Tauchen Ultimativ*. Cologne: BrunoMedia, 2006.

Hardt, Michael and Antonio Negri. *Multitude: War and Democracy in the Age of Empire*. New York: Penguin, 2004.

Harks, Tobias and Sebastian Vehlken, eds. *Neighborhood Technologies: Media and Mathematics of Dynamic Networks*. Zurich: Diaphanes, 2015.

Harlow, F. H. and B. D. Meixner. *The Particle-and-Force Computing Method for Fluid Dynamics*. Santa Fe: Los Alamos Scientific Laboratory, 1961.

Hartley, Ralph V. L. 'Transmission of Information.' *The Bell Technical Journal* 7 (1928): 535–563.

Hass, Hans. *Fotojagd am Meeresgrund*. Harzburg: Herring Verlag, 1942.

Hass, Hans and Werner Katzmann. *Der Hans Hass Tauchführer: Das Mittelmeer – Ein Ratgeber für Sporttaucher und Schnorchler.* Vienna: Molden, 1976.

Hayles, Catherine N. How We Became Posthuman: Virtual Bodies in Cybernetics, Literature, and Informatics, 4th ed. Chicago: University of Chicago Press, 2002.

Heider, Fritz. 'Ding und Medium.' *Symposion: Philosophische Zeitschrift für Forschung und Aussprache* 1 (1926): 109–157. Reprinted as *Ding und Medium.* Berlin: Kadmos, 2005.

Helbing, Dirk. 'Agent-Based Modeling.' In *Social Self-Organization: Agent-Based Simulations and Experiments to Study Emergent Social Behavior*, edited by Dirk Helbing, 25–70. Heidelberg: Springer, 2012.

———. *Social Self-Organization Agent-Based Simulations and Experiments to Study Emergent Social Behavior.* Berlin: Springer, 2012.

Helbing, Dirk and Anders Johansson. 'Pedestrian, Crowd, and Evacuation Dynamics.' In *Encyclopedia of Complexity and Systems Science*, edited by Robert A. Meyers, 6476–6495. New York: Springer, 2009.

Helbing, Dirk et al. 'The Dynamics of Crowd Disasters: An Empirical Study.' *Physical Review E* 75 (2007): 1–7.

Helbing, Dirk et al. 'Simulation Dynamical Features of Escape Panic.' *Nature* 407 (2000): 487–90.

Helbing, Dirk et al. 'Simulation of Pedestrian Crowds in Normal and Evacuation Situations.' In *Pedestrian and Evacuation Dynamics*, edited by Michael Schreckenberg and Som Deo Sharma, 21–58. New York: Springer, 2002.

Helm, Volker et al. 'Mobile Robotic Fabrication on Construction Sites: DimRob.' In *Proceedings of the 2012 IEEE/RSJ International Conference on Intelligent Robots and Systems*, 4335–4341. Piscataway, NJ: IEEE, 2012.

Helmreich, Stefan. *Silicon Second Nature: Culturing Artificial Life in a Digital World.* Berkeley: University of California Press, 1998.

Heppner, Frank. 'Avian Flight Formations.' *The Condor* 45 (1974): 160–170.

Heppner, Frank and Ulf Grenader. 'A Stochastic Nonlinear Model for Coordinated Bird Flocks.' In *The Ubiquity of Chaos*, edited by Saul Krasner, 233–238. Washington, DC: American Association for the Advancement of Science, 1990.

Hewstone, Miles et al., eds. *Introduction to Social Psychology: A European Perspective.* Oxford: Blackwell, 1988.

Hillis, W. Daniel. *The Connection Machine.* Cambridge, MA: MIT Press, 1985.

Hine, Thomas. 'One Architect Who Left His Mark on Cities.' *Philadelphia Inquirer* (9 February 1986): J01.

Hinske, Norbert. 'Die Aufklärung und die Schwärmer: Sinn und Funktion einer Kampfidee.' *Aufklärung* 3 (1988): 3–6.

Hoffmann, Christoph. *Unter Beobachtung: Naturforschung in der Zeit der Sinnesapparate.* Göttingen: Wallstein, 2006.

Horn, Eva. 'Das Leben ein Schwarm: Emergenz und Evolution in moderner Science Fiction.' In *Schwärme – Kollektive ohne Zentrum: Eine Wissensgeschichte zwischen Wissen und Information*, edited by Eva Horn and Lucas Marco Gisi, 102–124. Bielefeld: Transcript, 2009.

———. 'Schwärme – Kollektive ohne Zentrum: Einleitung.' In *Schwärme – Kollektive ohne Zentrum: Eine Wissensgeschichte zwischen Wissen und Information*, edited by Eva Horn and Lucas Marco Gisi, 13–26. Bielefeld: Transcript, 2009.

Hu, Xiaohui et al. 'Recent Advances in Particle Swarm.' In *Proceedings of the 2004 Congress on Evolutionary Computation*, Vol. 1, 90–97. Piscataway, NJ: IEEE, 2004.

Hunt, Graham et al. '3D Printing with Flying Robots.' In *Proceedings of 2014 IEEE International Conference on Robotics and Automation (ICRA)*, 4493–4499. Piscataway, NJ: IEEE, 2014.

Hunter, John R. 'Procedure for Analysis of Schooling Behavior.' *Journal of the Fisheries Research Board of Canada* 23 (1966): 547–562.

Husserl, Edmund. 'The Origins of Geometry.' Translated by David Carr. In *Edmund Husserl's Origin of Geometry: An Introduction*. By Jacques Derrida. Lincoln: University of Nebraska Press, 1989. 155–180.

Huth, Andreas and Christian Wissel. 'The Simulation of Fish Schools in Comparison with Experimental Data.' *Ecological Modelling* 75 (1994): 135–145.

———. 'The Simulation of the Movement of Fish Schools.' *Journal of Theoretical Biology* 156 (1992): 365–385.

Huxley, Thomas. 'The Struggle for Existence and its Bearing Upon Man.' *Nineteenth Century* 23 (1888): 161–180.

Inada, Yoshinubu. 'Steering Mechanism of Fish Schools.' *Complexity* 8 (2001): 1–9.

Inada, Yoshinubu and Keiji Kawachi. 'Order and Flexibility in the Motion of Fish Schools.' *Journal of Theoretical Biology* 214 (2002): 371–387.

Inagaki, Tadashi et al. 'Studies on the Schooling Behavior of Fish: Mathematical Modeling of Schooling Form Depending on the Intensity of Mutual Forces Between Individuals.' *Bulletin of the Japanese Society of Scientific Fisheries* 42/3 (1976): 265–270.

'Interview with Roland Snooks,' *suckerPUNCH* (25 April 2010): http://www.suckerpunchdaily.com/2010/04/25/interview-with-roland-snooks/.

Ioannou, C. C. et al. 'Predatory Fish Select for Coordinated Collective Motion in Virtual Prey.' *Science* 337 (2012): 1212–1215.

Jacob, François. *The Logic of Life: A History of Heredity*. Translated by Betty E. Spillmann. Princeton: Princeton University Press, 1973.

———. *Of Flies, Mice, and Men*. Translated by Giselle Weiss. Cambridge, MA: Harvard University Press, 1998.

———. *The Statue Within: An Autobiography*. Translated by Franklin Philip. Woodbury, NY: Cold Spring Harbor Laboratory Press, 1995.

Jaffe, Jules S. 'Methods for Three-Dimensional Sensing of Animals.' In *Animal Groups in Three Dimensions: How Species Aggregate*, edited by Julia K. Parrish et al., 17–35. Cambridge: Cambridge University Press, 1997.

Jahn, Ilse et al, eds. *Geschichte der Biologie: Theorien, Methoden, Institutionen, Kurzbiographien*, 3rd ed. Hamburg: Spektrum, 2000.

James, R. et al. 'Geometry for Mutualistic and Selfish Herds: The Limited Domain of Danger.' *Journal of Theoretical Biology* 228 (2004): 107–113.

Jammer, Max. *Concepts of Simultaneity: From Antiquity to Einstein and Beyond.* Baltimore: Johns Hopkins University Press, 2006.

Jennings, Nicholas R. 'On Agent-Based Software Engineering.' *Artificial Intelligence* 117 (2000): 277–296.

Jimenez-Pacheco, Alberto et al. 'Implementation of a Wireless Mesh Network of Ultra-Light MAVs with Dynamic Routing.' In *2012 IEEE Globecom Workshops*, 1591–1596. Piscataway, NJ: IEEE, 2012.

Johach, Eva. 'Andere Kanäle: Insektengesellschaften und die Suche nach den Medien des Sozialen.' *Zeitschrift für Medienwissenschaft* 4 (2011): 71–82.

———. 'Schwarm-Logiken: Genealogien sozialer Organisation in Insektengesellschaften.' In *Schwärme – Kollektive ohne Zentrum: Eine Wissensgeschichte zwischen Leben und Information*, edited by Eva Horn and Lucas Marco Gisi. Bielefeld: Transcript, 2009. 203–224.

Johannson, Anders et al. 'From Crowd Dynamics to Crowd Safety: A Video-Based Analysis.' *Advances in Complex Systems* 11 (2008): 497–527.

Johnson, Steven. *Emergence: The Connected Lives of Ants, Brains, Cities, and Software.* New York: Scribner, 2001.

Jordan, David Starr. *Fishes*. New York: Appleton, 1925.

Kaempfer, Engelbert. *The History of Japan, Together with a Description of the Kingdom of Siam*. 3 vols. London: H. Sloane, 1727, I.

Kalikhman, Igor L. and K. I. Yudanov. *Acoustic Fish Reconnaissance*. Boca Raton: Taylor & Francis, 2006.

Kampfwirth, Ralph. 'SchwarmStrom: Die Energie der Zukunft.' *LichtBlick: Die Zukunft der Energie* (2010): https://www.elke-seidel.de/sites/default/files/Dateien/101000_LichtBlick-Info_SchwarmStrom.pdf.

Karafyllis, Nicole. 'Bios und Zoe.' In *Naturphilosophische Grundbegriffe*, edited by Thomas Kirchhoff (2012). http://www.naturphilosophie.org/bios-und-zoe/.

Kassung, Christian, ed. *Die Unordnung der Dinge: Eine Wissens- und Mediengeschichte des Unfalls*. Bielefeld: Transcript, 2009.

Kaufmann, Stefan. 'Soldatische Subjekte: Disziplinierung in militärischen Netzwerken.' In *Vernetzte Steuerung: Soziale Prozesse im Zeitalter technischer Netzwerke*, edited by Stefan Kaufmann, 145–158. Zurich: Chronos, 2007.

Kazadi, Sanza T. *Swarm Engineering*. Doctoral Diss.: California Institute of Technology, 2000.

Keating, J. P. 'The Myth of Panic.' *Fire Journal*, 76/3 (1982): 57–61.

Keele, Kenneth D. *Leonardo Da Vinci's Elements of the Science of Man*. New York: Academic Press, 1983.

Keller, Evelyn Fox. *Making Sense of Life: Explaining Biological Development with Models, Metaphors, and Machines*. Cambridge, MA: Harvard University Press, 2002.

———. 'Models of and Models for: Theory and Practice in Contemporary Biology.' *Philosophy of Science* 67 (2000): 72–82.

———. 'Models, Simulations, and 'Computer Experiments'.' In *The Philosophy of Scientific Experimentation*, edited by Hans Radder, 198–215. Pittsburgh: University of Pittsburgh Press, 2003.

Kelley, Harold et al. 'Collective Behaviour in a Simulated Panic Situation.' *Journal of Experimental Social Psychology* 1 (1965): 20–54.

Kelly, Kevin. *Out of Control: The New Biology of Machines, Social Systems, and the Economic World*. London: Fourth Estate, 1994.

Kelty, Christopher and Hannah Landecker. 'A Theory of Animation: Cells, L-Systems, and Film.' *Grey Room* 17 (2004): 30–63.

Kennedy, James and R. C. Eberhart. 'Particle Swarm Optimization.' In *Proceedings of the IEEE International Conference on Neural Networks*, 1942–1948. Piscataway, NJ: IEEE Neural Networks Council, 1995.

———. *Swarm Intelligence: Collective, Adaptive*. San Francisco: Morgan Kaufman, 2001.

Kessel, J. F. 'Flocking Habits of the California Valley Quail.' *The Condor* 23 (1921): 167–168.

Khoshnevis, Behrokh. 'Automated Construction by Contour Crafting – Related Robotics and Information Technologies.' *Automation in Construction* 13 (2004): 5–19.

Kieser, R. et al. 'Bias Correction of Rockfish School Cross-Section Widths from Digitized Echosounder Data.' *Canadian Journal of Fisheries and Aquatic Sciences* 50 (1993): 1801–11.

Kimura, K. 'On the Detection of Fish-Groups by an Acoustic Method.' *Journal of the Imperial Fisheries Institute Tokyo* 24 (1929): 41–45.

Kipnis, Jeffrey. '(Architecture) After Geometry – An Anthology of Mysteries: Case Notes to the Mystery of the School of Fish.' In *Architecture After Geometry*, edited by Peter Davidson and Donald L. Bates, 43–47. London: Architectural Design, 1997.

Kittler, Friedrich. 'Computer Graphics: A Semi-Technical Introduction.' Translated by Sara Ogger. *Grey Room* 2 (2001): 30–45.

———. *Gramophone, Film, Typewriter.* Translated by Geoffrey Winthrop-Young and Michael Wutz. Stanford: Stanford University Press, 1999.

———. 'Signal-to-Noise Ratio.' In *The Truth of the Technological World: Essays on the Genealogy of Presence.* Translated by Erik Butler, 165–177. Stanford: Stanford University Press, 2013.

Klarreich, Erica. 'The Mind of the Swarm: Math Explains How Group Behavior Is More Than the Sum of Its Parts.' *Science News Online* 170/22 (November 2006). http://www.sciencenews.org/articles/20061125/bob10.asp.

Kleinrock, Leonard. 'Distributed Systems.' *Communications of the ACM* 28 (1985): 1200–1213.

Klüver, Jürgen. *The Dynamics and Evolution of Social Systems: New Foundations of a Mathematical Sociology.* Dortrecht: Springer, 2000.

Kocaman, Ibrahim. *Distributed Beamforming in a Swarm UAV Network.* MS Thesis: Naval Postgraduate School, 2008.

Kracauer, Siegfried. 'The Mass Ornament.' In *The Mass Ornament: Weimar Essays.* Translated by Thomas Y. Levin, 75–88. Cambridge, MA: Harvard University Press, 1995.

Kreutzer, Wolfgang. 'Grundkonzepte und Werkzeugsysteme objektorientierter Systementwicklung.' *Wirtschaftsinformatik* 32/3 (1990): 211–227.

Kropotkin, Piotr. *Mutual Aid: A Factor of Evolution.* Harmondsworth: Pelican, 1939.

Kruzelecki, Karol. 'Flying Ad-Hoc Networks,' *SMAVNET II Project Website* (2015): http://smavnet.epfl.ch/.

Kümmel, Albert and Erhard Schüttpelz. 'Medientheorie der Störung / Störungstheorie der Medien: Eine Fibel.' In *Signale der Störung,* edited by Albert Kümmel and Erhard Schüttpelz, 9–13. Munich: Fink, 2003.

———, eds. *Signale der Störung.* Munich: Fink, 2003.

Kunz, Hanspeter and Charlotte K. Hemelrijk. 'Artificial Fish Schools: Collective Effects of School Size, Body Size, and Body Form.' *Artificial Life* 9 (2003): 237–253.

Küppers, Günter and Johannes Lenhard. 'The Controversial Status of Simulations.' In *Networked Simulations and Simulated Networks: 18th European Simulation Multiconference,* edited by Graham Horton, 271–275. Erlangen: SCS, 2004.

Lamothe, Dan. 'Watch Perdix, the Secretive Pentagon Program Dropping Tiny Drones from Jets,' *The Washington Post* (8 March 2016): https://www.washingtonpost.com/news/checkpoint/wp/2016/03/08/watch-perdix-the-secretive-pentagon-program-dropping-tiny-drones-from jets/?noredirect=on&utmterm=.ce4952071168.

Landgraf, Tim et al. 'RoboFish: Increased Acceptance of Interactive Robotic Fish with Realistic Eyes and Natural Motion Patterns by Live Trinidadian Guppies.' *Bioinspiration & Biomimetics* 11 (2016): 1–14.

Lane, Nicholas et al. 'A Survey of Mobile Phone Sensing.' *IEEE Communication Magazine* 48/9 (2010): 140–150.

Langelaan, Jack W. and Nicholas Roy. 'Enabling New Missions for Robotic Aircraft.' *Science* 326 (2009): 1642–1644.

Langton, Christopher G. 'Artificial Life.' In *Artificial Life: The Proceedings of an International Workshop on the Synthesis and Simulation of Living Systems*, edited by Christopher G. Langton, 1–47. Redwood City: Addison-Wesley, 1989.

———. *Artificial Life: An Overview.* Cambridge, MA: MIT Press, 1995.

———. 'Life at the Edge of Chaos.' In *Artificial Life II: Proceedings of the Workshop on Artificial Life*. Redwood City: Addison-Wesley, 1991. 41–91.

Latour, Bruno. *Science in Action: How to Follow Scientists and Engineers through Society*. Cambridge, MA: Harvard University Press, 1987.

Laurent, Philip. 'The Supposed Synchronal Flashing of Fireflies.' *Science* 45 (1917): 44.

Lause, Markus and Peter Wippermann. *Leben im Schwarm: Die Spielregeln der Netzwerkökonomie*. Reutlingen: Red Indians, 2012.

Lautman, Albert. *Les schemas de structure*. Paris: Hermann, 1938.

Lawson, Brian. *How Designers Think: The Design Process Demystified*, 4th ed. Oxford: Butterworth Architecture, 2005.

Leach, Neil. 'Swarm Urbanism.' *Architectural Design* 79 (2009): 56–63.

Le Bon, Gustave. *The Crowd: A Study of the Popular Mind*. New York: Macmillan, 1896.

———. *Psychology of Crowds*. Edited by Anna Cuffaro and David Kauders. Southampton: Sparkling, 2009.

Leidenberger, Frank et al. 'Wie kämpfen die Landstreitkräfte künftig?' *Internal Thesis Paper of the Bundeswehr* (2017). https://www.gemhflatr.de/contrexx/media/archive1/dokumente/Thesenpapier_LSK_.pdf.

Lem, Stanisław. *The Invincible*. Translated by Wendayne Ackerman. New York: Ace Books, 1973.

———. 'The Upside-Down Evolution.' In *One Human Minute*. Translated by Catherine S. Leach, 37–68. New York: Harcourt Brace & Company, 1986.

Lemke, Thomas et al. 'Gouvernementalität, Neoliberalismus und Selbsttechnologien: Eine Einführung.' In *Gouvernementalität der Gegenwart: Studien zur Ökonomisierung des Sozialen*, edited by Ulrich Bröckling et al., 7–40. Frankfurt am Main: Suhrkamp, 2000.

Lettvin, Jerôme Y. et al. 'What the Frog's Eye Tells the Frog's Brain.' In *The Mind: Biological Approaches to Its Functions*, edited by William C. Corning and Martin Balaban, 233–258. New York: Interscience, 1968.

Levenez, Jean-Jacques et al. 'Acoustics in Halieutic Research.' *Les dossiers thématiques de l'IRD* (n.d.). http://www.suds-en-ligne.ird.fr/ecosys/ang_ecosys/pdf/acoustic.pdf.

Levin, Simon A. 'Conceptual and Methodological Issues in the Modeling of Biological Aggregations.' In *Animal Groups in Three Dimensions: How Species Aggregate*, edited by Julia K. Parrish et al., 245–256. Cambridge: Cambridge University Press, 1997.

Levy, Steven. *Artificial Life: A Report from the Frontier Where Computers Meet Biology.* New York: Vintage Books, 1993.
Li, Yanchao. *Oriented Particles for Scientific Visualization.* MCS Thesis: University of New Brunswick, 1997.
Licklider, J. C. R. 'Man-Computer Symbiosis.' *IRE Transactions on Human Factors in Electronics* 1 (1960): 4–11.
Lloyd, Morgan C. *Emergent Evolution.* London: Williams and Norgate, 1923.
Long, William J. *How Animals Talk, and Other Pleasant Studies of Birds and Beasts.* New York: Harper & Brothers, 1919.
Lorenz, Konrad. *On Aggression.* Translated by Marjorie Kerr Wilson. New York: Routledge, 2002.
Lovink, Geert. Zero Comments: Blogging and Critical Internet Culture. New York: Routledge, 2007.
Luhmann, Niklas. *Die Politik der Gesellschaft.* Frankfurt am Main: Suhrkamp, 2000.
Lynn, Greg. 'Body Matters.' In *Folds, Bodies & Blobs: Collected Essays*, by Greg Lynn, 135–153. Brussels: La lettre volée, 1998.
———. 'Why Tectonics is Square and Topology is Groovy.' In *Folds, Bodies & Blobs: Collected Essays*, by Greg Lynn, 169–186. Brussels: La lettre volée, 1998.
Maaler, Josua. *Die teütsch Spraach.* Zurich, 1561.
Macal, Charles M. and Michael J. North. 'Tutorial on Agent-Based Modeling and Simulation – Part 2: How to Model with Agents.' In *Proceedings of the 2006 Winter Simulation Conference*, edited by L. F. Perrone, 73–83. Piscataway, NJ: IEEE, 2006.
Macavinta, Courtney. 'Digital Actors in *Rings* Can Think.' *Wired Magazine* (13 December 2002): https://www.wired.com/2002/12/digital-actors-in-rings-can-think/.
MacCurdy, Edward, ed. *The Notebooks of Leonardo da Vinci.* Garden City: Garden City Publishing, 1942.
Maes, Patty, ed. *Designing Autonomous Agents: Theory and Practice from Biology to Engineering and Back.* Cambridge, MA: MIT Press, 1994.
Maeterlinck, Maurice. *The Life of the Bee.* Translated by Alfred Sutro. London: George Allen, 1901.
Magnenat, Stéphane et al. 'Autonomous Construction Using Scarce Resources in Unknown Environments.' *Autonomous Robots* 33 (2012): 467–485.
Maier, N. R. F. and Theodore C. Schneirla. *Principles of Animal Psychology.* New York: McGraw-Hill, 1935.
Major, Peter F. and Lawrence M. Dill. 'The 3D Structure of Airborne Bird Flocks.' *Behavioral Ecology and Sociobiology* 4 (1978): 111–122.
Mandelbrot, Benoit B. *The Fractal Geometry of Nature.* New York: W. H. Freeman, 1987.
Mann, Steve. '"Sousveillance": Inverse Surveillance in Multimedia Imaging.' In *Proceedings of the Twelfth Annual ACM International Conference on Multimedia*, 620–627. New York: ACM, 2004.

Marey, Étienne-Jules. *La méthode graphique dan les sciences expérimentales et particulièrement en physiologie et en médecine*. Paris: G. Masson, 1878.

Marey, Étienne-Jules. 'Preface.' In *La photographie animée*. By Eugène Trutat. Paris: Gauthier-Villars, 1889.

Marks, Paul. '3D Printing Takes Off with the World's First Printed Plane.' *New Scientist* 211 (2011): 17–18.

Matuda, Ko and Nobuo Sannomiya. 'Computer Simulation of Fish Behavior in Relation to Fishing Gear: Mathematical Model of Fish Behavior.' *Bulletin of the Japanese Society of Scientific Fisheries* 46 (1980): 689–697.

Mauss, Marcel. 'Techniques of the Body.' *Economy and Society* 2 (1973): 70–88.

Maye, Harun. 'Was ist eine Kulturtechnik?' *Zeitschrift für Medien- und Kulturforschung* 1 (2010): 121–136.

Mayer, Larry et al. '3D Visualization for Pelagic Fisheries Research and Assessment.' *ICES Journal of Marine Science* 59 (2002): 216–25.

Mayer, Ruth and Brigitte Weingart, eds. *Virus! Mutationen einer Metapher*. Bielefeld: Transcript, 2004.

McDougall, Marina. 'Introduction: Hybrid Roots.' In *Science is Fiction: The Films of Jean Painlevé*, edited by Andy Masaki Bellows et al., xiv–xviii. Cambridge, MA: MIT Press, 2000.

McLaughlin, Peter. 'Der neue Experimentalismus in der Wissenschaftstheorie.' In *Die Experimentalisierung des Lebens: Experimentalsysteme in den biologischen Wissenschaften 1850/1950*, edited by Hans-Jörg Rheinberger and Michael Hagner. Berlin: Akademie, 1993. 207–218.

McLuhan, Marshall and Quentin Fiore. *The Medium is the Massage: An Inventory of Effects*. New York: Penguin, 1967.

McLuhan, Marshall and Quentin Fiore. *War and Peace in the Global Village: An Inventory of Some of the Current Spastic Situations that Could Be Eliminated by More Feedforward*. New York: McGraw-Hill, 1968; repr. Corte Madera: Gingko Press, 2001.

McPhail, Clark. *The Myth of the Madding Crowd*. New York: Aldine de Gruyter, 1991.

Mecklenburg, Sebastian. 'Digitale Ork-Massen: Im Fantasy-Spektakel 'Die zwei Türme' schlägt der Rechner die Schlachten.' *C't: Magazin für Computertechnik* 26 (2002): 34.

Medosch, Armin. *Freie Netze: Geschichte, Politik und Kultur offener WLAN-Netze*. Hanover: Heise, 2004.

———. 'Meshing the Future: The Free Configuration of Everything and Everyone with Hive Networks.' *Mutatis Mutandis: The Node* (25 February 2006): http://www.nettime.org/Lists-Archives/nettime-l-0602/msg00076.html.

Menschen unter Haien. Directed by Hans Hass. Vienna: Hans Hass-Filmproduktion, 1947.

Mersch, Dieter. *Medientheorien zur Einführung*. Hamburg: Junius, 2006.
Mesyatsev, I. I. 'Structure of Shoals of Schooling Fish.' *Investiya AN SSSR: Seriya Biologicheskaya* 3 (1937): 737.
Metzger, Hans-Joachim. 'Genesis in Silico: Zur digitalen Biosynthese.' In *HyperKult: Geschichte, Theorie und Kontext digitaler Medien*, edited by Martin Warnke et al., 461–510. Bielefeld: Transcript, 1997.
Michelet, Jules. *The Sea*. Translated by W. H. D. Adams. London: T. Nelson, 1875.
Middtun, L. and G. Sætersdal. 'On the Use of Echo-Sounder Observations for Estimating Fish Abundance.' *Special Publication of the International Commission of North West Atlantic Fisheries* 11 (1957): 260–266.
'Military Robots Are Getting Smaller and More Capable.' *The Economist* (14 December 2017): https://www.economist.com/science-and-technology/2017/12/14/military-robots-are-getting-smaller-and-more-capable.
Miller, J. Abbott et al., eds. *Swarm*. Philadelphia: Fabric Workshop and Museum, 2005.
Miller, Peter. *The Smart Swarm: How Understanding Flocks, Schools, and Colonies Can Make Us Better at Communicating, Decision Making, and Getting Things Done*. New York: Avery, 2010.
Miller, Robert C. 'The Mind of the Flock.' *The Condor* 23 (1921): 183–186.
Millonas, Mark M. 'Swarms, Phase Transitions, and Collective Intelligence.' In *Computational Intelligence: A Dynamic System Perspective*, edited by M. Palaniswani et al., 137–151. Piscataway, NJ: IEEE, 1994.
Minar, Nelson et al. 'Hive: Distributed Agents for Networking Things.' In *Proceedings of the First and Third International Symposium on Agent Systems Applications and Mobile Agents*, 118–129. Piscataway, NJ: IEEE, 1999.
Minar, Nelson et al. 'The Swarm Simulation System: A Toolkit for Building Multi-Agent Simulations.' *Santa Fe Institute: Working Paper 96-06-042* (1996): https://pdfs.semanticscholar.org/3df5/d4c65795f85259c7c175c08cac58c53f203b.pdf.
Minati, Gianfranco and Eliano Pessa. *Collective Beings*. New York: Springer, 2006.
Mintz, A. 'Non-Adaptive Group Behavior.' *Journal of Abnormal Social Psychology* 46 (1951): 150–159.
Misund, Ole Arve. 'Underwater Acoustics in Marine Fisheries and Fisheries Research.' *Reviews in Fish Biology and Fisheries* 7 (1997): 1–34.
Mitson, R. B. and R. J. Wood. 'An Automatic Method of Counting Fish Echoes.' *ICES Journal of Marine Science* 26/3 (1961): 281–291.
Mogilner, Alexander and Leah Edelstein-Keshet. 'A Non-Local Model for a Swarm.' *Journal of Mathematical Biology* 38 (1999): 534–570.
Moorstedt, Tobias. 'Düstere Entscheidungen.' *Süddeutsche Zeitung Online* (1 March 2010): http://www.sueddeutsche.de/digital/videospiel-heavy-rain-duestere-entscheidungen-1.15821.

Moreira, André A. et al. 'Efficient System-Wide Coordination in Noisy Environments.' *Proceedings of the National Academy of Sciences of the Unites States of America* 101/33 (2004): 12085–12090.

Morrison, Margaret and Mary S. Morgan. 'Models as Mediating Instruments.' In *Models as Mediators: Perspectives on Natural and Social Science*, edited by Margaret Morrison and Mary S. Morgan, 10–38. Cambridge: Cambridge University Press, 1999.

Musse, Soraia Raupp et al. 'Groups and Crowd Simulation.' In *Handbook of Virtual Humans*, edited by N. Magnenat-Thalmann and Dirk Thalmann, 323–352. New York: Wiley, 2004.

Müller, Johannes. *Zur vergleichenden Physiologie des Gesichtssinnes des Menschen und der Thiere nebst einem Versuch über den menschlichen Blick*. Leipzig: C. Cnobloch, 1826.

Münker, Stefan. 'Information.' In *Grundbegriffe der Medientheorie*, edited by Alexander Roesler and Bernd Stiegler, 95–105. Munich: Fink, 2005.

Muscovici, Serge. *The Age of the Crowd: A Historical Treatise on Mass Psychology*. Translated by J. C. Whitehouse. Cambridge: Cambridge University Press, 1985.

Napp, Nils and Radhika Nagpal. 'Distributed Amorphous Ramp Construction in Unstructured Environments.' *Robotica* 32 (2014): 279–290.

Napp, Nils et al. 'Materials and Mechanisms for Amorphous Robotic Construction.' In *Proceedings of the 2012 IEEE/RSJ International Conference on Intelligent Robots and Systems*, 4879–4885. Piscataway, NJ: IEEE, 2012.

Neef, Andreas and Klaus Burmeister. 'Swarm Organization: A New Paradigm for the E-Enterprise of the Future.' In *The Practical Real-Time Enterprise: Facts and Perspectives*, edited by Bernd Kuhlin and Heinz Thielmann, 509–517. Berlin: Springer, 2005.

Nembrini, Julian et al. 'Mascarillons: Flying Swarm Intelligence for Architectural Research.' In *Proceedings of the 2005 IEEE Swarm Intelligence Symposium*, 225–232. Piscataway, NJ: IEEE, 2005.

Newland, C. Bingham. *What Is Instinct? Thoughts on Telepathy and Subconsciousness in Animals*. London: John Murray, 1916.

Nyhart, Lynn K. 'Natural History and the "New" Biology.' In *Cultures of Natural History*, edited by Nicholas Jardine et al., 426–443. Cambridge: Cambridge University Press, 1996.

Ofak, Ana and Philipp von Hilgers. 'Einleitung.' In *Rekursionen: Von Faltungen des Wissens*, edited by Ana Ofak and Philipp von Hilgers, 7–21. Munich: Fink, 2010.

———, eds. *Rekursionen: Von Faltungen des Wissens*. Munich: Fink, 2010.

Oosterhuis, Kas. *Hyperbodies: Towards an E-Motive Architecture*. Basel: Birkhäuser, 2003.

———. 'Swarm Architecture.' (2006) http://vhpark.hyperbody.nl/images/4/48/Swarm_Architecture_II.pdf

Oosterhuis, Kas and Lukas Feireiss, eds. *Game, Set, and Match II: On Computer Games, Advanced Geometries, and Digital Technologies*. Rotterdam: Episode, 2006.

Oreskes, Naomi. *Science and Technology in the Global Cold War*. Cambridge, MA: MIT Press, 2014.

Osborn, Jon. 'Analytical and Digital Photogrammetry.' In *Animal Groups in Three Dimensions: How Species Aggregate*, edited by Julia K. Parrish et al., 36–60. Cambridge: Cambridge University Press, 1997.

Page, Daniel and Howard Tripp. 'The Perfect Swarm.' *Aerospace International* (July 2012): https://www.aerosociety.com/news/the-perfect-swarm/.

Painlevé, Jean. 'Feet in the Water.' In *Science is Fiction: The Films of Jean Painlevé*, edited by Andy Masaki Bellows et al., 130–139. Cambridge, MA: MIT Press, 2000.

Paramo, Jorge et al. 'A Three-Dimensional Approach to School Typology Using Vertical Scanning Multibeam Sonar.' *Fisheries Research* 84 (2007): 171–179.

Parikka, Jussi. *Insect Media: An Archaeology of Animals and Technology*. Minneapolis: University of Minnesota Press, 2010.

———. 'Politics of Swarms: Translations Between Entomology and Biopolitics.' *Parallax* 14/3 (2008): 112–124.

Pariser, Eli. *The Filter Bubble. What the Internet is Hiding from You*. New York: Penguin, 2011.

Parker, George Howard. 'The Function of the Lateral-Line Organ in Fishes.' *Bulletin of the U.S. Bureau of Fisheries* 24 (1904): 185–207.

Parr, Albert Eide. 'A Contribution to the Theoretical Analysis of the Schooling Behavior of Fishes.' *Occasional Papers of the Bingham Oceanographic Collection* 1 (1929): 1–32.

Parrish, Julia K. and Leah Edelstein-Keshet. 'Complexity, Pattern, and Evolutionary Trade-Offs in Animal Aggregation.' *Science* 284 (1999): 99–101.

Parrish, Julia K. and Steven V. Viscido. 'Traffic Rules of Fish Schools: A Review of Agent-Based Approaches.' In *Self-Organization and Evolution of Social Systems*, edited by Charlotte Hemelrijk, 50–80. Cambridge: Cambridge University Press, 2005.

Parrish, Julia K. et al. 'Introduction – From Individuals to Aggregations: Unifying Properties, Global Framework, and the Holy Grails of Congregation.' In *Animal Groups in Three Dimensions: How Species Aggregate*, edited by Julia K. Parrish et al., 1–14. Cambridge: Cambridge University Press, 1997.

Parrish, Julia K. et al. 'Self-Organized Fish Schools: An Examination of Emergent Properties.' *Biological Bulletin* 202 (2002): 296–305.

Partridge, Brian L. and Michael J. Cullen. 'Computer Technology: A Low-Cost Interactive Coordinate Plotter.' *Behavior Research Methods & Implementation* 9 (1977): 473–479.

Partridge, Brian L. et al. 'The Three-Dimensional Structure of Fish Schools.' *Behavioral Ecology and Sociobiology* 6 (1980): 277–288.

Perella, Stephen. 'Bernard Cache/Objectile: Topological Architecture and the Ambiguous Sign.' *Architectural Design* 68 (1998): 66–69.

Petersen, Kirstin H. *Collective Construction by Termite-Inspired Robots*. Doctoral Diss.: Harvard University, 2014.

Petersen, Kirsin H. et al. 'TERMES: An Autonomous Robotic System for Three-Dimensional Collective Construction.' In *Robotics: Science and Systems VII*, edited by Hugh F. Durrant-Whyte et al., 257–264. Cambridge, MA: MIT Press, 2011.

Pflüger, Jörg. 'Writing, Building, Growing: Leitvorstellungen der Programmiergeschichte.' In *Geschichten der Informatik: Visionen, Paradigmen, Leitmotive*, edited by Hans Dieter Hellige, 275–319. Berlin: Springer, 2004.

Pias, Claus. '"Bat Men Begin": Die Fledermaus und die Erfindung von Abschreckung.' In *Politische Zoologie: Das Tier – Symbol, Modell, Funktion*, edited by Anne von der Heiden and Joseph Vogl, 305–316. Zurich: Diaphanes, 2007.

———. *Computer Game Worlds*. Translated by Valentine A. Pakis. Zurich: Diaphanes, 2017.

———. *Details zählen: Zur Epistemologie der Computersimulation*. Unpublished manuscript, 2008.

———. 'Das digitale Bild gibt es nicht: Über das (Nicht-)Wissen der Bilder und die informatische Illusion.' *Zeitenblicke* 2/1 (2003). http://www.zeitenblicke.de/2003/01/pias/index.html.

———. 'Synthetic History.' In *Mediale Historiographien*, edited by Lorenz Engell and Joseph Vogl, 171–183. Weimar: Universitätsverlag, 2001.

———. 'Zeit der Kybernetik: Zur Einführung.' In *Kursbuch Medienkultur: Die maßgeblichen Theorien von Brecht bis Baudrillard*, edited by Joseph Vogl et al., 427–431. Stuttgart: DVA, 1999.

Pickering, Andrew. 'A Gallery of Monsters: Cybernetics and Self-Organization, 1940–1970.' In *Mechanical Bodies, Computational Minds: Artificial Intelligence from Automata to Cyborgs*, edited by Stefano Franchi and Güven Güzeldere, 229–248. Cambridge, MA: MIT Press, 2005.

Pitcher, Tony J. 'The Three-Dimensional Structure of Schools in the Minnow.' *Animal Behaviour* 21 (1973): 673–686.

Pluta, Werner. 'Crowd Management: Smartphone soll Massenpanik verhindern.' *Golem.de: IT-News für Profis* (4 September 2012): https://www.golem.de/news/crowd-management-smartphone-soll-massenpanik-verhindern-1209-94331.html.

Poli, Riccardo et al. 'Particle Swarm Optimization.' *Swarm Intelligence* 1 (2007): 33–57.

Polverino, G. et al. 'Zebrafish Response to Robotic Fish: Preference Experiments on Isolated Individuals and Small Shoals.' *Bioinspiration & Biomimetics* 7 (2012): 1–13.

Pomeroy, Harold and Frank Heppner. 'Structure of Turning in Airborne Rock Dove (*Columba Livia*) Flocks.' *The Auk* 109 (1992): 256–267.

Popper, Karl. *The Logic of Scientific Discovery*. London: Routledge, 2002.

Potel, Michael J. and R. E. Sayre. 'Interacting with the GALATEA Film Analysis System.' In *SIGGRAPH: Proceedings of the 3rd Annual Conference on Computer Graphics, Interactive Techniques, and Image Processing*, edited by Udo W. Pooch. New York: ACM Press, 1976. 52–59.

Potel, Michael J. and Richard J. Wassersug. 'Computer Tools for the Analysis of Schooling.' *Environmental Biology and Fisheries* 6 (1981): 15–19.

Potts, Wayne R. 'The Chorus-Line Hypothesis of Manoeuvre Coordination in Avian Flocks.' *Nature* 309 (1984): 344–345.

Pritchard, Helen and Jennifer Gabrys. 'From Citizen Sensing to Collective Monitoring: Working through the Perceptive and Affective Problematics of Environmental Pollution.' *GeoHumanities* 2 (2016): 354–371.

Quarantelli, E. L. 'The Behaviour of Panic Participants.' *Sociology and Social Research* 41 (1957): 187–194.

Querner, Hans. 'Die Methodenfrage in der Biologie des 19. Jahrhunderts: Beobachtung oder Experiment?' In *Geschichte der Biologie: Theorien, Methoden, Institutionen, Kurzbiographien*, 3rd ed. Edited by Ilse Jahn et al., 420–430. Hamburg: Spektrum, 2000.

Radakov, Dimitri. *Schooling in the Ecology of Fish*. New York: Wiley, 1973.

Reeves, William T. 'Particle Systems: A Technique for Modeling a Class of Fuzzy Objects.' *ACM Transactions on Graphics* 2/2 (1983): 91–108.

Reichenbach, Hans. *Experience and Prediction: An Analysis of the Foundations and the Structure of Knowledge*. Chicago: University of Chicago Press, 1938; repr. 1961.

Reichert, Ramón. 'Huschen, Schwärmen, Verführen.' *Kunsttexte* 3 (2004). https://edoc.hu-berlin.de/bitstream/handle/18452/8065/reichert.pdf?sequence=1&isAllowed=y.

Reuter, Hauke and Broder Breckling. 'Selforganisation of Fish Schools: An Object-Oriented Model.' *Ecological Modelling* 75 (1994): 147–159.

Reynolds, Craig W. 'Flocks, Herds, and Schools: A Distributed Behavioral Model.' *Computer Graphics* 21 (1987): 25–34.

Rheinberger, Hans-Jörg. *Iterationen*. Berlin: Merve, 1994.

———. 'Nachwort.' In *Die Logik des Lebenden: Eine Geschichte der Vererbung*. By François Jacob. Frankfurt am Main: Fischer, 2002. 345–54.

———. 'Objekt und Repräsentation.' In *Mit dem Auge Denken: Strategien der Sichtbarmachung in wissenschaftlichen und virtuellen Welten*, edited by Bettina Heintz and Jörg Huber, 55–61. Zurich: Springer, 2001.

———. *Toward a History of Epistemic Things: Synthesizing Proteins in the Test Tube*. Stanford: Stanford University Press, 1997.

Rheinberger, Hans-Jörg and Michael Hagner. 'Experimentalsysteme.' In *Die Experimentalisierung des Lebens: Experimentalsysteme in den biologischen*

Wissenschaften 1850/1950, edited by Hans-Jörg Rheinberger and Michael Hagner, 7–27. Berlin: Akademie, 1996.

Rheingold, Howard. *Smart Mobs: The Next Social Revolution*. Cambridge, MA: Basic Books, 2002.

'Riders on a Swarm.' *The Economist* (12 August 2010): https://www.economist.com/science-and-technology/2010/08/12/riders-on-a-swarm.

Riedl, Rupert. 'Die Tauchmethode, ihre Aufgaben und Leistungen bei der Erforschung des Litorals: Eine kritische Untersuchung.' *Helgoländer wissenschaftliche Meeresuntersuchungen* 15 (1967): 295–352.

Rieger, Stefan. 'Der Frosch – ein Medium?' In *Was ist ein Medium?*, edited by Stefan Münker and Alexander Roesler, 285–303. Frankfurt am Main: Suhrkamp, 2008.

———. *Kybernetische Anthropologie: Eine Geschichte der Virtualität*. Frankfurt am Main: Suhrkamp, 2003.

Riemann, Bernhard. *On the Hypotheses Which Lie at the Bases of Geometry*. Translated by William Clifford. Basel: Birkhäuser, 2016.

Riley, Dave. 'Reginald Aubrey Fessenden (1866–1932): Some Biographical Notes.' *Soundscapes* 2 (1999). http://www.icce.rug.nl/~soundscapes/VOLUME02/Reginald_Aubrey_Fessenden.- shtml.

Rogers, David F. *An Introduction to NURBS: With Historical Perspective*. San Francisco: Morgan Kaufmann, 2001.

Roitman, Haggai et al. 'Harnessing the Crowds for Smart City Sensing.' In *Proceedings of the 1st International Workshop on Multimodal Crowd Sensing*, 17–18. New York: ACM, 2012.

Roux, Wilhelm. 'Aufgaben der Entwicklungsmechanik der Organismen.' *Aufgaben für Entwicklungsmechanik* 1 (1895): 75.

Rubenstein, Michael et al. 'Kilobot: A Low-Cost Scalable Robot System for Collective Behaviors.' In *Proceedings of 2012 IEEE International Conference on Robotics and Automation*, 3293–3298. Piscataway, NJ: IEEE, 2012.

Rugoff, Ralph. 'Fluid Mechanics.' In *Science is Fiction: The Films of Jean Painlevé*, edited by Andy Masaki Bellows et al., 48–57. Cambridge, MA: MIT Press, 2000.

Runnstrøm, S. 'A Review of the Norwegian Herring Investigations in Recent Years.' *ICES Journal of Marine Science* 12 (1937): 123–143.

Saenz, Aaron. 'Japan's Nuclear Woes Give Rise to Crowd-Sourced Radiation Maps in Asia and US.' *Singularity Hub* (24 March 2011): https://singularityhub.com/2011/03/24/japans-nuclear-woes-give-rise-to-crowd-sourced-radiation-maps-in-asia-and-us/#sm.00010etb051ei2fb2w3zrgmgipcdi.

Şahin, Erol. 'Swarm Robotics: From Sources of Inspiration to Domains of Application.' In *Swarm Robotics*, edited by Erol Şahin and William M. Spears, 10–20. New York: Springer, 2008.

Saidi, Kamel S. et al. 'Robotics in Construction.' In *Springer Handbook of Robotics*, edited by Siciliano Bruno and Oussama Khatib, 1079–1099. Berlin: Springer, 2008.

Sakai, Sumiko. 'A Model for Group Structure and Its Behavior.' *Biophysics (Japan)* 13/2 (1973): 82–90.

Sauvage, Léo. 'The Institute in the Cellar.' In *Science is Fiction: The Films of Jean Painlevé*, edited by Andy Masaki Bellows et al., 124–128. Cambridge, MA: MIT Press, 2000.

Schäfer, Armin and Joseph Vogl. 'Feuer und Flamme: Über ein Ereignis des 19. Jahrhunderts.' In *Kultur im Experiment*, edited by Henning Schmidgen et al., 191–211. Berlin: Kadmos, 2004.

Schaffer, Simon. 'Fish and Ships: Models in the Age of Reason.' In *Models: The Third Dimension of Science*, edited by Soraya de Chadarevian and Nick Hopwood, 71–108. Stanford: Stanford University Press, 2004.

Schäffner, Wolfgang. 'Nicht-Wissen um 1800: Buchführung und Statistik.' In *Poetologien des Wissens um 1800*, edited by Joseph Vogl, 123–144. Munich: Fink, 1999.

Scharre, Paul. *Robotics on the Battlefield Part II: The Coming Swarm*. Washington, DC: Center for a New American Security, 2014.

Schelling, Thomas C. 'Dynamic Models of Segregation.' *Journal of Mathematical Sociology* 1 (1971): 143–186.

———. 'Models of Segregation.' *American Economic Review* 59 (1969): 488–93.

Schieritz, Nadine and Peter Milling. 'Modeling the Forest or Modeling the Trees: A Comparison of System Dynamics and Agent-Based Simulation.' In *Proceedings of the 21st International Conference of the System Dynamics Society*, edited by Robert L. Eberlein et al., n.p. Albany: System Dynamics Society, 2003.

Schiff, Joel L. *Cellular Automata: A Discrete View of the World*. Hoboken: Wiley Interscience, 2008.

Schilt, Carl R. and Kenneth S. Norris. 'Perspectives on Sensory Integration Systems: Problems, Opportunities, and Predictions.' In *Animal Groups in Three Dimensions: How Species Aggregate*, edited by Julia K. Parrish and William M. Hammer, 225–244. Cambridge: Cambridge University Press, 1997.

Schmidt, Karl Patterson. 'Biographical Memoir of Warder Clyde Allee, 1885–1965.' In *National Academy of Sciences of the United States of America: Biographical Memoirs*, 1–40. New York: Columbia University Press, 1957.

Schneirla, Theodore C. 'Social Organization in Insects, as Related to Individual Function.' *Psychological Review* 48 (1941): 465–486.

Schreckenberg, Michael and Som Deo Sharma, eds. *Pedestrian and Evacuation Dynamics*. Berlin: Springer, 2002.

Schubbach, Arno. '"... A Display (Not a Representation) ...": Zur Sichtbarmachung von Daten.' *Navigationen: Zeitschrift für Medien- und Kulturwissenschaft* 7/2 (2007): 13–27.
Schumacher, Patrik. 'Parametricism: A New Global Style for Architecture and Urban Design.' *Architectural Design* 79 (2009): 14–23.
Schüttpelz, Erhard. 'Die Frage nach der Frage, auf die das Medium eine Antwort ist.' In *Signale der Störung*, edited by Albert Kümmel and Erhard Schüttpelz, 15–29. Munich: Fink, 2003.
Seeßlen, Georg and Markus Metz. *Blödmaschinen: Die Fabrikation der Stupidität*. Frankfurt am Main: Suhrkamp, 2011.
Seitter, Walter. *Menschenfassungen: Studien zur Erkenntnispolitikwissenschaft*. Munich: Boer, 1985.
Selous, Edmund. *Bird Life Glimpses*. London: G. Allen, 1905.
———. *Bird Watching*. London: J. M. Dent, 1901.
———. 'An Observational Diary of the Habits – Mostly Domestic – of the Great Crested Grebe (*Podicipes cristatus*).' *Zoologist* 5 (1901): 161–183.
———. Thought Transference (Or What?) in Birds. London: Constable, 1931.
Senebier, Jean. *L'art de observer*. Geneva: Cl. Philibert & Bart. Chirol., 1775.
Serres, Michel. *Genesis*. Translated by Geneviève James and James Nielson. Ann Arbor: University of Michigan Press, 1995.
———. *Hermès IV: La distribution*. Paris: Minuit, 1977.
———. *Hermes: Literature, Science, Philosophy*. Edited by Josué V. Harari and David F. Bell. Baltimore: Johns Hopkins University Press, 1982.
———. *Hermès ou la a communication*. Paris: Minuit, 1968.
———. 'Introduction.' In *A History of Scientific Thought: Elements of a History of Science*, edited by Michel Serres, 1 –16. Oxford: Blackwell, 1995.
———. *The Parasite*. Translated by Lawrence R. Schehr. Baltimore: Johns Hopkins University Press, 1982.
———. *Statues: The Second Book of Foundations*. Translated by Randolph Burks. New York: Bloomsbury Academic, 2015.
Serres, Michel and Nayla Farouki, eds. *Thesaurus der exakten Wissenschaften*, 3rd ed. Frankfurt am Main: Zweitausendeins, 2004.
Seyfried, Jörg et al. 'The I-SWARM Project: Intelligent Small World Autonomous Robots for Micro-Manipulation.' In *Swarm Robotics*, edited by Erol Şahin and William M. Spears, 70–83. New York: Springer, 2005.
Shannon, Claude E. 'A Mathematical Theory of Communication.' *The Bell System Technical Journal* 27 (1948): 379–423, 623–656.
———. 'The Redundancy of English.' In *Cybernetics: The Macy Conferences, 1946–1953*. 2 vols, edited by Claus Pias, I, 248–272. Zurich: Diaphanes, 2004.

Shao, Wei and Demetri Terzopoulos. 'Autonomous Pedestrians.' In *Eurographics/ ACM SIGGRAPH Symposium on Computer Animation*, edited by K. Anjiyo and P. Faloutsos, 246–274. New York: ACM, 2005.

Shanken, Edward A. 'Life as We Know It and/or Life as It Could Be: Epistemology and the Ontology/Ontogeny of Artificial Life.' *Leonardo* 31 (1998): 383.

Shaw, Evelyn. 'The Schooling of Fishes.' *Science* 206 (1962): 128–138.

Shiga, John. 'Ping and the Material Meanings of Ocean Sound.' In *Sustainable Media: Critical Approaches to Media and Environment*, edited by N. Starosielski and J. Walker, 128–145. New York: Routledge., 2015.

———. 'Sonar and the Channelization of the Ocean.' In *Living Stereo: Histories and Cultures of Multichannel Sound*, edited by P. Théberge et al., 85–106. New York: Bloomsbury 2015.

———. 'Sonar: Empire, Media, and the Politics of Underwater Sound.' *Canadian Journal of Communication*, 38 (2013): 357–378.

Siegert, Bernhard. *Cultural Techniques: Grids, Filters, Doors, and Other Articulations of the Real*. Translated by Geoffrey Winthrop-Young. New York: Fordham University Press, 2014.

———. 'Die Geburt der Literatur aus dem Rauschen der Kanäle: Zur Poetik der phatischen Funktion.' In *Electric Laokoon: Zeichen und Medien, von der Lochkarte zur Grammatologie*, edited by Michael Franz et al., 5–41. Berlin: Akademie, 2007.

———. 'Kakographie oder Kommunikation.' In *Mediale Historiographien*, edited by Lorenz Engell and Joseph Vogl, 87–99. Weimar: Universitätsverlag, 2001.

———. 'Der Nomos des Meeres: Zur Imagination des Politischen und ihre Grenzen.' In *Politiken der Medien*, edited by Daniel Gethmann and Markus Stauff, 39–56. Berlin: Diaphanes, 2005.

Siegfried, André. *Aspekte des 20. Jahrhunderts*. Munich: Langen-Müller 1956.

Sighele, Scipio. *La folla delinquente: Studio di psicologia collettiva*. Turin: Fratelli Bocca, 1891.

Sime, Jonathan D. 'Crowd Psychology and Engineering.' *Safety Science* 21 (1995): 1–14.

Simmel, Georg. 'Bridge and Door.' Translated by Mark Ritter. *Theory, Culture & Society* 11 (1994): 5–10.

Simmonds, John and David MacLennan. *Fisheries Acoustics: Theory and Practice*, 2nd ed. Oxford: Blackwell, 2005.

Simon, Herbert A. *The Sciences of the Artificial*, 3rd ed. Cambridge, MA: MIT Press, 1996.

Sims, Karl. 'Particle Animation and Rendering Using Data Parallel Computation.' In *SIGGRAPH 1990: Proceedings of the 17th Annual Conference on Computer Graphics and Interactive Techniques*, 405–413. New York: ACM Press, 1990.

Sismondo, Sergio. 'Models, Simulations, and Their Objects.' *Science in Context* 12 (1999): 247–260.

Soria, Marc et al. 'Effects of External Factors (Environment and Survey Vessel) on Fish School Characteristics Observed by Echosounder and Multibeam Sonar in the Mediterranean Sea.' *Aquatic Living Resources* 16 (2003): 145–157.

Soleymani, Touraj et al. 'Autonomous Construction with Compliant Building Material.' *Intelligent Autonomous Systems* 13 (2015): 1371–1388.

Southward, A. J. and E. K. Roberts. 'The Marine Biological Association 1884–1984: One Hundred Years of Marine Research.' *Report and Transactions of the Devonshire Association for the Advancement of Science* 116 (1984): 155–199.

Snyder, Joel. 'Visualization and Visibility.' In *Picturing Science, Producing Art*, edited by Caroline A. Jones and Peter Galison, 379–399. New York: Routledge, 1998.

Spooner, Guy M. 'Some Observations on Schooling in Fish.' *Journal of the Marine Biological Association of the United Kingdom* 17 (1931): 421–448.

Stäheli, Urs. 'Protokybernetik in der Massenpsychologie.' In *Transformationen des Humanen: Beiträge zur Kulturgeschichte der Kybernetik*, edited by Michael Hagner and Erich Hörl, 299–325. Frankfurt am Main: Suhrkamp, 2007.

Stauff, Markus. 'Zur Gouvernementalität der Medien: Fernsehen als 'Problem' und 'Instrument'.' In *Politiken der Medien*, edited by Daniel Gethmann and Markus Stauff, 89–110. Berlin: Diaphanes, 2005.

Steele, Luc. 'Towards a Theory of Emergent Functionality.' In *From Animals to Animats: Proceedings of the First International Conference on Simulation and Adaptive Behavior*, edited by Jean-Arcady Meyer and Stewart W. Wilson, 451–461. Cambridge, MA: MIT Press, 1990.

Steinbeck, John. *The Log from the Sea of Cortez*. New York: Viking, 1951.

Stephan, Achim. 'Emergente Eigentschaften.' In *Philosophie der Biologie: Eine Einführung*, edited by Ulrich Krohs and Georg Toepfer, 88–105. Frankfurt am Main: Suhrkamp, 2005.

Strogatz, Steven. *Sync: How Order Emerges from Chaos in the Universe, Nature, and Daily Life*. New York: Hyperion, 2003.

Stroud, Marion Boulton. 'Approaching Swarms.' In *Swarm*, edited by J. Abbott Miller et al., 7–11. Philadelphia: Fabric Workshop and Museum, 2005.

Stroupe, A. et al. 'Behavior-Based Multi-Robot Collaboration for Autonomous Construction Tasks.' In *Proceedings of the 2005 IEEE/RSJ International Conference on Intelligent Robots and Systems*, 1495–1500. Piscataway, NJ: IEEE, 2005.

Sumpter, David J. T. 'The Principles of Collective Animal Behaviour.' *Philosophical Transactions of the Royal Society: Biological Sciences* 361 (2006): 5–22.

Sund, Oscar. 'Echo Sounding in Fishery Research.' *Nature* 135 (1935): 935.

Surowiecki, James. *The Wisdom of Crowds: Why the Many Are Smarter than the Few*. New York: Doubleday, 2004.

Sutherland, Ivan E. *Sketchpad: A Man-Machine Graphical Communication System*. Doctoral Diss.: MIT, 1963.

———. 'Three-Dimensional Data Input by Tablet.' *Proceedings of the IEEE* 62 (1974): 453–461.

Tembrock, Günter. *Angst: Naturgeschichte einer psychobiologischen Phänomen*. Darmstadt: Wissenschaftliche Buchgesellschaft, 2000.

Terzopoulos, Demetri. 'Artificial Life for Computer Graphics.' *Communications of the ACM* 42/8 (1999): 33–42.

Terzopoulos, Demetri et al. 'Artificial Fishes: Autonomous Locomotion, Perception, Behavior, and Learning in a Simulated Physical World.' *Artificial Life* 1 (1994): 327–351.

Thacker, Eugene. *Biomedia*. Minneapolis: University of Minnesota Press, 2004.

———. 'Networks, Swarms, Multitudes.' *CTheory* (18 May 2004). http://www.ctheory.net/articles.aspx?id=423.

Tien, Joseph H. et al. 'Dynamics of Fish Shoals: Identifying Key Decision Rules.' *Evolutionary Ecology Research* 6 (2004): 555–565.

Tiqqun. 'The Cybernetic Hypothesis' (2001): https://theanarchistlibrary.org/library/tiqqun-the-cybernetic-hypothesis.

Towler, Richard H. et al. 'Visualizing Fish Movement, Behavior, and Acoustic Backscatter.' *Aquatic Living Resources* 16 (2003): 277–282.

Truffaut, François. *Hitchcock/Truffaut*. New York: Simon & Schuster, 1983.

Tu, Xiaoyuan and Demetri Terzopoulos. 'Artificial Fishes: Physics, Locomotion, Perception, Behavior.' In *SIGGRAPH 94: Proceedings of the 21st Annual Conference on Computer Graphics and interactive Technologies*, 43–50. New York: ACM Press, 1994.

Turner, Ralph H. and Lewis M. Killian. *Collective Behaviour*. Englewood Cliffs, NJ: Prentice-Hall, 1975.

Ulam, Stanislaw. 'On Some Mathematical Problems Connected with Patterns of Growth of Figures.' *Proceedings of the Symposium of Applied Mathematics* 14 (1962): 219–231.

Vabø, Rune and Leif Nøttestad. 'An Individual Based Model of Fish School Reactions: Predicting Antipredator Behavior as Observed in Nature.' *Fisheries Oceanography* 6/3 (1997): 155–171.

Vagt, Christina. 'Zeitkritische Bilder: Bergson zwischen Topologie und Fernsehen.' In *Zeitkritische Medien*, edited by Axel Volmar et al., 105–126. Berlin: Kadmos, 2009.

Van Olst, J. C. and J. R. Hunter. 'Some Aspects of the Organization of Fish Schools.' *Journal of the Fisheries Research Board of Canada* 27 (1970): 1225–1238.

Vásárhelyi, Gábor et al. 'Outdoor Flocking and Formation Flight with Autonomous Aerial Robots.' In *Proceedings of the 2014 IEEE/RSJ International Conference on Intelligent Robots and Systems*, 3866–3873. Piscataway, NJ: IEEE, 2014.

Vehlken, Sebastian. 'After Affects. Zealous Zombies, Panic Prevention, Crowd Simulation.' In *The Timing of Affect*, edited by Angerer, Marie-Luise et al., 303–320. Berlin, Zurich: Diaphanes 2014.

———. 'Angsthasen: Schwärme als Transformationsungestalten zwischen Tierpsychologie und Bewegungsphysik.' *Zeitschrift für Kultur- und Medienforschung* 1 (2009): 133–147.

———. 'Gesichter im Sand: Schwärme zwischen Nicht-Identität und Pattern Recognition.' In *Im Netz der Eindeutigkeiten: Unbestimmte Figuren und die Irritation von Identität*, edited by Michael Andreas and Natascha Frankenberg, 165–188. Bielefeld: Transcript, 2013.

———. 'Pervasive Intelligence. The Tempo-Spatiality of Drone Swarms.' *Digital Culture and Society* 4/1 (2018): Rethinking AI: Neural Networks, Biometrics and the New Artificial Intelligence. Eds. Mathias Fuchs and Ramòn Reichert, 99–124

———. 'Swarming. A Novel Cultural Technique for Generative Architecture.' *Footprint* 15 (2014) (= Special Issue Data-Driven Design). Eds. Henriette Bier, Terry Knight: 9–17.

Vicsek, Tamás et al. 'Novel Type of Phase Transition in a System of Self-Driven Particles.' *Physical Review Letters* 75 (1995): 1226–1229.

Vicsek, Tamás et al. 'Spontaneously Ordered Motion of Self-Propelled Particles.' *Journal of Physics A: Mathematical and General* 30 (1997): 1375–1385.

Vines, Gail. 'Psychic Birds (Or What?).' *New Scientist* 182 (2004): 48–49.

Viscido, Steven V. et al. 'The Effect of Population Size and Number of Influential Neighbors on the Emergent Properties of Fish Schools.' *Ecological Modelling* 183 (2005): 347–363.

Viscido, Steven V. et al. 'Individual Behavior and Emergent Properties of Fish Schools: A Comparison of Observation and Theory.' *Marine Ecology Progress Series* 273 (2004): 239–249.

Vitruvius. *The Ten Books on Architecture*. Translated by Morris H. Morgan. New York: Dover, 1960.

Vogl, Joseph. 'Becoming-Media: Galileo's Telescope.' Translated by Brian Hanrahan. *Grey Room* 29 (2007): 14–25. Originally 'Medien-Werden: Galileis Fernrohr.' In *Mediale Historiographien*, edited by Lorenz Engell and Joseph Vogl. Weimar: Universitätsverlag, 2001. 115–123.

———. 'Gefieder, Gewölk.' In *Synaesthetics: Konturen einer physiologischen Medienästhetik*, edited by Christian Filk et al. Cologne: Herbert von Halem, 2004. 140–149.

———. 'Über soziale Fassungslosigkeit.' In *Kollektive Gespenster: Die Masse, der Zeitgeist und andere unfassbare Körper*, edited by Michael Gamper and Peter Schnyder. Berlin: Rombach, 2006. 171–189.

Von der Heiden, Anne and Joseph Vogl, eds. *Politische Zoologie*. Zurich: Diaphanes, 2007.

Von Frisch, Karl. *Bees: Their Vision, Chemical Senses, and Language*. Ithaca, NY: Cornell University Press, 1950.

———. *Erinnerungen eines Biologen*. 3rd ed. Berlin: Springer, 1973.

———. 'Zur Psychologie des Fischschwarms.' *Die Naturwissenschaften* 37 (1938): 601–606.

Von Mammen, Sebastian and Christian Jacob. 'Swarm-Driven Idea Models: From Insect Nests to Modern Architecture.' *WIT Transactions on Ecology and Environment* 113 (2008): 117–126.

Von Mammen, Sebastian et al. 'Evolving Swarms that Build 3D Structures.' In *Proceedings of the 2005 IEEE Congress on Evolutionary Computation*, 1434–41. Piscataway, NJ: IEEE, 2005.

Von Neumann, John. 'Probabilistic Logics and the Synthesis of Reliable Organisms from Unreliable Components.' In *Collected Works* (6 vols.) edited by Abraham H. Taub, V, 329–378. Oxford: Pergamon Press, 1976,

———. *Theory of Self-Reproducing Automata*. Edited by Arthur W. Burks. Urbana: University of Illinois Press, 1966.

Von Uexküll, Jakob. 'Die Bedeutung der Planmäßigkeit für die Fragestellung in der Biologie.' In *Kompositionslehre der Natur: Biologie als undogmatische Naturwissenschaft. Ausgewählte Schriften*, edited by Thure von Uexküll, 213–217. Frankfurt am Main: Ullstein, 1980.

Wawerla, Jens et al. 'Collective Construction with Multiple Robots.' In *Proceedings of the 2002 IEEE/RSJ International Conference on Intelligent Robots and Systems*, 2696–2701. Piscataway, NJ: IEEE, 2002.

Weaver, Warren. 'Recent Contributions to the Mathematical Theory of Communication.' In *The Mathematical Theory of Communication*, by Claude E. Shannon and Warren Weaver, 1–28. Urbana: University of Illinois Press, 1949; repr. 1964.

Weihs, David. 'A Hydrodynamical Analysis of Fish Turning Maneouvers.' *Proceedings of the Royal Society of London – Series B: Biological Sciences* 182 (1972): 59–72.

Weill, Alan et al. 'MOVIES-B: An Acoustic Detection Description Software – Application to Shoal Species' Classification.' *Aquatic Living Resources* 6 (1993): 255–267.

Werber, Niels. *Ameisengesellschaften: Eine Faszinationsgeschichte*. Frankfurt am Main: S. Fischer, 2013.

Werfel, Justin et al. 'Collective Construction of Environmentally-Adaptive Structures.' In *Proceedings of the 2007 IEEE/RSJ International Conference on Intelligent Robots and Systems*, 2345–2352. Piscataway, NJ: IEEE, 2007.

Werfel, Justin et al. 'Designing Collective Behavior in a Termite-Inspired Robot Construction Team.' *Science* 343 (2014): 754–758.

Werfel, Justin et al. 'Distributed Construction by Mobile Robots with Enhanced Building Blocks.' In *Proceedings of the 2006 IEEE International Conference on Robotics and Automation*, 2787–2794. Piscataway, NJ: IEEE, 2006.

Wheeler, William M. 'The Ant-Colony as an Organism.' *Journal of Morphology* 21 (1911): 307–325.

——— 'On the Founding of Colonies by Queen Ants, with Special Reference to the Parasitic and Slave-Making Species.' *Bulletin of the American Museum of Natural History* 22/4 (1906): 33–105.

——— *Social Life Among the Insects*. London: Constable & Co., 1922.

Whitman, Charles Otis. 'Animal Behavior.' In *Biological Lectures Delivered at the Marine Biological Laboratory of Woods Hole, Mass., 1898*, by Whitman, 285–338.. Boston: Ginn and Company, 1899.

Wiener, Norbert. *Cybernetics, or Control and Communication in the Animal and the Machine*, 2nd ed. Cambridge, MA: MIT Press, 1961.

Wiener, Norbert. *The Human Use of Human Beings: Cybernetics and Society*. Boston: Houghton Mifflin, 1950.

———. *Nonlinear Problems in Random Theory*. Cambridge, MA: MIT Press, 1958.

Willmann, Jan et al. 'Aerial Robotic Construction: Towards a New Field of Architectural Research.' *International Journal of Architectural Computing* 10 (2012): 439–459.

Wilson, E. B. *The Cell in Development and Heredity*, 3rd ed. New York: Macmillan, 1925.

Wilson, Edward O. *Sociobiology: The New Synthesis*. Cambridge, MA: Harvard University Press, 1978.

Winkler, Hartmut. 'Rekursion: Über Programmierbarkeit, Wiederholung, Verdichtung und Schema.' *C't: Magazin für Computertechnik* 9 (1999): 234–240.

Winsberg, Eric. 'Simulations, Models, and Theories: Complex Physical Systems and Their Representations.' *Philosophy of Science* 68 (2001): 442–454.

Winter, Philipp. 'Science is Fiction: The Films of Jean Painlevé.' *Electric Sheep: A Deviant View of Cinema* (30 August 2007). http://www.electricsheepmagazine.co.uk/reviews/2007/08/30/science-is-fiction-the-films-of-jean-painleve/.

Winthrop-Young, Geoffrey. 'Cultural Techniques: Preliminary Remarks.' *Theory, Culture & Society* 30 (2013): 3–19.

Wirz, Martin et al. 'Inferring Crowd Conditions from Pedestrians' Location Traces for Real-Time Crowd Monitoring During City-Scale Mass Gatherings.' In *Proceedings of the 21st IEEE International WETICE Conference*, edited by Sumitra Reddy and Khalil Drira, 367–372. Piscataway, NJ: IEEE, 2012.

Wittkower, Rudolf. *Architectural Principles in the Age of Humanism*. New York: W. W. Norton, 1962.

Wolfram, Stephen. *A New Kind of Science*. Champaign, IL: Wolfram Media, 2002.

Wood, A. B. et al. 'A Magnetostriction Echo Depth-Recorder.' *Journal of the Institution of Electrical Engineers* 76 (1935): 550–566.

Wuketits, Franz. *Die Entdeckung des Verhaltens: Eine Geschichte der Verhaltensforschung*. Darmstadt: Wissenschaftliche Buchgesellschaft, 1995.

Zeng, Yifeng et al. 'SwarmArchitect: A Swarm Framework for Collaborative Construction.' In *Proceedings of the 9th Annual Conference on Genetic and Evolutionary Computation*, 186. New York: ACM, 2007.

Ziegler, Cai. 'Von Tieren lernen: Optimierungsprobleme lösen mit Schwarmintelligenz.' C't: *Magazin für Computertechnik* 3 (2008): 188–191.

For Product Safety Concerns and Information please contact our EU representative GPSR@taylorandfrancis.com Taylor & Francis Verlag GmbH, Kaufingerstraße 24, 80331 München, Germany

Batch number: 10725973

Printed by Printforce, the Netherlands